Laser-Matter Interaction for Radiation and Energy

Laser-Matter Interaction for Radiation and Energy

Hitendra K. Malik

CRC Press is an imprint of the
Taylor & Francis Group, an **informa** business

First edition published 2021
by CRC Press
6000 Broken Sound Parkway NW, Suite 300, Boca Raton, FL 33487-2742

and by CRC Press
2 Park Square, Milton Park, Abingdon, Oxon, OX14 4RN

CRC Press is an imprint of Taylor & Francis Group, LLC

ISBN: [978-1-138-03203-3] (hbk)
ISBN: [978-1-315-39602-6] (ebk)

Typeset in Palatino
by KnowledgeWorks Global Ltd.

Dedicated to

Mammi Shrimati Rajbala

&

Pitaji Shri Surendra Singh MALIK

For

Always showering their blessings

&

Inspiring at each step of life

Contents

Foreword

It gives me immense pleasure to see the present textbook on *Laser-Matter Interaction for Radiation and Energy,* which covers key topics on lasers, electromagnetic wave propagation in different media, terahertz (THz) radiation generation, THz detection techniques, surface plasmon resonance, techniques for particle acceleration, X-ray lasers, high harmonic generation, attosecond laser generation, fusion by lasers, controlled fusion devices, and so forth. These topics are important not only to researchers but also to the students at postgraduate and undergraduate levels at different universities and institutions all over the world who are eager to dive deeper into the subject and want to visualize various physical phenomena. I complement the author and appreciate his efforts in presenting this wonderful book. The text is comprehensive with respect to the introduction, methodology and state of the art. The explanation of each topic the author has given is commendable. I am very happy to note that this book carries all the elements including high-quality solved examples, making the book an excellent textbook.

After going through many of the topics, I recalled the days when my group at Utsunomiya University, Japan used to produce high-quality research and we pioneered many of the areas of laser/microwave-plasma interaction including solitons, particle acceleration, and radiation. Particle acceleration has been the subject of our key interest in view of laser accelerators and fusion-related research. The *laser-plasma-based high-energy particle accelerator concept* was originally demonstrated in our laboratory in the middle of the 1980s. Wakefield excitation is a key technique for exciting ultrahigh fields in the accelerators, and the physics of electron acceleration using different shapes of laser pulses is a very interesting and significant trial in this field. The author had proved this based on his theoretical research during visits to Utsunomiya University, where I worked as vice president from 2004 to 2007. Another most important aspect of high-power laser and matter interaction is the laser thermonuclear fusion control. In the case of inertial confinement fusion, the resonance frequency of solid-state density plasmas is close to the optical frequencies, demonstrating that infrared, optical, and UV radiations are strongly absorbed by such plasmas. So the coherence of the lasers, both spatial and temporal, make them appropriate to heat these dense plasmas up to high temperatures. In the case of the large density approach, the lasers created important tools for controlling thermonuclear fusion. When we talk about radiation, the THz radiation stands first in view of its diverse applications and the recent interest of researchers throughout the globe. The author has demonstrated many techniques to generate this radiation (in addition to X-ray lasers and attosecond lasers), which employ semiconducting materials, nanostructures, gases, and plasmas. The author's concept of tuning their frequency, power, and focus based on the application of a magnetic field and tailoring the laser pulses is well received globally. The concept of surface plasmon resonance for the THz radiation is also remarkable and has been discussed in this book. Very systematically the author has introduced various techniques for the detection of these types of radiation, which are vital for making readers understand the whole journey of creating the radiation types and using them for the benefit of the society.

I am well aware of the rigorous and enriching teaching experience at Indian Institutes of Technology (IITs), which originates from the interaction with the best engineering and science students. In the IIT system, strong feedback from all kinds of students results in continuous evolution and refinement of the teachers. I am impressed to see this spirit of the author in the book during the comprehensive and in-depth handling of the recondite topics in a pellucid manner. I am happy to affirm that this textbook has been penned down by an IITian and a top-ranking researcher. I am sure that this book would serve as a very good textbook on the subject *laser-matter interaction*. This book is an important learning resource not only for the students but also for the teachers. I hope that the book is very well received in the academic world.

Yasushi Nishida

Yasushi Nishida, Ph.D.
Former Vice-President, Utsunomiya University, Japan
Former Director, Cooperative Research Centre, Utsunomiya University, Japan
Professor Emeritus with Utsunomiya University, Japan
Fellow of the American Physical Society,
Honorary Professor, Zhejiang University of Technology of China, Zhejiang, China
Honorary Professor, University of Electronic Science and Technology of China, Chengdu, China
Chair Professor of Biomedical Physics, Taipei Medical University, Taiwan
Former Professor, Lunghwa University of Science and Technology, Taiwan
Research Professor, National Cheng Kung University, Taiwan

Preface

Today, laser sources are available that allow the concentration of large amounts of energy inside a small volume and for a short period, thanks to the chirped pulse amplification (CPA) technique, fibre laser technology, combination of Mach-Zehnder technique with CPA, cascaded conversion compression and its combination with a large-scale pump laser, and others. High-power lasers interact with matter in an incoherent manner where nonlinearity is involved, which can lead to the generation of X-rays and γ-rays and opens a new way for high-energy elementary particle physics. Based on the interaction of lasers with semiconducting and nanomaterials we can obtain terahertz (THz) radiation, which is supposed to replace X-rays in medical applications. Time-domain THz spectroscopy has been very efficient for the characterization of materials. The THz radiation can generally be achieved by heating any object (thermal sources), and also these radiations can be efficiently obtained by waving or more appropriately moving electric charges (synchrotron and free-electron laser sources). The transitions between defined energy levels are the basis of the lasers, and these principles form the operation of the Ge-laser, the quantum cascade laser, and the molecular laser, which operate on THz frequencies.

When a short, intense laser pulse is focussed in a gas and partially ionizes the gas, higher-order harmonic generation may take place. At the intensities that are required to ionize the gas, the gas response is seen to be strongly nonlinear and it leads to the generation of new frequencies that can extend very far in the VUV (vacuum ultraviolet) to XUV (extreme ultraviolet) domain. The duration of XUV pulses can be controlled by modulating the polarization of the laser pulse. So the XUV radiation can be confined down to the isolated attosecond pulse levels, which are required to study the electronic rearrangement and ultrafast processes. Powerful femtosecond laser pulses propagate in an apparent form of filamentation in all transparent optical media. More new applications and phenomena related to the filamentation of such lasers have been discovered recently. These include slice-by-slice self-focussing, intensity clamping, white light laser generation (supercontinuum generation), background energy reservoir, and multiple filamentation competition. The dynamic balance between the Kerr self-focussing and defocussing by the self-generated plasma leads to an upper limit of the intensity at the self-focus. This clamped intensity has been found to be pressure independent. In the transverse pattern of the laser profile there is a tiny hot spot (accompanying the occurrence of a filament) squeezed by a wide weak background. This has recently been recognized as an important background energy reservoir for the whole filamentation process.

The most important aspect of the high-power laser and matter interaction is the laser thermonuclear fusion control. If we look at the inertial confinement fusion (ICF), we observe that the resonance frequency of solid-state density plasmas (density $\sim 10^{22}/\text{cm}^3$) is close to the optical frequencies. Thus, we may expect that UV, optical, and infrared radiations are strongly absorbed by such plasmas. In this context, the spatial and temporal coherence of the lasers make them appropriate to heat these dense plasmas to high temperatures, and they appear to be a very appropriate tool for controlling thermonuclear fusion in the case of a large density approach. The accelerated particles can also be used in fusion phenomenon, and people have tried to use lasers directly for accelerating the electrons and protons. There are other mechanisms for particle acceleration; for example, the wakefield by lasers in plasma. The energy gain to the particles can be enhanced by wisely chosen laser pulse shapes. The magnetic field also has proved to be an extra controlling parameter in the case of microwave application to the particle acceleration in plasma-filled waveguides.

As per linear theory, an electromagnetic wave is reflected when it is normally incident on a plasma and its frequency ω is less than or equal to the plasma frequency ω_p. However, for waves of higher frequency, $\omega > \omega_p$, the plasma behaves as an optically rarer medium with refractive index $\eta = \sqrt{\left(1 - \omega_p^2 / \omega^2\right)}$. The propagation of electromagnetic waves in an underdense plasma ($\omega_p < \omega$), absorption by collisions, reflection at critical density ($\omega_p = \omega$), and evanescence in overdense plasma ($\omega_p > \omega$) are interesting phenomena. Other interesting phenomena include beat wave excitation, induced transparency, and excitation of surface plasmon resonance (SPR), which are observed to be quite useful for THz radiation generation. On the other hand, lasers can tunnel ionize the gas, i.e. they produce plasma, and cause the electrons/dipoles to oscillate at the THz frequencies, which leads to the emission of the THz radiation. Researchers are concerned about attaining such radiations with tunable frequency and power along with a controllable focus of these radiations for their diverse applications. The use of an external magnetic field and density ripples in the plasma has proved to be an effective means for attaining such properties of THz radiation.

This reference book discusses the abovementioned interesting topics that are realized when a high-power laser interacts with matter. We start with a brief review of electromagnetic wave propagation in vacuums, dielectrics, conductors, and plasmas in chapter 1, where we also discuss the continuous wave (CW) and pulsed lasers, and the methods for achieving such laser output and their polarization. In chapter 2, we discuss the interaction of lasers with the semiconductors / nanostructures in view of THz radiation generation by different mechanisms. Further, in chapter 3 we briefly discuss the excitation of SPR and its application to THz radiation generation. Other methods of THz radiation generation using gases/plasmas, for example, tunnel ionization and beat wave, are detailed in chapter 4. The THz radiation detection technologies are discussed in chapter 5. The details of plasma-based particle acceleration technologies are included in chapter 6, where generation of wakefields in plasma by lasers is discussed in depth. Then we move to other interesting topics including X-ray lasers (chapter 7), harmonic generation (chapter 8), and attosecond lasers (chapter 9). Finally, we focus on the application or interaction of high-power lasers with matter to thermonuclear controlled fusion in chapters 10 and 11. In every chapter, we present an abstract and list the conclusions and future perspectives of the discussed topics for the benefit of the researchers and readers. In view of the nature of the topics and their level of discussion, this book is expected to have great potential for researchers and postgraduate and undergraduate students all over the world.

Acknowledgments

It has been a 20-year journey since I embarked upon the field of laser and microwave interaction with plasma at Utsunomiya University, Japan through the Japan Society for the Promotion of Science (JSPS) Fellowship and under my host professor, Professor Yasushi Nishida, a world-renowned figure in the field of plasma physics who pioneered the fields of solitons, particle acceleration, radiation and many more. He was actually an experimentalist but used to catch mistakes in the theoretical calculations done by researchers. He could tell the outcome based on his long experience and intuition! That was when I discovered physics embedded within the complex equations and started visualizing most of the physical phenomena. With my learning and in-depth discussions with him and other colleagues, I could impress him. I could see myself pondering any research problem his way and getting new problems to solve and finally cherishing the outcome. The 2-year (2000–2002) association with him was a golden time in my life, when all my dreams came true and I joined as Assistant Professor of Physics at the Indian Institute of Technology (IIT) Delhi. Professor Nishida is a real teacher for me who not only taught me the way to do research but also taught me the art of living. I remember his quote 'excellent people excel in all the fields – not only the science but also at different roles in life too!'. The credit for excelling in research and teaching goes to him. My collaboration with him is still alive, and I have visited Utsunomiya University several times after joining IIT Delhi in the year 2002.

At IIT Delhi I had to stand on my own without any godfather and compete with otherwise advantaged colleagues. This was a blessing in disguise, however, since I worked very hard and could develop new research areas and guided a large number of Ph.D. scholars in addition to undergraduate and postgraduate students. This strengthened me more and more, and the odd situation contributed a great deal to this strengthening process. On the other hand, rigorous teaching at the IIT and my love for teaching physics in a simple manner by citing the examples provided by nature (Almighty) inspired me to pen this textbook *Laser-Matter Interaction for Radiation and Energy* in addition to another book on *Engineering Physics* (McGraw Hill).

I could not complete this book alone; several students helped me at different stages and in different ways. Discussion with them on many occasions enhanced the quality of the text and its overall presentation. In particular, Rajat worked with complete sincerity and commitment in bringing out the book without mistakes. Dimple, Bharat, Ashu, Sheetal, Dhananjay Singh, Rajesh, Manish, Dhananjay, and Sonu devoted significant amounts of time. Also helpful were Tamanna, Subhajit, Yetendra, Sandeep, and Mohit.

It was impossible for me to complete such a big task without the support of my family. My wife Professor Anushree Malik and my son Lohit have stood with me through thick and thin. They allowed me to devote my all time to this book and to keep this work as my first priority!

I have always been inspired by my friends and close relatives, who contribute directly or indirectly to any challenge I take up and finally achieve my goal with full satisfaction.

Finally, I remember the Almighty God in the form of Hanuman ji, Shiv-Shankar ji, and Saraswati Maa!

Hitendra K. Malik

Acknowledgments

Author Biography

Born in a village Chhainsa of GB Nagar in Uttar Pradesh (India), **Hitendra K. Malik** is currently a professor of physics at the Indian Institute of Technology (IIT) Delhi, from where he earned a PhD degree in the field of plasma physics in 1995. He has been a merit scholarship holder throughout his academic career. Owing to his worldwide recognition, his name was included in *Marquis Who's Who* in 2011, which was published in the United States. Based on the survey conducted by ResearchGate in 2016, his scientific score was found within the top 5% of scientists and researchers all over the world. Professor Malik is highly cited in India and abroad for his research work and books with an h-index of 33 and an i10-index of 98. He has accomplished 14 sponsored research projects by arranging funds from various agencies such as Department of Science and Technology (DST), Council of Scientific and Industrial Research (CSIR), Defence Research and Development Organization (DRDO), All India Council for Technical Education (AICTE), Deutscher Akademischer Austausch Dienst (DAAD), Indo-French Centre for the Promotion of Advanced Research (IFCPAR / CEFIPRA), and so forth from the governments of India, Germany, and France. He has been on the editorial board of six reputed research journals, and is recently working with two journals from Springer and Taylor and Francis. In recognition of his outstanding research and teaching contributions, he has been asked to deliver more than 65 keynote and invited talks in India, Japan, South Korea, the United States, France, Germany, South Africa, and Turkey. Also, he has been chief guest at various universities, the mentor of faculty colleagues of engineering institutions, and a member of organizing and advisory committees of national and international conferences held in India and abroad. He has guided more than 115 theses including 29 at the Ph.D. level in the areas of laser / microwave plasma interactions, particle acceleration, solitons, terahertz radiation, Hall thrusters, plasma material interaction, and nanotechnology. He has published more than 440 scientific papers in very reputed journals and conferences, including more than 20 independent articles. He has been a reviewer for 86 journals of international repute, several sponsored research projects (Indian and foreign agencies), and 36 Ph.D. theses. He is an expert member of academic and administrative bodies of 14 different universities and institutions from 8 states of India including University Grants Commission (UGC).

Professor Malik is the recipient of the prestigious 2018 Who's Who in the World Top 3%, in the United States; a Career Award from AICTE, Government of India (for his teaching and research); Outstanding Scientist Award from VIF, India (for his contributions to science); and the 2017 Albert Nelson Marquis Lifetime Achievement Award from the United States. In addition, he received the prestigious Erasmus Mundus Visiting Fellowship from the European Union (Germany and France), the JSPS Fellowship (two times) from Japan, the FRD Fellowship from South Africa, the DAAD Fellowship from Germany, and many others.

Apart from this textbook on *Laser-Matter Interaction for Radiation and Energy* published by Taylor and Francis (CRC Press), he has also authored a textbook on *Engineering Physics* (McGraw Hill), three chapters in the books *Wave Propagation*, InTechOpen Science, Croatia (featured as a highly downloaded chapter); *Society, Sustainability and Environment*, Shivalik Prakashan, New Delhi; and *Plasma Science and Nanotechnology*, Apple Academic Press, exclusive worldwide distribution by CRC Press, a Taylor & Francis Group.

1

Electromagnetic Waves and Lasers

1.1 Maxwell's Equations

The pioneering work of James Clark Maxwell brought a revolutionary discovery in the field of electrodynamics by providing the world with four major equations that governed the electromagnetic (EM) fields. These four Maxwell's equations are a set of equations which can be represented in differential and integral forms. These equations explain how the electric and the magnetic fields would evolve with respect to space and time (Malik and Singh 2018). These can also speak about how the evolution of EM fields would affect the sources of the fields like charges and current densities.

The spatial rates of change of the fields are represented in the form of divergences and curls of the electric field $\vec{E}$ and the magnetic field $\vec{B}$, i.e. div $\vec{E}$, div $\vec{B}$, curl $\vec{E}$ and curl $\vec{B}$. The time rates of the change of the fields are coupled with their spatial rates of change. Overall, these equations are of utmost importance as these can be applied to all charges and current densities whether static or time-dependent.

The differential form of the Maxwell's first equation is represented as follows:

$$\vec{\nabla} \cdot \vec{E} = \frac{\rho}{\varepsilon_0} \tag{1.1}$$

Here, ρ is the volume charge density. This is simply Gauss's law of electrostatics, which says that the spreading of the electric field out from a closed surface is determined by the density of charges in the volume enclosed by the surface. This also means that the divergence of the electric field is more if the enclosed charge carries larger density.

The differential form of the Maxwell's second equation is represented as follows:

$$\vec{\nabla} \cdot \vec{B} = 0 \tag{1.2}$$

This is simply Gauss's law of magnetostatics, revealing the non-existence of magnetic monopoles, because the zero divergence of the magnetic field $\vec{B}$ is only possible for the case of the two opposite poles.

The differential form of the Maxwell's third equation is simply the relation of the curl of field $\vec{E}$ and the time rate of change of field $\vec{B}$ as,

$$\vec{\nabla} \times \vec{E} = -\frac{\partial \vec{B}}{\partial t} \tag{1.3}$$

This is also known as Faraday's law of electromagnetic induction, saying that the changing magnetic field induces electric field; the minus sign, depicting the oppose to the change, is in accordance with the Lenz's law.

The fourth Maxwell's equation is the most powerful equation, which reads

$$\vec{\nabla} \times \vec{B} = \mu_0 \vec{J} + \mu_0 \varepsilon_0 \frac{\partial \vec{E}}{\partial t} \tag{1.4}$$

Here, $\vec{J}$ is the current density, which generally produces the magnetic field. However, Maxwell's important contribution is through the addition of a second term on the right-hand side (RHS), carrying the time variation of the electric field. This equation is the modified form of the Ampere's law. Actually, the time-variation of the electric field $\vec{E}$ induces the magnetic field $\vec{B}$, which in conjunction with time-variation of field $\vec{B}$ through Eq. (1.3) establishes the evolution of EM field.

The applications of Maxwell's equations are indispensable. All the fields, which rely on the basic principle of EM theory, depend on Maxwell's equations that form the basis for their applications. For example, their use is vital from medical to optoelectronics to security through magnetic resonance imaging (MRI), X-rays and so forth, in hospitals to waveguides, antennae, transmission lines to radars and security cameras.

1.2 Electromagnetic Wave Propagation in Vacuum and Dielectric Medium

Electromagnetic (EM) waves are non-mechanical waves and, hence, can travel through the vacuum. They do not need any material medium to propagate, and transport their energy from one point to another. However, during the process of energy transport, the absorption and the re-emission of wave energy takes place while interacting with the atoms of the material medium. When an EM wave strikes the surface of target material, the energy is absorbed due to which electrons inside the atoms undergo vibrations. Then these vibrating electrons emit a new EM wave of the same frequency as that of the original wave. The EM wave travels with the speed of light (c) through the vacuum, but in the medium the speed is less than c due to the absorption and re-emission of the wave in the material medium. The amount of decrease in the speed depends on the optical density of the medium. Because the high (low) density materials cause more (less) absorption of light, the speed of light will be different in low and high density materials.

In vacuum, the charge density is $\rho = 0$ and the current density is $\vec{J} = 0$. Then Maxwell's equations read

$$\vec{\nabla} \cdot \vec{E} = 0, \vec{\nabla} \cdot \vec{B} = 0, \vec{\nabla} \times \vec{E} = -\frac{\partial \vec{B}}{\partial t}, \vec{\nabla} \times \vec{B} = \frac{1}{c^2} \frac{\partial \vec{E}}{\partial t}, \text{together with } c^2 = \frac{1}{\mu_0 \varepsilon_0}$$

The curl of Eq. (1.3) along with the use of Eq. (1.4) yields

$$\vec{\nabla} \times \left(\vec{\nabla} \times \vec{E} \right) = -\frac{1}{c^2} \frac{\partial^2 \vec{E}}{\partial t^2} \tag{1.5}$$

Using the vector identity $\vec{\nabla} \times \left(\vec{\nabla} \times \vec{E} \right) = \vec{\nabla} \left(\vec{\nabla} \cdot \vec{E} \right) - \nabla^2 \vec{E}$ and realizing that $\vec{\nabla} \cdot \vec{E} = 0$, from Maxwell's first equation, we obtain

$$\nabla^2 \vec{E} = \frac{1}{c^2} \frac{\partial^2 \vec{E}}{\partial t^2} \tag{1.6}$$

This is the wave equation for the electric field vector $\vec{E}$ in vacuum. Its plane wave solution can be expressed as, $\vec{E} = \vec{E}_0 e^{i(\vec{k}.\vec{r}-\omega t)}$. This solution will satisfy wave Eq. (1.6) only if

$(\omega^2 - c^2k^2)\vec{E} = 0$. Since $\vec{E} \neq 0$, the required condition reads

$$\omega^2 = c^2k^2 \tag{1.7}$$

This relation [Eq. (1.7)] between ω and k is known as the dispersion relation. From this, we can find the wavelength $\lambda (= 2\pi/k)$ and the phase velocity $v_\phi (= \omega/k)$ of the EM wave in the vacuum. On the other hand, the magnetic field $\vec{B}$ of the wave can be obtained from the Maxwell's fourth equation, i.e. $\vec{\nabla} \times \vec{B} = \frac{1}{c^2}\frac{\partial \vec{E}}{\partial t}$, once we know the expression of the electric field vector.

In a dielectric medium also, $\rho = 0$ and $\vec{J} = 0$. However, in the presence of the electric field of the wave, the medium gets polarized and this effect of polarization is introduced through the electric displacement vector $\vec{D} = \varepsilon_0\vec{E} + \vec{P} = \varepsilon_0\vec{E} + \varepsilon_0\chi_e\vec{E} = \varepsilon_0(1+\chi_e)\vec{E}$ or, $\vec{D} = \varepsilon\vec{E}$ together with $\varepsilon = \varepsilon_0(1+\chi_e)$ for linear dielectric medium. Similarly, the effect of magnetization is introduced through the vector $\vec{H}$, and the relation between $\vec{B}$ and $\vec{H}$ is given by $\vec{B} = \mu\vec{H}$, together with $\mu = \mu_0(1+\chi_m)$; χ_e and χ_m represent the electric and magnetic susceptibilities, respectively. Hence, Maxwell's equations for a linear dielectric medium read

$$\vec{\nabla}\cdot\vec{E} = 0, \vec{\nabla}\cdot\vec{B} = 0, \vec{\nabla}\times\vec{E} = -\frac{\mu\partial\vec{H}}{\partial t}, \vec{\nabla}\times\vec{H} = \varepsilon\frac{\partial\vec{E}}{\partial t}$$

Similar mathematical treatment of these equations yields the following equation for the electric field vector $\vec{E}$ in a linear dielectric medium

$$\nabla^2\vec{E} = \mu\varepsilon\frac{\partial^2\vec{E}}{\partial t^2} \tag{1.8}$$

The plane wave solution, $\vec{E} = \vec{E}_0\, e^{i(\vec{k}\cdot\vec{r}-\omega t)}$, of this equation gives rise to the following condition

$$\omega^2\mu\varepsilon = k^2 \tag{1.9}$$

Since $\mu = \mu_0\mu_r$ and $\varepsilon = \varepsilon_0\varepsilon_r$ in terms of relative permeability μ_r and relative permittivity ε_r, this condition reads,

$$\omega^2\mu_r\varepsilon_r = c^2k^2 \tag{1.10}$$

This is the dispersion relation for the EM wave propagating in a linear dielectric medium. The phase velocity $v_\phi = \omega / k = c / \sqrt{\mu_r\varepsilon_r}$ reveals that the wave will propagate in the dielectric medium with the speed less than the speed of light. For a non-magnetic medium, $\mu_r = 1$, the phase velocity $v_\phi = c / \sqrt{\varepsilon_r}$. Since the ratio c / v_ϕ represents the index of refraction, the refractive index $\tilde{n}$ of a linear, non-magnetic dielectric medium is simply given by

$$\tilde{n} = \sqrt{\varepsilon_r} = \sqrt{\text{dielectric constant}} \tag{1.11}$$

The EM wave in a medium, whose refractive index is more than 1, would propagate at the speed less than c. However, the phase velocity would exceed c in a medium whose refractive index $\tilde{n} < 1$. This situation arises in plasma.

1.3 Electromagnetic Wave Propagation in Conductor

A conductor consists of free charges moving around in response to EM fields. Any charge supplied to the conductor reaches quickly to the surface of the conductor and the electric field inside the conductor does not survive. In the cases of vacuum and dielectric medium or insulators, current density $\vec{J}$ was neglected in Maxwell's equations because the electric conductivity of the medium is zero. However, a conducting medium has large conductivity σ due to which the conduction current density $\vec{J}$ cannot be neglected. Actually the flow of charges in conductors is not independently controlled and the current density in general cannot be neglected. On the other hand, free charge (ρ_{free}) supplied to a conductor gets dissipated because of which Maxwell's first equation still reads $\vec{\nabla}\cdot\vec{E} = 0$ even in the conducting medium. This can be understood as follows:

$$\vec{\nabla}\cdot\vec{J}_{\text{free}} = -\frac{\partial \rho_{\text{free}}}{\partial t} \text{ and } \vec{J}_{\text{free}} = \sigma\vec{E}$$

These equations form the following relation while using together ($\rho_{\text{free}} = \rho$)

$$\vec{\nabla}\cdot\vec{E} = -\frac{1}{\sigma}\frac{\partial \rho}{\partial t} \tag{1.12}$$

This is true for isotropic medium, whose conductivity does not change spatially. Since we want to establish a relation between ρ and t, we eliminate $\vec{E}$ by using Gauss's law of electrostatics, i.e. $\vec{\nabla}\cdot\vec{E} = \frac{\rho}{\varepsilon}$. This gives rise to

$$\frac{1}{\rho}\frac{\partial \rho}{\partial t} = -\frac{\sigma}{\varepsilon}$$

Hence, on integration this gives

$$\rho(t) = \rho_0 e^{-\frac{\sigma t}{\varepsilon}} \tag{1.13}$$

where, ρ_0 is the initial free charge density.

This Eq. (1.13) shows that charges do not reside inside the conductor; rather, they will flow out to the surface of conductor in characteristic time $\tau_f = \frac{\varepsilon}{\sigma}$. For a perfect conductor, this characteristic time $\tau_f = 0$ as $\sigma = \infty$. For a good conductor, τ_f will be very small.

With the above description, now we can write Maxwell's equations in the case of an isotropic and linear conductor as

$$\vec{\nabla}\cdot\vec{E} = 0 \tag{1.14}$$

$$\vec{\nabla}\cdot\vec{H} = 0 \tag{1.15}$$

$$\vec{\nabla} \times \vec{E} = -\mu \frac{\partial \vec{H}}{\partial t} \tag{1.16}$$

$$\vec{\nabla} \times \vec{H} = \sigma \vec{E} + \varepsilon \frac{\partial \vec{E}}{\partial t} \tag{1.17}$$

Similar mathematical treatment of these equations yields the following wave equation for the electric field in a conducting medium

$$\nabla^2 \vec{E} = \mu\sigma \frac{\partial \vec{E}}{\partial t} + \mu\varepsilon \frac{\partial^2 \vec{E}}{\partial t^2} \tag{1.18}$$

The wave equation in terms of the magnetic field vector $\vec{H}$ is obtained as

$$\nabla^2 \vec{H} = \mu\sigma \frac{\partial \vec{H}}{\partial t} + \mu\varepsilon \frac{\partial^2 \vec{H}}{\partial t^2} \tag{1.19}$$

In general, we can write

$$\nabla^2 \vec{f} = \mu\sigma \frac{\partial \vec{f}}{\partial t} + \mu\varepsilon \frac{\partial^2 \vec{f}}{\partial t^2} \tag{1.20}$$

Comparing these equations with the wave equations, obtained for the cases of vacuum and dielectric medium, we find that an additional term appears because of the presence of the conductivity σ. This term is known as the dissipative term, since it allows the current to flow through the medium.

The general Eq. (1.20) in the one-dimensional case can be written as

$$\frac{\partial^2 \vec{f}}{\partial z^2} = \mu\sigma \frac{\partial \vec{f}}{\partial t} + \mu\varepsilon \frac{\partial^2 \vec{f}}{\partial t^2} \tag{1.21}$$

The plane wave solution of this equation is given by

$$\vec{f}(z,t) = \vec{f}_0 e^{i(kz-\omega t)}$$

Putting this solution in Eq. (1.21) gives

$$k^2 = \mu\varepsilon\omega^2 + i\mu\sigma\omega \tag{1.22}$$

This equation follows that k is a complex quantity. Writing $k = k_r + ik_i$ and using it in this equation gives

$$k_r = \omega\sqrt{\frac{\mu\varepsilon}{2}\left[\sqrt{1+(\sigma/\omega\varepsilon)^2}+1\right]} \text{ and } k_i = \omega\sqrt{\frac{\mu\varepsilon}{2}\left[\sqrt{1+(\sigma/\omega\varepsilon)^2}-1\right]}$$

The real part k_r determines the wave propagation characteristics, i.e. wavelength $\lambda = 2\pi/k_r$, phase velocity $v_\phi = \omega/k_r$, and refractive index $\tilde{n} = c/v_\phi = ck_r/\omega$. On the other hand, the imaginary part k_i is attached to the amplitude f_0 as the solution for the complex wave number k. The solution $\vec{f} = \vec{f}_0 e^{i(kz-\omega t)}$ reads

$$\vec{f}(z,t) = \vec{f}_0\, e^{-k_i z} e^{i(k_r z-\omega t)} \tag{1.23}$$

It is inferred from the above equation that when an EM wave propagates through the conductor, attenuation of its amplitude takes place, leading to the exponential decay in amplitude with distance. The distance through which the amplitude falls to the value $1/e$ times of its original value is known as skin depth. An EM wave can travel through a few skin depths only. The skin depth (say δ) can be calculated based on the fact that the amplitude is $\frac{1}{e} f_0$ at $z = \delta$, i.e.

$$\frac{f_0}{e} = f_0 e^{(-k_i\delta)}. \text{ This gives } \delta = \frac{1}{|k_i|}$$

For good conductors, $\sigma \gg \omega\varepsilon$ and hence, $k_i \approx \sqrt{\pi f \mu \sigma}$. So, the skin depth reads

$$\delta = \frac{1}{\sqrt{\pi f \mu \sigma}} \tag{1.24}$$

The skin depth is small at the high frequencies. The skin effect causes the effective resistance of the conductor to increase at higher frequencies where the skin depth is smaller, thus reducing the effective cross-section of the conductor. For a typical metallic conductor such as copper, whose electrical conductivity at room temperature is about 6×10^7 $(\Omega\text{-m})^{-1}$, the skin depth is given by

$$\delta \approx \frac{6}{\sqrt{f(Hz)}} \text{ cm}$$

It follows that the skin depth is about 6 cm at frequency of 1 Hz, but it is only about 2 mm at 1 kHz. At the frequency of 10^{16} Hz, the skin depth is about 0.6 nm. Hence, by the *skin effect,* an oscillating EM signal of increasing frequency transmitted along a copper wire is confined to an increasingly narrower layer on the surface of the wire.

1.4 Electromagnetic Wave Propagation in Plasma

The average charge neutrality of plasmas can be attributed to electrons' inclination to spend greater time in the vicinity of positive ions (or atoms) than away from them. Any perturbation to electrons from a uniform background of ions (mass of ions is quite large, so their motion is neglected) trigger the buildup of an electric field that will tend to restore the charge neutrality of the plasma by pulling the electrons back to their original positions. The electrons, however, overshoot the equilibrium position due to inertia, resulting in their oscillation around the equilibrium position with a characteristic frequency known as 'plasma frequency'. The plasma frequency in terms of plasma density n_0 can be given as:

$$\omega_P = \left(\frac{n_0 e^2}{\varepsilon_0 m}\right)^{1/2} \text{ rad/s}$$

Here, m and e are, respectively, the mass and charge of an electron.

The magnitude of plasma frequency is usually very high owing to the smallness of electron mass. We can use the following approximate formula for plasma frequency f_p for numerical purpose

$$f_p = \frac{\omega_p}{2\pi} \approx 9\sqrt{n_0} \text{ Hz, with } n_0 \text{ in m}^{-3}$$

Plasma waves concerning the oscillations of the electrons can be excited by different means. When a high frequency EM wave passes through the plasma, only the electrons respond to the oscillatory field of the wave. The plasma density or the number density of electrons plays a major role in allowing the propagation of EM wave of frequency ω (Malik and Aria 2010; Tomar and Malik 2013; Malik 2013, 2014). An incident EM wave having a frequency lesser than the plasma frequency ($\omega < \omega_p$) will be unable to propagate through the plasma and will be totally reflected. This cutoff is modified when an external magnetic field is applied and that too depends on the directions of applied magnetic field.

The propagating and non-propagating situations can be explained based on the refractive index or dielectric constant of the plasma that shows a dependence on the frequency of the incident wave. The imaginary nature of refractive index also implies that the associated energy flux will be zero which will prove a total reflection of any low frequency wave incident on plasma. Nonlinear effects are also associated with the nature of refractive index. Plasma is inherently a nonlinear medium and these nonlinear effects also occur when a large amplitude plasma wave is excited by an external means; for example, by an application of oscillating potential to a grid in plasma or external injection of electron beams.

If we represent the oscillating electric field by $\vec{E}'$, and oscillating magnetic field by $\vec{B}'$, then based on the vanishing and non-vanishing $\vec{B}'$ and the external magnetic field by $\vec{B}_0$, we can analyze the nature of the wave (Chen 1984). For example, the wave is electrostatic when the oscillating magnetic field $\vec{B}'$ is zero and EM otherwise. This can be shown based on Maxwell's third equation, i.e. $\vec{\nabla} \times \vec{E} = -\frac{\partial \vec{B}}{\partial t}$, for the field oscillations of frequency ω and wave number k behaving as $\sim \exp i(\vec{k} \cdot \vec{r} - \omega t)$ or $\exp i(\omega t - \vec{k} \cdot \vec{r})$. Then, we have

$$\vec{k} \times \vec{E}' = \omega \vec{B}' \tag{1.25}$$

For a longitudinal wave $\vec{k} \parallel \vec{E}'$ and $\vec{k} \times \vec{E}'$ vanishes, and the wave is also said to be electrostatic. If the wave is transverse, $\vec{k} \times \vec{E}'$ is finite, and the oscillating magnetic field $\vec{B}'$ is finite and the wave is EM in nature. The terms 'parallel' and 'perpendicular' are used to describe the direction of $\vec{k}$ with respect to that of the undisturbed magnetic field $\vec{B}_0$. The terms 'longitudinal' and 'transverse' are used to describe the direction of the oscillating electric field $\vec{E}'$ with respect to that of $\vec{k}$. 'R' and 'L' are used for denoting right-hand polarization and left-hand polarization, respectively, and the corresponding waves are called R wave and L wave. An electric field vector for the R wave will be rotating clockwise in time as viewed along the direction of $\vec{B}_0$ and anticlockwise for the L wave. The term 'extraordinary wave' is used to describe EM waves propagating in the presence of a magnetic field applied perpendicular to the direction of oscillatory field $\vec{E}'$, and 'ordinary wave' is the wave that remains unaffected by the external magnetic field.

In case of 'transverse waves' with $\vec{k} \perp \vec{E}'$ the wave could be an ordinary wave, i.e. $\vec{E}' \parallel \vec{B}_0$ or extraordinary wave where $\vec{E}' \perp \vec{B}_0$. In case of ordinary waves, the electric field causes the electrons to move parallel to $\vec{B}_0$, i.e. $\vec{v} \times \vec{B}_0$ vanishes and the wave shows the same behavior as if it is propagating in the absence of magnetic field. In contrast, in the case of extraordinary waves, the electron motion will be affected by $\vec{B}_0$, thus changing the dispersion relation. These waves tend to be elliptically polarized instead of being plane polarized. The electric vector will be elliptically polarized due to an additional component generated by the electrons' motion in the direction of propagation. So this wave is partly transverse and partly longitudinal. It loses its EM character at resonance and the wave turns to have only the electrostatic oscillations.

The motion of the ions in response to the electric field of the wave is ignored for simplicity, while studying the electron waves. In the presence of an external magnetic field, the waves concerning the low frequency ion oscillations are of two types, namely, the hydromagnetic wave (also known as the Alfven wave) propagating along $\vec{B}_0$, and the magnetosonic wave propagating across $\vec{B}_0$.

1.4.1 Absence of Magnetic Field $\vec{B}_0$: Ordinary Waves

In plasma, the current with density $\vec{J}$ is generated because of the electric field of the EM wave. The fourth Maxwell's equation reads for the oscillating fields $\vec{E}'$and $\vec{B}'$and the current density $\vec{J}'$

$$c^2\vec{\nabla}\times\vec{B}' = \frac{\vec{J}'}{\epsilon_0} + \frac{\partial \vec{E}'}{\partial t}$$

For the oscillations of kind $e^{i(\vec{k}\cdot\vec{r}-\omega t)}$, we obtain the following wave equation with the use of Maxwell's equations

$$-\vec{k}\left(\vec{k}\cdot\vec{E}'\right) + k^2\vec{E}' = \frac{i\omega}{\epsilon_0 c^2}\vec{J}' + \frac{\omega^2}{c^2}\vec{E}' \tag{1.26}$$

In case of the transverse waves, $\vec{k}\cdot\vec{E}' = 0$, and for light waves or the microwaves where frequency of the incident EM waves is very high, the ions can be considered as fixed, and the current density $\vec{J}'$ is contributed entirely by the motion of the electrons. This can be written as

$$\vec{J}' = -n_0 e\vec{v}_e'$$

Substituting the value of $\vec{v}_e'$ from the linearized equation of motion for the electrons, $m\frac{\partial \vec{v}_e'}{\partial t} = -e\vec{E}'$, i.e. $\vec{v}_e' = \frac{e\vec{E}'}{i\omega m}$ in $\vec{J}'$ and then substitute the expression of $\vec{J}'$ in Eq. (1.26), we obtain the following dispersion relation of the wave

$$\omega^2 = \omega_p^2 + k^2c^2 \tag{1.27}$$

The physics behind this dispersion relation can be understood as follows. If we increase the plasma density n_0, the plasma frequency ω_p will increase (wave number k will decrease in order to satisfy the dispersion relation) and a situation may arise at a particular density n_c where $\omega_p = \omega$. Under this situation, $k = 0$ and $\lambda = \frac{2\pi}{k} = \infty$. This is called cutoff for the wave and the wave is reflected back. The wave propagates only when k is real, i.e. when $\omega > \omega_p$. For the density $n_0 > n_c$, the wave number k becomes imaginary and hence, the wave amplitude is exponentially attenuated. In this situation, we can talk about the skin depth, i.e. the distance travelled by the wave when its amplitude reaches a value equal to $1/e$ times of its original value. This particular density below which the wave propagation is possible and above which the wave is attenuated is called the critical density.

1.4.2 Perpendicular Magnetic Field ($\vec{k} \perp \vec{B}_0$): Extraordinary Waves

In the case of the extraordinary waves, the oscillating electric field, $\vec{E}'$ is perpendicular to the external magnetic field $\vec{B}_0$. The magnetic field $\vec{B}_0$ also influences the electron motion in the directions perpendicular to the electric field $\vec{E}'$ and $\vec{B}_0$. So, a component of electric field perpendicular to $\vec{E}'$ is also generated. If we take $\vec{B}_0 = B_0\hat{x}$ and $\vec{k} = k\hat{y}$, then the field $\vec{E}'$ will have y- as well as z-components. The linearized equation of motion of electrons

$m\left(\frac{d\vec{v}'}{dt}\right) = -e\vec{E}' - e\vec{v}' \times \vec{B}_0$, under this configuration shall give

$$v_y = \frac{e}{m\omega}\left(-iE_y + \frac{\omega_c}{\omega}E_z\right) \Big/ \left(1 - \frac{\omega_c^2}{\omega^2}\right)$$

$$v_z = \frac{e}{m\omega}\left(-iE_z - \frac{\omega_c}{\omega}E_y\right) \Big/ \left(1 - \frac{\omega_c^2}{\omega^2}\right)$$

together with $\omega_c = eB_0/m$ as the electron cyclotron frequency. Keeping the longitudinal term, the wave equation is written as:

$$\left(\omega^2 - c^2k^2\right)\vec{E}' + c^2kE_z\vec{k} = -\frac{i\omega}{\epsilon_0}\vec{J}' \quad (1.28)$$

The current density $\vec{J}'$ will have y- and z-components in view of the velocity components v_y and v_z. Separating out the wave Eq. (1.28) into its y- and z- components, we get

$$\omega^2 E_z = \frac{i\omega n_0 e}{\epsilon_0}\frac{e}{m\omega}\left(-iE_z - \frac{\omega_c}{\omega}E_y\right) \Big/ \left(1 - \frac{\omega_c^2}{\omega^2}\right)$$

$$(\omega^2 - c^2k^2)E_y = -\frac{i\omega n_0 e}{\epsilon_0}\frac{e}{m\omega}\left(iE_y - \frac{\omega_c}{\omega}E_z\right) \Big/ \left(1 - \frac{\omega_c^2}{\omega^2}\right)$$

These two simultaneous equations for E_y and E_z become compatible only if the determinant of the coefficients vanishes. This condition finally gives

$$\frac{c^2k^2}{\omega^2} = \frac{\omega^2 - \omega_h^2 - \left[\left(\omega_p^2\omega_c/\omega\right)^2 / \left(\omega^2 - \omega_h^2\right)\right]}{\omega^2 - \omega_c^2} \quad (1.29)$$

Here we have defined upper hybrid frequency ω_h as $\omega_h^2 = \omega_p^2 + \omega_c^2$. This equation yields the dispersion relation for extraordinary wave as

$$\frac{c^2k^2}{\omega^2} = \frac{c^2}{v_\varphi^2} = 1 - \frac{\omega_p^2}{\omega^2}\frac{\omega^2 - \omega_p^2}{\omega^2 - \omega_h^2} \quad (1.30)$$

As the wave propagates through a region in which ω_p and ω_c are changing, it may encounter cutoffs and resonances. The reflection of the wave takes place at the cutoff $(k = 0)$ and it is absorbed at the resonance $(k = \infty)$. In case of extraordinary wave, the resonance is found at a point in the plasma, where $\omega_h = \omega$. In the situation, the wavelength of the wave and hence, its phase velocity becomes zero and the refractive index becomes infinite as $\tilde{n} = ck/\omega$ in the plasma. So, the wave energy gets absorbed and the upper hybrid oscillations are excited in the plasma. On the other hand, at the cut-off the wavelength becomes infinite and the refractive index approaches zero.

At $k = 0$ in Eq. (1.30), we obtain two different kinds of cutoff frequencies. We shall call these as ω_R and ω_L in view of their respective higher and lower magnitudes. These are known as right-hand and left-hand cutoffs, respectively. The following expressions are obtained for these frequencies

$$\omega_R = \frac{1}{2}\left[\omega_c + \sqrt{(\omega_c^2 + 4\omega_p^2)}\right] \tag{1.31}$$

$$\omega_L = \frac{1}{2}\left[-\omega_c + \sqrt{(\omega_c^2 + 4\omega_p^2)}\right] \tag{1.32}$$

Based on the magnitudes of various frequencies, the dispersion diagram is divided into regions of non-propagation and propagation by the cutoff and resonance frequencies. The frequencies in their increasing magnitudes read as $\omega_L, \omega_P, \omega_h$ and ω_R. There is no propagation possible between $\omega = \omega_R$ and $\omega = \omega_h$ since ω^2/k^2 is negative. Between $\omega = \omega_h$ and $\omega = \omega_L$, the propagation is again possible. The resonance is at $\omega = \omega_h$, where the phase velocity v_ϕ approaches zero. In the region of wave propagation, the velocity will be faster or slower than c depending on whether ω is smaller or larger than ω_p. The wave will propagate at the speed of light c at $\omega = \omega_p$, i.e. the velocity is less than c for $\omega > \omega_p$, but it is larger than c for $\omega < \omega_p$, and the region of propagation for $\omega > \omega_R$. There is another region of non-propagation for $\omega < \omega_L$. The wave propagates at the speed of light when its frequency is much higher, i.e. when $\omega \gg \omega_R$.

1.4.3 Parallel Magnetic Field ($\vec{k} \parallel \vec{B}_0$): R and L Waves

Now let $\vec{k}$ and $\vec{B}_0$ lie along the x-axis $\vec{k} = k\hat{x}$, $\vec{B} = B_0\hat{x}$ and allow $\vec{E}'$ to have both y-and z-transverse components. Similar mathematical treatment of the Maxwell's equations yields the wave equation whose y- and z-components are given as

$$\left(\omega^2 - c^2k^2\right)E_y = \frac{\omega_p^2}{1 - \omega_c^2/\omega^2}\left(E_y + \frac{i\omega_c}{\omega}E_z\right) \tag{1.33}$$

$$\left(\omega^2 - c^2k^2\right)E_z = \frac{\omega_p^2}{1 - \omega_c^2/\omega^2}\left(E_z - \frac{i\omega_c}{\omega}E_y\right) \tag{1.34}$$

Setting the determinant of the coefficients to zero, we have $\left(\omega^2 - c^2k^2\right) = \frac{\omega_p^2}{1 \pm \omega_c/\omega}$. Hence, there are two different waves that can propagate along $\vec{B}_0$. Their dispersion relations read as

$$(\text{R wave})\frac{c^2k^2}{\omega^2} = 1 - \frac{\omega_p^2/\omega^2}{1 - (\omega_c/\omega)} \tag{1.35}$$

$$(\text{L wave})\frac{c^2k^2}{\omega^2} = 1 - \frac{\omega_p^2/\omega^2}{1+(\omega_c/\omega)} \tag{1.36}$$

In case of an R wave, the wave is in resonance ($k = \infty$) with the cyclotron motion of the electrons at $\omega = \omega_c$. The direction of rotation of the plane of polarization and the direction of gyration of the electrons for this wave are the same, causing the wave to lose energy in continuous acceleration of the electrons and making it incapable of propagating. This wave has a cutoff at $\omega = \omega_R$ and a stop-band between ω_R and ω_c and below ω_c, i.e. when $\omega < \omega_c$ and it propagates at a velocity less than the speed of light c. The wave in this low frequency region is called the whistler mode. The whistler waves are extremely useful in the study of ionospheric phenomena. The L wave, on the other hand, has no resonance with the cyclotron motion of the electrons and has a lower cutoff frequency ω_L compared with the cutoff frequency ω_R of the R wave. This wave could have been in resonance with the motion of ions under the action of magnetic field B_0. Since we have neglected this motion, the term revealing the resonance is not appearing in the dispersion relation of the L wave.

1.4.4 Parallel Magnetic Field ($\vec{k} \parallel \vec{B}_0$): Alfven Waves

Alfven waves in plasma were first generated and detected by Allen, Baker, Pyle, and Wilcox at Berkeley, California, and by Jephcott in England in 1959. The Alfven wave has $\vec{k} \parallel \vec{B}_0$ (say, in y-direction in plane geometry), $\vec{E}'$ and $\vec{J}'$ are perpendicular to $\vec{B}_0$ where $\vec{E}'$ has components in the z-direction, $\vec{B}'$ and $\vec{v}'$ are perpendicular to both $\vec{B}_0$ and $\vec{E}'$. In this case, where low frequencies are considered and thermal motions are ignored, the current density $\vec{J}'$ is generated jointly by both the ions and the electrons. The Alfven velocity v_A is the constant velocity with which the hydromagnetic wave travels along $\vec{B}_0$. This is a characteristic velocity at which the perturbations of the magnetic field's lines travel. The interesting feature is that the whole plasma appears to be attached with the magnetic lines and oscillates with them. The Alfven velocity is given by

$$\vec{v}_A = \vec{B}_0 / (\mu_0 \rho)^{\frac{1}{2}} \tag{1.37}$$

where, $\rho = n_0 M$ is the mass density of the ions.

Since we are considering low frequencies, the current density $\vec{J}'$ will now have contributions from both the ions and the electrons. The z-component of the wave equation is nontrivial, as the electric field $\vec{E}'$ is along the z-axis. Thus, the wave equation reads

$$\varepsilon_0 \left(\omega^2 - c^2k^2\right) E_z' = -i\omega n_0 e \left(v_{iz} - v_{ez}\right) \tag{1.38}$$

Here v_{iz} and v_{ez} are the z-components of the velocities of the ions and the electrons.

Thermal motions are insignificant for this wave and the ion equation of motion without pressure gradient force may, therefore, be used. The components v_{iz} and v_{iy} are so obtained as

$$v_{iz} = \frac{ie}{M\omega} E_z' \Big/ \left(1 - \frac{\Omega_c^2}{\omega^2}\right)$$

$$v_{iy} = \frac{e}{M\omega} \frac{\Omega_c}{\omega} E_z' \Big/ \left(1 - \frac{\Omega_c^2}{\omega^2}\right)$$

where Ω_c is the ion cyclotron frequency given by eB_0/M. Based on these expressions, we can obtain the velocity components for the electrons by letting $M \to m,\ e \to -e,\ \Omega_c \to -\omega_c$ and then taking the limit $\omega_c^2 \gg \omega^2$. In this limit, the electrons are found to have simply an $\vec{E}' \times \vec{B}_0$ drift, i.e. $v_{ey} = -{}^{E_z'}/_{B_0}$. Hence, we obtain from Eq. (1.38)

$$\left(\omega^2 - c^2k^2\right)E_z' = \frac{\Omega_P^2}{\left(1 - \frac{\Omega_c^2}{\omega^2}\right)} E_z' \tag{1.39}$$

On making a further assumption of $\Omega_c^2 \gg \omega^2$, in view of the fact that the hydromagnetic waves have frequencies well below the frequency of ion cyclotron resonance. The above equation in this limit gives

$$\frac{\omega^2}{k^2} = \frac{c^2}{1 + \left(\rho\mu_0/B_0^2\right)c^2}$$

This is generally seen that $\frac{\rho\mu_0}{B_0^2} \gg 1$. Thus, the phase velocity of the Alfven wave or the Alfven velocity can be given as

$$v_A = \frac{\omega}{k} = \frac{B_0}{\sqrt{\mu_0\rho}} \tag{1.40}$$

1.4.5 Perpendicular Magnetic Field ($\vec{k} \perp \vec{B}_0$): Magnetosonic Waves

A magnetosonic wave is a wave in which plasma is compressed and released due to the oscillations produced by $\vec{E}' \times \vec{B}_0$ drifts in the direction of $\vec{k}$ but across $\vec{E}'$ and $\vec{B}_0$. In view of this, we may take $\vec{B}_0 = B_0\hat{x}$, $\vec{E}' = E'\hat{y}$ and $\vec{k} = k\hat{z}$, and also pressure gradient term $\left(\vec{\nabla}p\right)$ needs to be included in the equation of motion. We consider the ion temperature as T_i, the electron temperature as T_e and the specific heat ratios as γ_i and γ_e for the ions and the electrons. For the ions, we have

$$Mn_0\frac{\partial \vec{v}_i'}{\partial t} = en_0\left(\vec{E}' + \vec{v}_i' \times \vec{B}_0\right) - \gamma_i KT_i\vec{\nabla}n'$$

Keeping in mind the directions of $\vec{E}'$ and $\vec{B}_0$, we obtain from this equation

$$v_{iy} = \frac{\frac{ie}{M\omega}E_y}{\left(1 - \frac{\frac{\Omega_c^2}{\omega^2}}{1 - \frac{k^2}{\omega^2}\frac{\gamma_i KT_i}{M}}\right)} \tag{1.41}$$

In order to obtain v_{ey}, we make appropriate changes in v_{iy} and take the limit of small electron mass, so that $\omega^2 \ll \omega_c^2$ and $\omega^2 \ll k^2v_{\text{the}}^2$ together with $v_{\text{the}} = \sqrt{{}^{\gamma_e KT_e}/_m}$. This leads to the following expression for v_{ey}

$$v_{ey} = \frac{-ik^2\gamma_e KT_e}{\omega B_0^2 e}E_y \tag{1.42}$$

We make use of v_{iy} and v_{ey} in the wave equation $\varepsilon_0(\omega^2 - c^2k^2)E_y = -i\omega n_0 e(v_{iy} - v_{ey})$

and also assume $\omega^2 \ll \Omega_c^2$ to obtain the dispersion relation

$$\frac{\omega^2}{k^2} = c^2 \frac{v_S^2 + v_A^2}{c^2 + v_A^2} \tag{1.43}$$

Here v_s is the acoustic speed, given by

$$v_s = \sqrt{\frac{\gamma_e K T_e + \gamma_i K T_i}{M}}$$

In view of this, the magnetosonic wave is also called an acoustic wave. In the absence of magnetic field $\vec{B}_0$, $v_A = 0$ and the dispersion relation takes the form of that of acoustic wave and hence, the magnetosonic wave becomes an ordinary ion acoustic wave. However, in the absence of ion and electron temperatures, $v_s = 0$, and the dispersion relation carries only v_A terms along with c. Under this situation, $\omega/k > v_A$ and the wave is said to be a fast hydromagnetic wave.

1.5 Laser: A Source of EM Radiation

Existence of EM radiation can be perceived in nature and its effects can be realized and utilized in our daily life from short wavelength TV rays, microwaves in household to X-rays in medical industry. These all radiations fall under the EM spectrum, in which the most important one is the human visible range (i.e. the light) which falls between 380 nm (approximately violet) and 780 nm (approximately red color). Other rays such as gamma, ultraviolet (UV) and infrared (IR) also have their own applications in different areas of science and technology. For example, gamma rays are used to treat malignant tumors in radiotherapy. In hospitals, UV lamps are used to sterilize surgical equipment and the air in operating theatres. Suitable doses of UV rays cause the body to produce vitamin D. In view of their wavelength close to that of visible light, the applications of IR radiation can be somewhat similar to the visible light. Near-IR rays are used in electronic applications such as TV remote sensors and photography.

Laser light is somehow different form of light that covers a much wider portion of EM spectrum (Figure 1.1). The invention of laser has uncovered many phenomena which were not observable with the conventional sources of ordinary light of intensity that could hardly influence the electrons of the atoms. The revolutionary invention of laser by Maiman in 1960 led to the development of a new era of light which further paved the way to many fascinating phenomena and techniques. The term LASER originated as an acronym for 'light amplification by stimulated emission of radiation'. Laser is a highly intense, tunable and coherent source of light which spans almost the whole EM spectrum and hence, it has many applications in diversified areas depending on the wavelength of light emitted.

The operation of a laser involves two major phenomena, namely amplification and the stimulated emission of radiation / light.

When the energy is supplied to the atom, the atom goes to an excited state and then returns to its ground state via two processes, either 'spontaneous emission' or 'stimulated emission'. When an atom comes down to the ground state on its own (as the lifetime of a particle in the excited state is very small ~10^{-9} sec), the emission is known as spontaneous

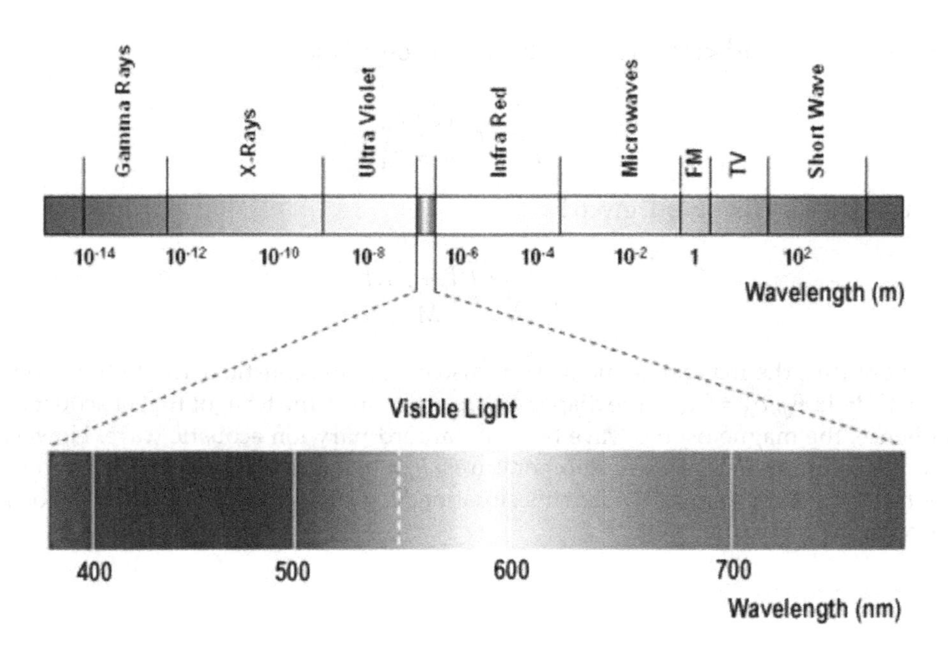

FIGURE 1.1
EM spectrum.

emission. The simplest example of such emission is the filament bulb at our homes, where the emitted light is due to the relaxation of excited atoms to the ground state via spontaneous emission. During this process the packets of optical energy are released in the form of photons. The photons so radiated are independent of each other and there is no definite phase relationship between them and hence, the light so generated is incoherent. This is one of the important characteristics which differentiates the ordinary light from the laser light.

Einstein proposed that excited atoms could also produce light by converting the stored energy into light through stimulated emission. The photon emitted during the spontaneous emission may interact with the excited atoms and stimulates them to emit photons. In such a way two photons are produced and these two photons may interact with other excited atoms to produce four photons. Hence, this process is multiplicative and most importantly the photons so emitted have identical wavelength, same direction, phase and state of polarization. With these characteristics, the light so produced is coherent, which keeps on amplifying if the sufficient numbers of excited atoms are present.

If somehow one is able to excite the large number of atoms to a high energy state from the ground state (known as a population inversion), then the specific photon (stimulus) may trigger an avalanche of stimulated photons that are all emitted in phase and in the same direction. In a laser, the stimulating radiation is generated as a result of feedback within a resonant optical cavity. A laser cavity practically, consists of two mirrors, one is a fully reflecting mirror and the other is a partially reflecting front mirror. It means the laser light can be emitted out from the partially reflecting mirror, which is providing the feedback required for laser action. When the laser is illuminated by an external energy source such as a flash lamp or diode laser, the spontaneous photon emission also takes place, such photons travel back and forth and get amplified depending on the configuration of the resonator and the optical characteristics of the active medium. This results in emission of photons taking place in the same direction and preserving the definite phase relationship.

1.6 Various Laser Operations

Lasers can be operated in different modes and are categorized on the basis of output of a laser. For example, continuous wave (CW) operation and pulsed operation are the two basic modes of operations.

1.6.1 CW Operation

In this mode of operation, laser emits an uninterrupted beam of light continuously with a stable output. The output wavelength emitted by laser depends on the characteristics of the active medium. The first CW laser invented was a helium-neon (He-Ne) gas laser operating at 1153 nm. Afterwards a number of CW lasers were developed like solid-state lasers (Nd:YAG laser, semiconductor lasers), dye lasers and other gas lasers. However, in many lasers, CW operation is difficult to achieve due to low gain. The output power of a CW laser remains constant for long durations of time, but power variations may occur due to mode beating, if single-frequency operation is not achieved. In this mode, the laser's output remains constant if a pumping source is properly maintained so that the population inversion required for lasing action is also continually maintained. Since the output is continuous, this mode of operation is also termed as free-running mode.

Many types of lasers can be made to operate in CW mode for their certain applications. Many of these lasers actually lase in several longitudinal modes at the same time, and beat between the slightly different optical frequencies of those oscillations. For example, Nd:YAG laser lase between 1047 nm and 1064 nm. Certain lasers are also termed as CW lasers due to the fact that their output power is constant when averaged over lengths of time periods.

The major requisite of the CW laser is that its power should not fluctuate and should remain stable for long time durations, i.e. for hours or weeks and also for short intervals of times, i.e. for milliseconds to microseconds, as per requirement of different applications. This stability is achieved by using microprocessor control units which even overcome changing environment conditions like surrounding temperature, stray vibration and so forth. Additional circuitry servers may be required to maintain the stable output, which may get disturbed by various factors like temperature and misalignment of the resonant mirrors. CW lasers are used in many applications that require high processing speeds (such as high-speed seam welding applications), as these can penetrate the target to desired depths.

Every laser is specific with a linewidth that depends on factors like gain bandwidth, framework of optical resonators and other elements such as filters and etalons, used to narrow the linewidth. A single laser line includes a range of wavelengths. The wavelength of the output beam can be determined by longitudinal modes of the cavity as per the relation

$$n\lambda = 2L \tag{1.44}$$

where, n is an integer known as mode number, λ is the wavelength and L is the length of the cavity. This is usually very large number as wavelength (λ) is a lot smaller than the cavity length. The physical meaning of this relation is that the round trip distance inside the cavity must be an exact multiple of λ. The wavelengths satisfying the above condition are called longitudinal cavity modes. The output of a CW laser is defined by the overlap

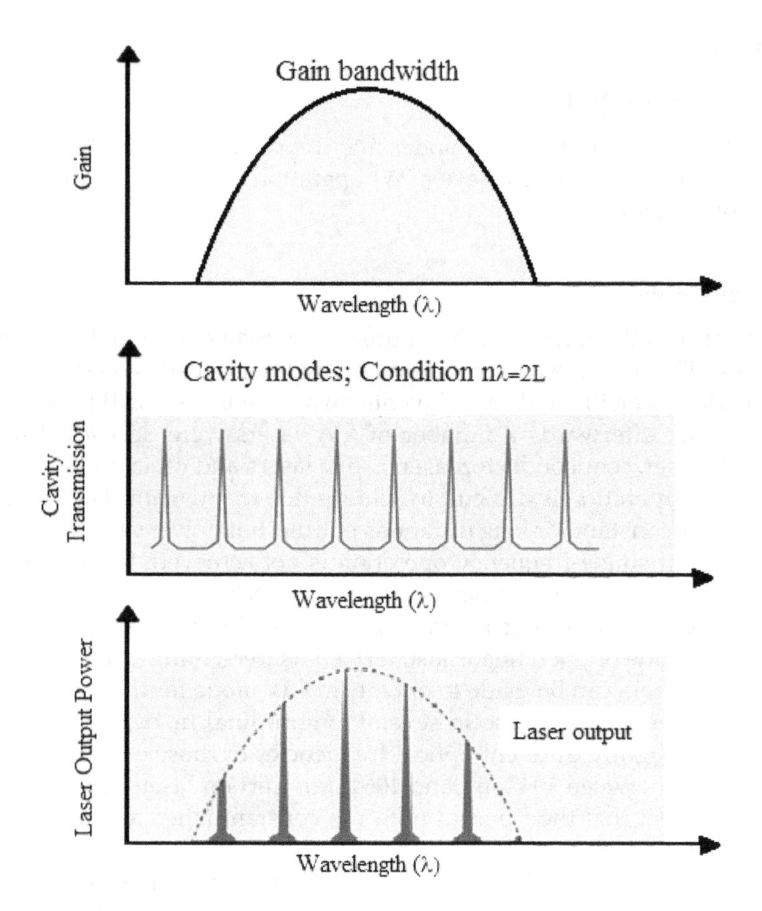

FIGURE 1.2
Gain bandwidth, cavity modes and output of the laser.

of the gain bandwidth and these resonant cavity modes, as shown in Figure 1.2. It means the output wavelength of the laser is the one that corresponds to the cavity modes falling within the gain bandwidth.

1.6.2 Quasi-Continuous Wave Operation

In some lasers, strong heating of the gain medium is caused in CW operation mode. Therefore, to minimize the heating of the active medium, another technique is used which is known as quasi-continuous wave operation. In this technique, the pump power is only switched on for short intervals of time. It means the pumping is done for certain time intervals, which are short enough to reduce heating effects efficiently, but intense and long enough to initiate and sustain laser process. Therefore, the laser is practically in the state of CW mode only (Figure 1.3), where the thermal effects such as thermal lensing and damage through overheating are reduced with the help of geometrical components like mirrors and filters of high damage threshold. Because of less thermal lensing effects, these are appropriate to provide high quality intense laser output. An important point is that the quasi-continuous wave operation is also a power saving mode, as it provides higher output peak power by consuming lower input average powers.

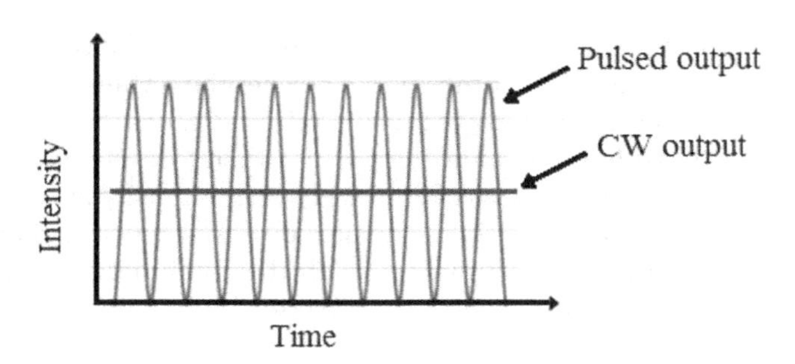

FIGURE 1.3
Pulsed and continuous wave lasers.

1.6.3 Pulsed Operation

In this mode of operation, the output of the laser is in the form of pulses that vary with time like sine and cosine waveforms (Figure 1.3). Here, all the energy is dumped out in a single pulse which may last for nanoseconds to picoseconds. Any operation of a laser which is not continuous may be called the pulsed operation of lasers. In other words, operation of a laser that does not provide continuous output and the output power appears in the form of pulses of some duration at a certain repetition rate is termed as pulsed operation mode. This mode of operation is utilized in numerous applications, covering a range of technologies. For example, in pulsed laser deposition (PLD) technique, short laser pulses are utilized to create plasma from targeted raw material and after tailoring the properties of this plasma it is transferred to the component to be coated. Excimer laser is a pulsed laser which is used in synthesizing the microelectromechanical systems (MEMS) as well as in high resolution photolithography.

PLD technique enables to perform high quality (with defects) materials to almost every substrate with high quality adhesive properties which do not break or delaminate during or after coating. Contrary to this is the laser ablation method, in which a material is removed from a solid surface with the help of a pulsed laser where the CW laser fails to accomplish the desired task. Excimer lasers are mainly used for both the techniques. They deliver the high output power and pulse energy in the UV regions. The average power output typically lies between 10 to 300 W. Pulsed lasers, like the Nd:YAG laser, are used for drilling, spot welding, remote sensing, surgery and for implantable medical devices or batteries. The CO_2 laser is used for industrial cutting, whereas the chemical lasers are used for airborne weapons.

Different kinds of modes are associated with pulse operation, which include Q-switching, mode locking, and pulsed pumping.

1.6.3.1 Q-switching

Q-switching is a method used for the production of giant / energetic pulses from lasers. It is a process in which a laser is prevented to produce small pulses for some duration. This is done by closing the cavity and allowing the energy to be built inside the cavity to a certain limit and thereby releasing at once so that all the concentrated energy gives rise to an extremely giant pulse of short duration as a pulsed output beam.

In simpler terms, using Q-switching in the laser assembly means disrupting the optical cavity of the laser and preventing the formation of laser beam by some means, which could be external or internal. As the disruptive action is stopped, the laser pulse is released. In a Q-switched laser, the cavity conditions are made unfavourable for lasing action by choosing an appropriate method and in the meantime, the population inversion is allowed to build up inside. When the stored pump energy reaches the certain desired level, then Q is adjusted to favourable conditions and the pulse is released. Since the cavity Q-factor in this technique is switched low to high value, the process, therefore, is known as Q-switching. With the help of this technique, pulses of extremely high intensity and peak power (about gigawatt) can be produced. Q-switching generates pulses of much higher energies, lower repetition rates, and longer pulse durations. With the combination of Q-switching and mode locking, a constant amplitude output of the laser, both in the form of CW or pulsed, can be achieved.

Q-switching can be done via the following two methods.

Active Q-switching

The mechanical devices like shutters, chopper wheels, spinning mirror or electro-optical devices like Pockels cells, Kerr cell or acousto-optic devices are used to modulate the losses. These devices are located externally, i.e. outside the laser cavity. For this reason it is also known as *external quenching*. Based on the energy stored in the gain medium during the time for which the laser remains Q-switched, we can tailor the output pulse energy and pulse duration.

Passive Q-switching

Passive Q-switching is also known as *self Q-switching*. Here the losses are modulated internally by filling the cavity with a suitable substance which becomes transparent for emission as the intensity of light increases beyond a particular threshold value. For example, Cr:YAG, is used for Q-switching of Nd:YAG lasers. Since passive switching is done by placing suitable saturable absorbers inside a cavity like dyes, this type of quenching is also known as *internal quenching*. Saturable absorber is selected wisely, as it is crucial that this should have a recovery time much longer than the pulse duration. If it is not so, then energy will be needed to keep the absorber saturated during the pulse. So the absorber recovery time should be many times longer than the round trip time of the resonator. In view of less circuitry requirement, this is a cost-effective and simple technique, though the pulse energies are quite lower but can be used where high pulse repetition rates are needed. In this, pulse energy and pulse duration are independent of the pump power.

1.6.3.2 Mode Locking

The technique of producing pulses of extremely short duration typically of the order of picoseconds to femtoseconds is called mode locking. Here laser is active for a very small time interval during which the pulses of picoseconds or femtoseconds durations are emitted. Pulses so produced are separated by the time interval that a pulse takes to complete one to-and-fro movement inside the resonator cavity. As these pulses have small temporal length they spread over wide wavelength range following the principle of uncertainty due to which the output of laser has broad gain profile. In this technique, we get several longitudinal modes lasing in a coherent manner with respect to each other. These modes would form a periodic pulse train emitted from the laser. The variation of laser intensity with time in perfectly locked and random-locked situations is shown in Figure 1.4(a) and (b), respectively. Ti:Sapphire laser used for research applications is a mode-locked tunable pulsed laser.

This is clear that in mode-locked lasers the resonant modes of the optical cavity affect the characteristics of the output beam. Here beating effect is realized due to the interference of different modes as the phases of these different frequency modes are synchronized,

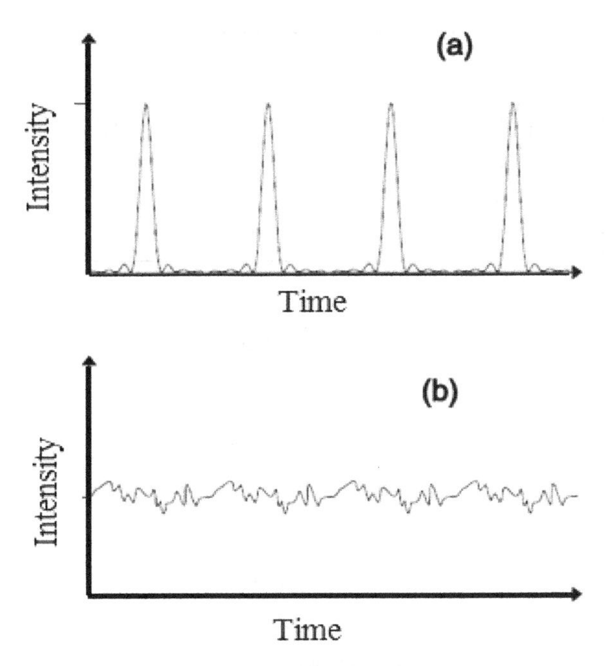

FIGURE 1.4
Variation of laser intensity with time in (a) perfectly locked and (b) random-locked situations.

and finally regularly spaced pulses are obtained as output. Sometimes mode locking technique is applied along with Q-switching. A mode-locked laser generates extremely high peak powers (of the order of 10^{12} W) than Q-switched laser operating under similar conditions. However, Q-switching alone is found to produce a single strong and short pulse of laser radiation, in contrast to the mode locking where we obtain a train of pulses.

Mode locking can also be either active or passive. Active mode locking produces modulated light as it depends on an external source used to stimulate oscillation in the resonant cavity, whereas an internal cavity source itself is used to instigate a change within the cavity in passive mode locking.

Active Mode Locking

Active mode locking technique involves the modulation of the resonator losses periodically or round-trip phase change is achieved with the help of an acousto-optic or electro-optic modulator, as shown by M in Figure 1.5(a). In this figure, GM represents the gain medium. The periodic modulation is synchronized with to-and-fro trips of light inside the cavity. This leads to the generation of ultrashort pulses of picosecond pulse durations. However, different effects such as the limited gain bandwidth sometimes contribute to the shortening or broadening of these pulses.

Temporal evolution of optical power and losses in an active mode locking is shown in Figure 1.5(b). Here it can be seen that a pulse with the 'correct' timing is allowed to pass through the modulator when the losses are at a minimum. However, the trails of these pulses may experience a little attenuation that leads to slight shortening of pulse in each round trip. This pulse shortening is compensated by other effects such as gain narrowing that tend to broaden the pulse. The strength of the modulator signal also decides the time duration of the pulses and this weak dependence arises from the fact that the pulse-shortening effect of the modulator becomes less effective for shorter

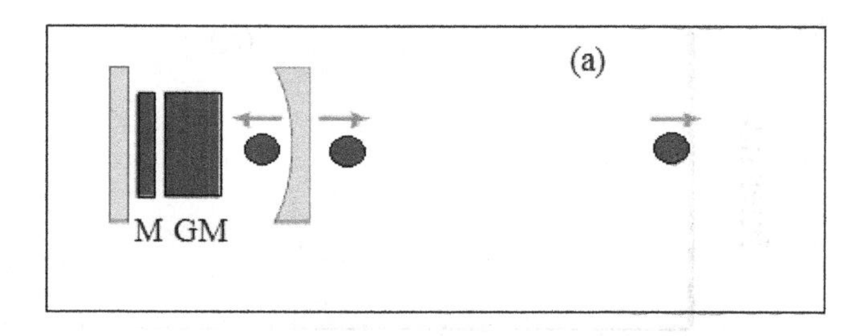

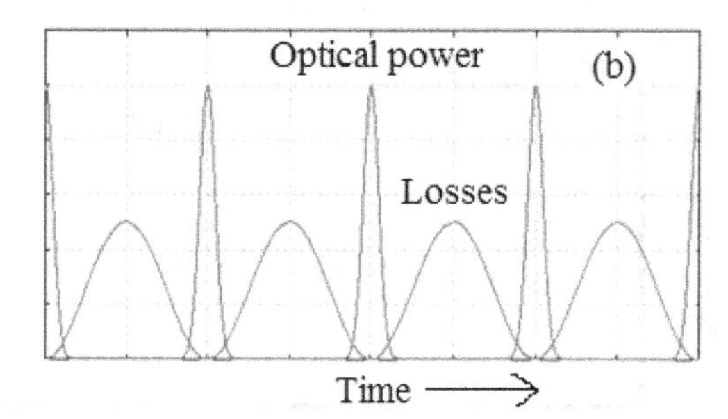

FIGURE 1.5
(a) Active mode locking technique. (b) Temporal evolution of optical power and losses.

pulse durations. In that situation, other effects such as chromatic dispersion become more effective and broaden the pulse.

Passive Mode Locking

Passive mode locking is done by using a suitable saturable absorber [shown by A in Figure 1.6(a)] inside the cavity that allows the generation of pulses of much shorter durations of the order of femtoseconds. This absorber when driven by short pulses can modulate the resonator losses at a much faster rate than an electronic modulator. Much faster modulation takes place, when the pulse becomes shorter provided the absorber has a sufficiently short recovery time. In this case, the pulse duration can be made much shorter than the recovery time of the absorber.

Let us consider that there is a single circulating fundamental frequency pulse inside the cavity and a fast saturable absorber. The pulses saturate the absorption every time when they interact with the saturable absorber and hence, reduce the losses temporarily [Figure 1.6(b)]. A steady state is reached when the laser gain is saturated to a level, which is just sufficient for compensating the losses for the circulating pulse. Since the absorber cannot be saturated by the light of lower intensity, such radiation which hits the absorber at other times experiences losses higher than the gain for the circulating pulse. This way the absorber can suppress any weaker pulses in addition to any continuous background light. Also, the absorber constantly attenuates the leading wing of the circulating pulse. If the absorber can recover sufficiently quickly, it can also attenuate the trailing wing of the pulse, tending to decrease the pulse duration. This effect in the steady state will balance other effects, which tend to broaden the pulse; e.g. chromatic dispersion.

Comparison of Active and Passive Mode Lockings

Active mode locking typically generates longer pulses than passive mode locking, and there is a disadvantage of the need for an optical modulator. However, active mode locking

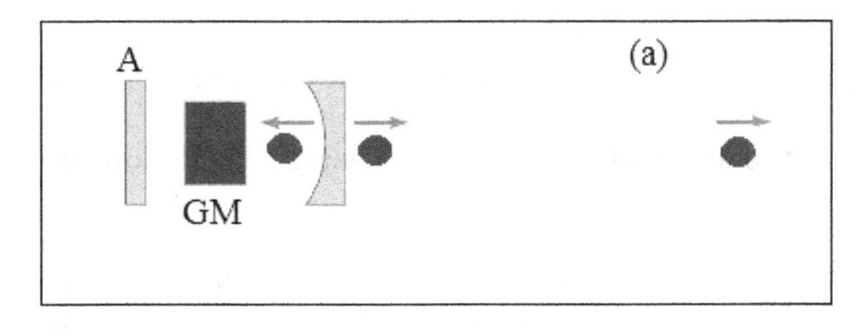

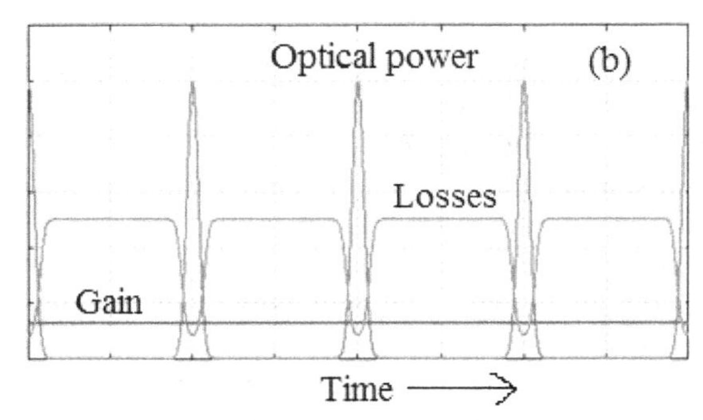

FIGURE 1.6
(a) Passive mode locking. (b) Temporal evolution of optical power and losses.

is a very useful tool for the situation where pulse trains are required synchronized or when many lasers need to be operated in synchronism. For such reasons, active mode locking is mostly used in the context of optical fiber communications.

On the other hand, passive mode locking technique allows the generation of much shorter pulses. This is because a saturable absorber can efficiently adjust the resonator losses at a much faster rate than any electronic modulator. Here loss modulation is realized quickly as the circulating pulse is already very small. This works well for the leading edge of the pulse, as the absorber is bleaching by the time. However, for trailing edge it may or may not work, as the absorber recovery may take a longer time.

1.6.3.3 Pulsed Pumping

Another technique to drive pulsed laser is pulsed pumping, where active material is pumped by a self-pumped source only. The self-pumped source could be a pulsed laser or flash pump producing flashes periodically. Dye laser is pumped by a pulsed laser only because the inverted population lifetime of a dye molecule is so short that a high energy fast pump is required. Hence, large capacitors are charged first and then switched to discharge through the flash lamp. The lasers, such as excimer laser and copper vapour lasers, utilizing the pulsed pumping can never be operated in CW mode.

1.6.4 Comparison of CW and Pulsed Lasers

The average power density for both the continuous (CW) and pulsed wave (PW) lasers is determined by the ratio of power of the beam P and the area A of the laser beam's spot. Hence

$$\text{Average power density} = P / A$$

In the case of CW lasers, P is the total power, whereas in the pulsed laser's case P is the average pulse peak power over the duration of an individual pulse. It means for the pulsed laser

$$\text{Average peak power} = \text{Pulse energy}/\text{Pulse duration}$$

Another parameter, known as the spatial peak power density, represents the maximum power density across the laser beam profile. In the case of top-hat profiles, the spatial peak power density is the same as the average power density. A comparison between a CW and individual pulses obtained with a PW laser provides major information about the interaction of a CW laser and a PW laser with the materials. In this context, the interaction time is the most useful parameter, which can be interpreted as the time at which a specific point located in the centre line of the beam is exposed to the laser beam.

The input energy in normal PW lasers is stored within the power supply of the laser. This energy when stored, like in capacitor, can reach very high levels and discharging this stored energy quickly into the laser gives rise to high intensity laser pulses. The peak power level of the pulses in PW lasers exceeds very much the power levels obtainable in CW lasers. For normal pulsed laser operation, it is common to produce peak power in the excess of several Joules of energy in a pulse lasting for up to 10 ms, though many pulsed lasers can produce pulses lasting for a much shorter time of say 100 μs. On the other hand, the pulse repetition time in PW lasers can be as short as 0.01 second, which amounts to a frequency of 100 pulses/sec. The repetition time depends on the specific design features of the laser. This is also worth noticing that a CW laser pulse gives the appearance of being pulsed when it is turned off rapidly.

1.7 Laser Field and Profile

Laser field is an electromagnetic (EM) field that may be allowed to acquire a particular profile for its specific application.

1.7.1 Ponderomotive Force

The nonlinear relationship between the displacement field vector $\vec{D}$ and the electric field vector $\vec{E}$ gives rise to nonlinearity in the refractive index of dielectric material. The nonlinearity can be explained by the following two mechanisms.

Electrostriction

This is the force exerted by a uniform electric field on a material medium. This force affects the density of the material which in turn modifies the dielectric constant. The electrorestrictive force is proportional to the gradient of the square of the electric field, i.e. the gradient of intensity.

Kerr Effect

This effect is felt when an intense light wave is incident on a liquid with molecules having an anisotropically polarized tensor. The incident light wave will tend to orient the anisotropically polarized molecules so as to align the direction of maximum polarizability along the direction of propagation.

Ponderomotive force is a nonlinear electrorestrictive force experienced by a charged particle in an inhomogeneous oscillating EM field. The mechanism of the ponderomotive force can be explained by considering the motion of a charge in an oscillating electric field. In a homogenous field, after one cycle of oscillation, the charge returns to its initial position, whereas in case of an inhomogeneous field, the force exerted on the charged particle is largely spent in a region with higher amplitude points in the direction where the field is weaker.

1.7.1.1 Nonrelativistic Regime

Consider an electric field $\vec{E} = \vec{A}(\vec{r})e^{-i\omega t}$, where $\vec{A}(\vec{r})$ is the amplitude that varies with space. The equation of motion of electron in non-relativistic regime will be

$$m\left[\frac{\partial \vec{v}}{\partial t} + (\vec{v}\cdot\vec{\nabla})\vec{v}\right] = -e\vec{E} - e\vec{v}\times\vec{B} \tag{1.45}$$

$$\text{Maxwell's third equation is } \vec{\nabla}\times\vec{E} = -\frac{\partial \vec{B}}{\partial t} \tag{1.46}$$

Since the variation of $\vec{E}$ with time is exponential $\left(\sim e^{-i\omega t}\right)$, $\vec{B}$ will have the same variation. Hence, we can respectively write Eqs. (1.46) and (1.45) as

$$\vec{\nabla}\times\vec{E} = i\omega\vec{B} \tag{1.47}$$

$$m\left(\frac{\partial \vec{v}}{\partial t}\right) = -e\vec{E} - e\vec{v}\times\vec{B} - m\left(\vec{v}\cdot\vec{\nabla}\right)\vec{v} \tag{1.48}$$

The last two terms of RHS of Eq. (1.48) are nonlinear and have the dimensions of a force called ponderomotive force. We use the iteration method, first to solve linear part to the zeroth order, i.e.

$$m\left(\frac{\partial \vec{v}}{\partial t}\right) = -e\vec{E}, \text{ that gives}$$

$$\vec{v} = \frac{e\vec{E}}{mi\omega} \tag{1.49}$$

Using this expression for velocity in ponderomotive force (say $\vec{F}_P$), we get:

$$\vec{F}_p = -e\vec{v}\times\vec{B} - m(\vec{v}\cdot\vec{\nabla})\vec{v}$$

$$= -\frac{e}{2}Re\left(\vec{v}\times\vec{B}^*\right) - \frac{m}{2}\mathrm{Re}\left[(\vec{v}\cdot\vec{\nabla})\vec{v}^*\right]$$

$$= -\frac{e^2}{2m\omega^2}\mathrm{Re}\left(\frac{\vec{E}\times\vec{B}^*}{i}\right) - \frac{e^2}{2m\omega^2}Re\left[\left(\vec{E}\cdot\vec{\nabla}\right)\vec{E}^*\right] \tag{1.50}$$

Using $\frac{\vec{\nabla}\times\vec{E}}{i\omega} = \vec{B}$ and $\frac{\vec{\nabla}\times\vec{B}}{i\omega} = -\vec{E}$ in Eq. (1.50), we get

$$\vec{F}_p = -\frac{e^2}{2m\omega^2}\mathrm{Re}[(\vec{E}\cdot\vec{\nabla})\vec{E}^*] - \frac{e^2}{2m\omega^2}Re(\vec{E}\times\vec{\nabla}\times\vec{E}^*)$$

We can write,

$$\vec{F}_p = -\frac{e^2}{4m\omega^2}\left[[(\vec{E}\cdot\vec{\nabla})\vec{E}^*] + (\vec{E}\times(\vec{\nabla}\times\vec{E}^*)) + (\vec{E}^*\times(\vec{\nabla}\times\vec{E})) + [(\vec{E}\cdot\vec{\nabla})\vec{E}^*]\right] \tag{1.51}$$

Here, all exponential terms get cancelled out, when we use the expression of $\vec{E}$. So

$$\vec{F}_p = -\frac{e^2}{4m\omega^2}\left[\left[(\vec{A}\cdot\vec{\nabla})\vec{A}^*\right] + (\vec{A}\times(\vec{\nabla}\times\vec{A}^*)) + (\vec{A}^*\times(\vec{\nabla}\times\vec{A})) + \left[(\vec{A}\cdot\vec{\nabla})\vec{A}^*\right]\right]$$

For simplicity, consider only x-component:

$$\vec{F}_{px} = -\frac{e^2}{4m\omega^2}\frac{\partial}{\partial x}\left[A_xA_x^* + A_yA_y^* + A_zA_z^*\right]$$

$$\vec{F}_{px} = -\frac{e^2}{4m\omega^2}\frac{\partial}{\partial x}|A|^2 \tag{1.52}$$

Based on this, we can generalize the expression as

$$\vec{F}_{pr} = -\frac{e^2}{4m\omega^2}\vec{\nabla}|E|^2 \tag{1.53}$$

This has the unit of Newton (in SI units). In its expression e is the electrical charge of the electron, m is its mass, ω is the angular frequency of oscillation of the field and E is the amplitude of the electric field. The physical meaning of the equation is that a charged particle in an inhomogeneous oscillating field not only oscillates at the frequency ω of the field, but is also accelerated by $\vec{F}_p$ towards the weak field direction. This is a rare case where the sign of the charge on the particle does not change the direction of the force because of the appearance of square of charge.

1.7.1.2 Relativistic Regime

At higher intensities, the magnetic field component in the Lorentz equation becomes strong enough to induce a significant change in the electron dynamic causing it to become non-linear. In the relativistic regime, the quiver momentum of the electron starts approaching or even exceeding the rest mass momentum of the electron. Hence, at higher intensities, in the relativistic regime, it is required to add the relativistic correction in electron mass. The relativistic factor is given as

$$\gamma = \frac{1}{\sqrt{1-\frac{v^2}{c^2}}}$$

The relativistic generalization of the ponderomotive force is given by

$$\vec{F}_p = -\frac{e^2}{4\gamma m_0 \omega^2}\vec{\nabla}|E|^2 \tag{1.54}$$

Alternatively, in relativistic regime, total energy $E = K.E. + P.E.$ gives rise to $mc^2 = K.E. + m_0c^2$, together with $m = \gamma m_0$.

Hence, $K.E. = mc^2 - m_0c^2$

When laser gets switched off, the total energy gain by electron can be written as $m_0c^2(\gamma - 1)$. From this we get (for $K.E.$ as U)

$$\vec{F}_p = -\vec{\nabla}U \text{ or}$$

$$\vec{F}_p = -m\vec{\nabla}(\gamma - 1) \tag{1.55}$$

1.8 Laser Interaction: Basic Facts

In the case of focussed intensity of a laser pulse larger than 10^{18} W/cm^2, it can be seen that the quiver velocity of an electron is close to the speed of light, in such a high EM field. Then various nonlinear phenomena are caused by the relativistic effect of the electron motion, for example, self-focusing, high harmonic generation, and so on. These phenomena are well known in nonlinear optics, which have also been observed in laser-plasma interactions. An ultrashort intense laser pulse propagating in plasma can excite a plasma wave by exerting a nonlinear force, i.e. ponderomotive force (Khazanov and Krivorutsky 2013). This way, the longitudinal electric field is formed by the plasma wave, which can accelerate the electrons trapped in the potential of the plasma wave. Such a high accelerating field has enabled the realization of compact electron accelerators and / or obtaining extremely high energy electrons.

Considering the motion of an electron in an EM field through the Lorentz equation, we write

$$\frac{d\vec{p}}{dt} = -e\left(\vec{E} + \vec{v} \times \vec{B}\right) \tag{1.56}$$

where $\vec{p}(=\gamma m\vec{v})$, $\vec{v}$, $\vec{E}$ and $\vec{B}$ are the momentum of the electron, velocity of the electron, electric field and magnetic field, respectively. Here, m and γ are the electron mass and the relativistic Lorentz factor. In the case of a weak EM field, the $\vec{v} \times \vec{B}$ component is negligible, hence Eq. (1.56) can be simplified to $m\frac{d\vec{v}}{dt} = -e\left(\vec{E}\right)$. For a linearly polarized laser field, the electron quiver velocity $\vec{v}_q$ is given by $\vec{v}_q = e\vec{E}_L / m\omega_L$, where $\vec{E}_L$ and ω_L are the amplitude and frequency of the laser field. The ratio of the electron quiver velocity to the speed of light c is defined by

$$a_0 = \frac{eE_L}{m\omega_L c} = 8.5 \times 10^{-10}\lambda_L\ [\mu\text{m}]\ \sqrt{(I_L[W / cm^2])} \tag{1.57}$$

where λ_L and I_L are the wavelength and intensity of the laser. This is called the normalized vector potential a_0, whose magnitude demarcates the nonrelativistic and relativistic regimes.

For example, the relativistic effects are brought about in an electron motion in a laser field yielding $a_0 \geq 1$. This region is called the relativistic region. As the intensity of 2.2×10^{18} W/cm^2 gives $a_0 = 1$ for 800 nm laser light, so it satisfies for the relativistic case. By averaging $(mv_q^2)/2$ over one oscillation period of a field, we can obtain the electron quiver energy

$$U_p = \frac{1}{2}\left(m\langle v_q^2\rangle\right) = \mathrm{m}\left(eE_L / 2m\omega_L\right)^2 \tag{1.58}$$

This is an expression for the ponderomotive potential U_p. The ponderomotive potential results in a force $\vec{F}_p = -\vec{\nabla} U_p$. This ponderomotive force is directed along the intensity gradient of a laser pulse envelope and is perpendicular to the direction of propagation of the laser. This force becomes the driving force for exciting a plasma wave by driving electrons out of the region of the laser-plasma interaction. If the velocity of electrons trapped into a plasma wave is closed to the phase velocity of plasma wave ($\sim c$), the electrons can be accelerated very efficiently. The energy gain $W_{\max}$ is simply given by the product of the accelerating field and the acceleration length (Kupersztych et al. 2004). However, the acceleration length is limited by the dephasing (phase slippage) of trapped electrons in the plasma wave. The dephasing is the difference between the phase velocity of the plasma wave and the velocity of the accelerated electrons. The accelerated phase of the plasma wave is outrun by the accelerated electrons in the plasma wave. This causes the electrons to enter into the deceleration phase. The dephasing length L_d in the linear region is given by

$$L_d = \frac{n_c}{n_e} \lambda_p \tag{1.59}$$

Here n_c is the critical density, and λ_p is the plasma wavelength. Finally, the maximum energy gain is given by

$$W_{\max} = 4\frac{mn_c c^2}{n_e} \tag{1.60}$$

For the case of a laser with a wavelength of 800 nm and electron density of 10^{19}cm^{-3}, the dephasing length and the maximum energy gain can be found as 1.8 mm and 350 MeV, respectively. It means the electrons with an energy of 350 MeV can be accelerated in only a 1.8 mm length.

When an intense ultrashort laser pulse is impinged onto a solid target, there is a rapid occurrence of multiple ionizations without significant ablation. It has been shown that the laser excitation of surface plasma waves can lead to significant modifications in the energy spectrum of electrons, thus increasing the laser absorption in the framework of multi-photon photoelectric emission of metals. This also leads to enhanced electron production. The excitation of surface plasmon waves by the p-polarized laser wave has been found to accelerate the electrons.

1.9 Dielectric Tensor of Plasma

For an isotropic medium, such as glass, the light faces the same structure in every direction while travelling in the medium. In contrast, for an anisotropic medium, the nature of light will be dependent on the direction of its propagation, i.e. light behaves differently in every direction due to the variation in the refractive index of the medium in different directions. With the application of appropriate external electric field, we can induce anisotropy in some media like liquid crystals.

Starting with the general approach of Maxwell in a linear medium, we write Maxwell's third and fourth equations as

$$\vec{\nabla}\times\vec{E} = -\frac{\partial\vec{B}}{\partial t} \text{ and } \vec{\nabla}\times\vec{B} = \mu_0\vec{J} + \frac{1}{c^2}\frac{\partial\vec{E}}{\partial t}$$

As the plasma is uniform, the Fourier analyzed expressions in space and time are given as:

$$i\vec{k}\times\vec{E} = \omega\vec{B} \text{ and, } \vec{k}\times\vec{B} = \mu_0\vec{J} + \left(\frac{-i\omega\vec{E}}{c^2}\right)$$

Eliminating $\vec{B}$ in the above equations, we get

$$i\vec{k}\times\left(\vec{k}\times\vec{E}\right) = \omega\mu_0\vec{J} - \frac{i\omega^2\vec{E}}{c^2}$$

$$\text{or } \vec{k}\times(\vec{k}\times\vec{E}) + \frac{\omega^2\vec{E}}{c^2} + i\omega\mu_0\vec{J} = 0 \tag{1.61}$$

Now, to build the appropriate relationship between $\vec{J}$ and $\vec{E}(\vec{k},\omega)$, we must solve the plasma equations first; but in order to keep it simple we pick the linear general relationship as: $\vec{J} = \vec{\sigma}\cdot\vec{E}$; where $\vec{\sigma}$ is the 'conductivity tensor'. Hence

$$\begin{pmatrix} J_x \\ J_y \\ J_z \end{pmatrix} = \begin{pmatrix} \sigma_{xx} & \sigma_{xy} & \sigma_{xz} \\ \sigma_{yx} & \sigma_{yy} & \sigma_{yz} \\ \sigma_{zx} & \sigma_{zy} & \sigma_{zz} \end{pmatrix}\begin{pmatrix} E_x \\ E_y \\ E_z \end{pmatrix} \tag{1.62}$$

This is the general form of Ohm's law. If the plasma is isotropic in nature then only the diagonal σ will survive. Hence, $\vec{J} = \vec{\sigma}\cdot\vec{E}$ applies.

Now from Eq. (1.61), we can write,

$$\vec{k}\,(\vec{k}\cdot\vec{E}) - k^2\vec{E} + \frac{\omega^2\vec{E}}{c^2} + i\omega\mu_0\vec{\sigma}\cdot\vec{E} = 0 \tag{1.63}$$

Modification of atom (or molecules) causes the 'polarization' in medium; dielectric medium will be discussed in terms of a dielectric constant as

$$\underbrace{\vec{D}}_{\text{Displacement}} = \varepsilon_0\vec{E} + \underbrace{\vec{P}}_{\text{Polarization}} \text{ and } \vec{\nabla}\cdot\vec{D} = \underbrace{\rho_{\text{ext}}}_{\text{external charge density}}$$

We can also write for the polarization $\vec{P} = \underbrace{\chi}_{\text{susceptiibility}} \varepsilon_0\vec{E}$

We consider here the response of dielectric medium in terms of the current density $\vec{J}$ (not in terms of polarization, $\vec{P}$) by using Ampere's law or Maxwell's fourth equation with the dielectric constant $\varepsilon_r = 1+\chi$. It means

$$\frac{1}{\mu_0}\vec{\nabla}\times\vec{B} = \frac{\partial\vec{D}}{\partial t} = \frac{\partial\left(\varepsilon_0\varepsilon_r\vec{E}\right)}{\partial t}, \text{ if } \vec{J}_{\text{external}} = 0 \tag{1.64}$$

Thus, the polarization current can be expressed explicitly in the form of an equivalent dielectric expression if

$$\vec{J}+\varepsilon_0\frac{\partial\vec{E}}{\partial t}=\ddot{\sigma}\cdot\vec{E}+\varepsilon_0\frac{\partial\vec{E}}{\partial t}=\frac{\partial(\varepsilon_0\ddot{\varepsilon}_r\cdot\vec{E})}{\partial t}$$

$$\text{Or } \varepsilon_r=1+\left(\frac{\sigma}{-i\omega\varepsilon_0}\right)$$

We can notice here the tensor nature of dielectric constant because of the anisotropic nature of the medium. If we take current implicitly in terms of a dielectric medium we would write this as $\frac{\partial(\varepsilon_0\ddot{\varepsilon}_r\cdot\vec{E})}{\partial t}$; where ε_r is the dielectric constant. Now we redefine the above wave Eq. (1.63) using $\varepsilon_r = 1+\frac{\sigma}{-i\omega\varepsilon_0} = 1+\frac{i\mu_0 c^2\sigma}{\omega}$. That is

$$\vec{k}(\vec{k}\cdot\vec{E})-k^2\vec{E}+\frac{\omega^2\ddot{\varepsilon}_r\cdot\vec{E}}{c^2}=0 \tag{1.65}$$

along with $\ddot{\varepsilon}\,(\vec{k},\omega)$ as the dielectric tensor.

1.9.1 Magnetic Field along Direction of Wave Propagation

Let us assume all the particles to move together and obey Newton's second law of motion as follows

$$m\frac{d\vec{v}}{dt}=q\left(\vec{E}+\vec{v}\times\vec{B}\right) \tag{1.66}$$

We consider $\vec{E}_0 = 0$, $\vec{B} = \vec{B}_0$ and zero velocity here due to which all the motion is contributed by the wave. As the speed of particles stays small, the magnetic field of wave can be ignored. Hence, Eq. (1.66) reads

$$m\frac{d\vec{v}}{dt}=q\left(\vec{E}+\vec{v}\times\vec{B}_0\right) \tag{1.67}$$

Opting axes such that $\vec{B}_0 = B_0\ (0,0,1)$, depicting direction of the magnetic field along the direction of propagation (i.e. z-axis), we have from Eq. (1.67)

$$-i\omega m v_x=q\left(E_x+v_yB_0\right) \tag{1.68}$$

$$-i\omega m v_y=q\left(E_y-v_xB_0\right) \tag{1.69}$$

$$-i\omega m v_z=qE_z \tag{1.70}$$

Solving the velocity components in terms of electric field components

$$v_x=\frac{q}{m}\left(\frac{i\omega E_x-\Omega E_y}{\omega^2-\Omega^2}\right) \tag{1.71}$$

$$v_y = \frac{q}{m}\left(\frac{i\omega E_y + \Omega E_x}{\omega^2 - \Omega^2}\right) \tag{1.72}$$

$$v_z = \frac{qi}{m\omega} E_z \tag{1.73}$$

The gyrofrequency is defined here as $\Omega = \frac{qB_0}{m}$, where the charge on the particles taking under consideration decides the sign. Using the current $\vec{J} = q\vec{v}n = \vec{\sigma} \cdot \vec{E}$ and making use of the above equations we can write the conductivity tensor as

$$\sigma_j = \begin{pmatrix} \frac{q_j^2 n_j}{m_j} \frac{i\omega}{\omega^2 - \Omega_j^2} & \frac{-q_j^2 n_j}{m_j} \frac{\Omega_j}{\omega^2 - \Omega_j^2} & 0 \\ \frac{q_j^2 n_j}{m_j} \frac{\Omega_j}{\omega^2 - \Omega_j^2} & \frac{q_j^2 n_j}{m_j} \frac{i\omega}{\omega^2 - \Omega_j^2} & 0 \\ 0 & 0 & \frac{iq_j^2 n_j}{m\omega} \end{pmatrix} \tag{1.74}$$

Here j specifies the particles, for example it is 'i' for +ve ions and 'e' for the electrons. The total conductivity of the system contributed by all the species is defined as the sum of the individual conductivities $\sigma = \sum_j \sigma_j$. So, we have

$$\sigma_{xx} = \sigma_{yy} = \sum_j \frac{q_j^2 n_j}{m_j} \frac{i\omega}{\omega^2 - \Omega_j^2} \tag{1.75}$$

$$\sigma_{xy} = -\sigma_{yx} = -\sum_j \frac{q_j^2 n_j}{m_j} \frac{\Omega_j}{\omega^2 - \Omega_j^2} \tag{1.76}$$

$$\sigma_{zz} = \sum_j \frac{iq_j^2 n_j}{m\omega} \tag{1.77}$$

Substituting $\chi = -\frac{1}{i\omega\epsilon_0}\vec{\sigma}$, ε_r can be written as

$$\varepsilon_r = \begin{bmatrix} \epsilon_{xx} & \epsilon_{xy} & 0 \\ \epsilon_{yx} & \epsilon_{yy} & 0 \\ 0 & 0 & \epsilon_{zz} \end{bmatrix} = \begin{bmatrix} S & -iD & 0 \\ iD & S & 0 \\ 0 & 0 & P \end{bmatrix} \tag{1.78}$$

$$\epsilon_{xx} = \epsilon_{yy} = S = 1 - \sum_j \frac{\omega_{pj}^2}{\omega^2 - \Omega_j^2} \tag{1.79}$$

$$i\epsilon_{xy} = -i\epsilon_{yx} = D = \sum_j \frac{\omega_{pj}^2}{\omega^2 - \Omega_j^2} \frac{\Omega_j}{\omega} \tag{1.80}$$

$$\epsilon_{zz} = P = 1 - \sum_j \frac{\omega_{pj}^2}{\omega^2} \tag{1.81}$$

1.9.2 Magnetic Field Perpendicular to Direction of Wave Propagation

When an applied magnetic field is perpendicular to the direction of wave propagation, i.e. $\vec{B} = \vec{B}_0(1,0,0)$, we can use the same mathematical treatment and obtain the following form of the dielectric tensor

$$\epsilon = \begin{bmatrix} \epsilon_{xx} & 0 & 0 \\ 0 & \epsilon_{yy} & \epsilon_{yz} \\ 0 & \epsilon_{zy} & \epsilon_{zz} \end{bmatrix} = \begin{bmatrix} P & 0 & 0 \\ 0 & S & -iD \\ 0 & iD & S \end{bmatrix}, \text{ together with}$$

$$\epsilon_{zz} = \epsilon_{yy} = S = 1 - \sum_j \frac{\omega_{pj}^2}{\omega^2 - \Omega_j^2} \tag{1.82}$$

$$i\epsilon_{zy} = -i\epsilon_{yz} = D = \sum_j \frac{\omega_{pj}^2}{\omega^2 - \Omega_j^2} \frac{\Omega_j}{\omega} \tag{1.83}$$

$$\epsilon_{xx} = P = 1 - \sum_j \frac{\omega_{pj}^2}{\omega^2} \tag{1.84}$$

1.9.3 Physical Significance

Polarization field can be defined as the response of medium when an external electric field is applied. In an anisotropic medium, it is not required that this electric field be aligned in the direction of the polarizing field. It can be thought that the dipoles induced in the medium by the electric field have certain preferred directions related to the physical structure of the crystal. In this case, the electric susceptibility does not behave as a number but as a rank 2 tensor. With the relation of dielectric tensor to polarization, we can conclude dielectric tensor as a quantity that measures the ability of a substance to store electrical energy in different electric field directions. In non-magnetic and transparent materials, electric susceptibility tensor is real and symmetric. However, when magnetic field is applied externally, few materials can have a dielectric tensor which is complex-Hermitian and the principal axes are complex-valued vectors.

1.10 Polarization of Laser Light

Light is an EM wave having oscillating electric field ($\vec{E}$) and oscillating magnetic field ($\vec{B}$) coupled in perpendicular directions to the direction of wave propagation; primes on these quantities have been dropped for the sake of simplicity. In view of its transverse nature, the components of vibrations are present in all the possible directions perpendicular to

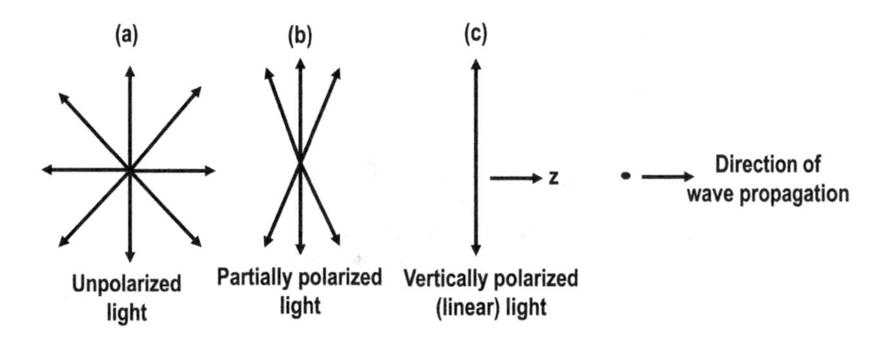

FIGURE 1.7
Different situations of the (a) unpolarized, (b) partially polarized and (c) totally polarized light.

the direction of propagation (say $\vec{k}$). Light from the ordinary sources like bulb in the room, flash lamps or candle flame, contains electric field vector oscillating in all the possible directions perpendicular to the direction of propagation. Such type of light is called unpolarized light. The light in which the electric field is restricted to oscillate in a specific direction is called polarized light. Convention is that the polarization of an EM wave is taken in the direction of the oscillating electric field vector. Just to mention the magnetic field vector oscillates in a direction transverse to both the directions of the electric field vector and the wave propagation. Different situations of the unpolarized, partially polarized and totally polarized light are shown in Figure 1.7. The plane determined by the field $\vec{E}$ and the wave vector $\vec{k}$ is generally referred to as the plane of polarization; this is true particularly for the linearly or plane polarized wave. Then the plane containing vectors $\vec{B}$ and $\vec{k}$ is called plane of vibration. However, sometimes the term plane of polarization is also applied to the plane containing $\vec{B}$ and $\vec{k}$. In the cases of elliptical and circular polarizations, the field $\vec{E}$ has two components in the directions perpendicular to $\vec{k}$ and hence, plane of vibration and plane of polarization are the same.

Depending on the orientation of the electric field components, polarized light can have three types of polarizations.

1.10.1 Linearly Polarized Light

When the electric field vector components of light are restricted along one particular direction transverse to the direction of propagation, light is said to be linearly polarized. This type of light is shown in Figure 1.8, where the electric field is restricted to oscillate only in the x-direction and the wave propagation is along the z-direction.

1.10.2 Circularly Polarized Light

If the electric field vector has two mutually perpendicular components of equal amplitudes with a phase difference of $\pi/2$, then the resulting electric field so produced rotates in a circular manner about the direction of propagation of the wave. This type of light is called circularly polarized light. Depending on the rotation either clockwise or anti-clockwise (when viewed in the direction of wave propagation) this light is called right-handed or left-handed circularly polarized light.

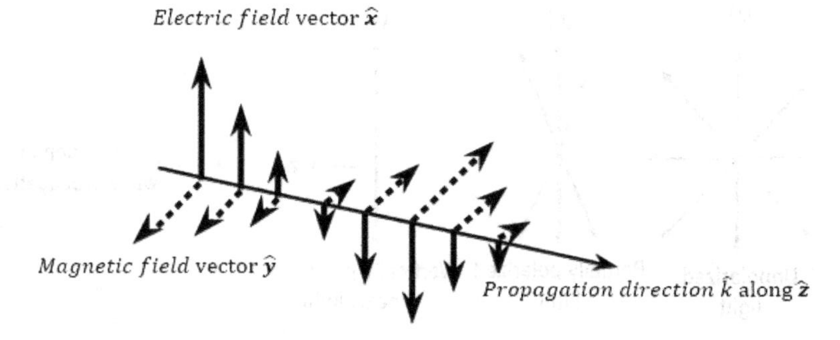

FIGURE 1.8
Linearly polarized light.

1.10.3 Elliptically Polarized Light

When the electric field vector oscillates with two linear components having different amplitudes and phase difference other than $\pi/2$, the resulting electric field traces an ellipse about the direction of wave propagation. This type of light is called elliptically polarized light. Hence, this is clear that circularly and linearly polarized lights can be regarded as special cases of elliptically polarized light.

1.10.4 The s and p Designations for Laser Light

In order to understand the s-polarized or p-polarized wave, first we need to talk about the plane of incidence. The plane of incidence is the plane made by the wave propagation vector $\vec{k}$ and the vector $\hat{n}$ which is normal to the plane of an interface. It means the plane of incidence is the plane in which the wave or light travels before and after its reflection or refraction. Now there are two possibilities of the direction of electric field of the wave, as it could be parallel or perpendicular to the plane of incidence. The first geometry is called p-like (parallel) and the second one is called s-like (perpendicular). In view of this, the polarized light with its electric field along the plane of incidence is termed as p-polarized light, whereas the light whose electric field is normal to the plane of incidence is called s-polarized light. The p-polarization is commonly referred to as transverse-magnetic (TM); it is also given the names pi-polarized or tangential plane polarized. On the other hand, the s-polarization is called transverse-electric (TE), as well as sigma-polarized or sagittal plane polarized.

1.11 Production of Polarized Laser Light

Polarized light can be produced by a number of methods.

1.11.1 Interaction with Matter

Light can be polarized on interaction with a certain media. For example, an unpolarized light on entering a certain anisotropic medium splits up into two perpendicular components, polarized in the perpendicular directions. Such uneven response of certain media may cause absorption, reflection and transmission of light depending on the medium.

Anisotropic Absorption - Dichroism

Polarized light is also obtained by using polarizers, as the polarizers have the ability to block one of the components of the electric field more effectively and transmits the other components. Such type of ability is known as dichroism. Actually a dichroic polarizer selectively absorbs the electric field oscillations in a specific direction and transmits electric field oscillation in the perpendicular direction, i.e. along the transmission axis. The material of which such polarizers are made up of are known as dichroic materials. Such type of polarization is a result of chemical composition of the material, which consists of long chains of molecules aligned in one particular direction. While manufacturing, these long chains are stretched along the entire length of polarizer to give them a polarization axis. Then the oscillations parallel to this axis are transmitted and the other ones are blocked.

Reflection

When an unpolarized light is reflected from surface of a transparent dielectric medium, i.e. a non-metallic medium, the reflected light is partially polarized in TM mode. Here, the degree of polarization depends on the angle of incidence and at a certain angle of incidence the reflected light is completely polarized. That angle of incidence is known as the angle of polarization or polarizing angle i_p. This angle is different for different media which means this angle depends on the nature of the reflecting surface. The refractive index μ is related to the polarizing angle i_p through the following law, called Brewster's law

$$\mu = \tan i_p$$

This is interesting that the sum of polarization angle and the angle of refraction amounts to 90°.

1.12 Visualization of Laser's Polarization

The simplest technique to analyze the laser beam's polarization is with the help of a microscope slide. Here we have to hold the slide into the beam but oriented at Brewster's angle and then rotate it around the incident beam axis. We keep on doing this until the reflected ray disappears. The direction in which the reflected ray disappears is the direction of the beam's polarization. However, the major flaw of this technique is of stray laser beams which are extremely harmful for people working in the laboratory.

There are other alternate ways to visualize the laser's polarization. For example, we can insert a thin layer of photosensitive material between two glass plates. This layer is illuminated, and the transmitted and scattered lights are observed carefully. The scattered light is seen when it undergoes total internal reflection at the interface of the glass and the air. On analysis, the pattern of the scattered light reveals that whether the light is linearly polarized or not, and in which direction it is polarized. This is depicted in Figure 1.9, where the diagram in the bottom of each photograph indicates the polarization of the laser beam.

In this phenomenon what happens is as follows. The incident laser beam stimulates polymerization of the photosensitive material. Hence, microscopic particles are formed that

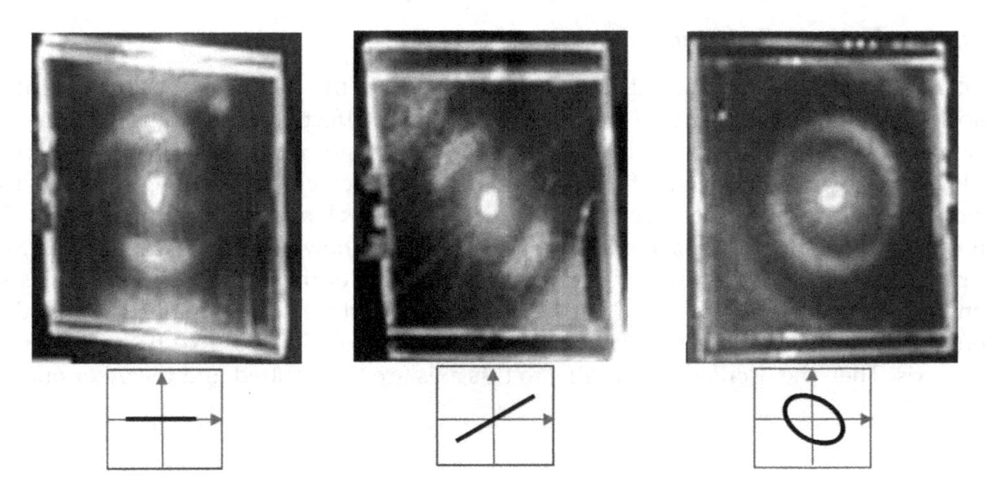

FIGURE 1.9
Visualization of polarization of light with the help of microscope slides.

cause the scattering of light. As per classical laws of scattering, the major part of scattering comes out in a perpendicular direction to the beam's polarization when the light is scattered from a small particle. This scattered beam is guided by total internal reflection at the glass-air interface and through successive reflections it comes out from the other end.

The next part is to analyze the light coming out. For this, a rotating board having several coloured light-emitting diodes (LEDs) is placed in front of the polarizer. A detector placed in the middle of the board controls which LEDs are illuminated. Since the board is rotating, different color LEDs generate circles of different colors. The light after coming out of the polarizer is made to fall on the detector. It is noticed that the LED glows more when the beam is more intense (Figure 1.10). Since polarized light falls on the board, the angle / orientation of glowing LEDs can reveal the state of polarization of the light. Various states of polarization are shown in Figure 1.11. However, the diagram obtained for circular polarization would be the same as for the unpolarized light.

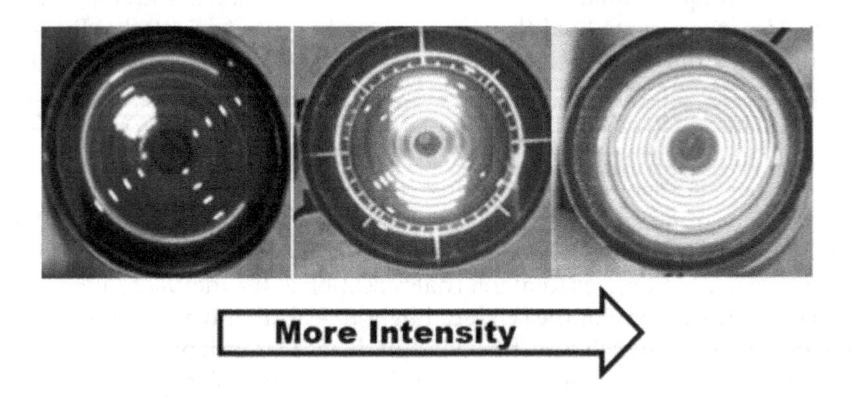

FIGURE 1.10
LED is shown to glow more for the case of more intensity of the beam.

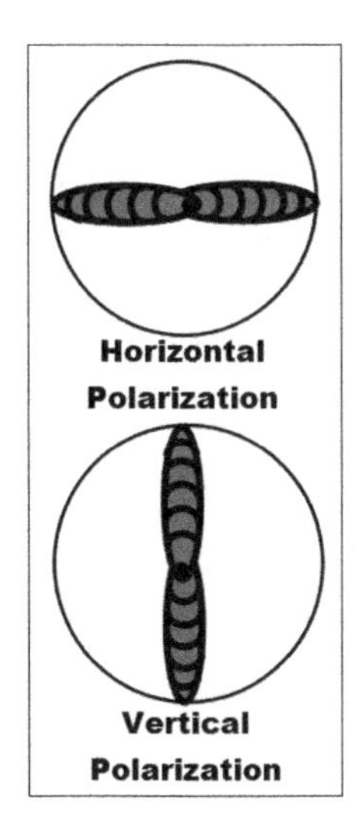

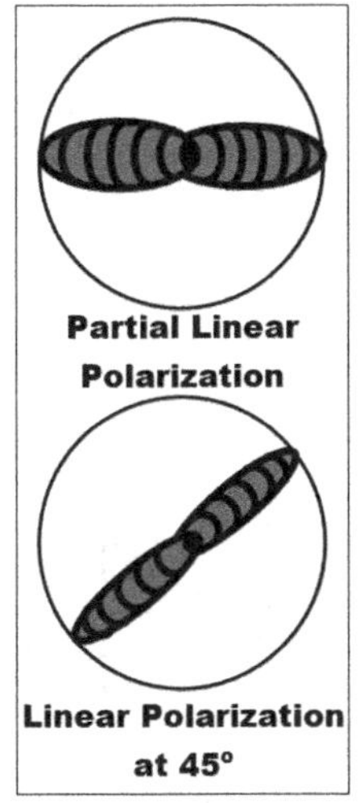

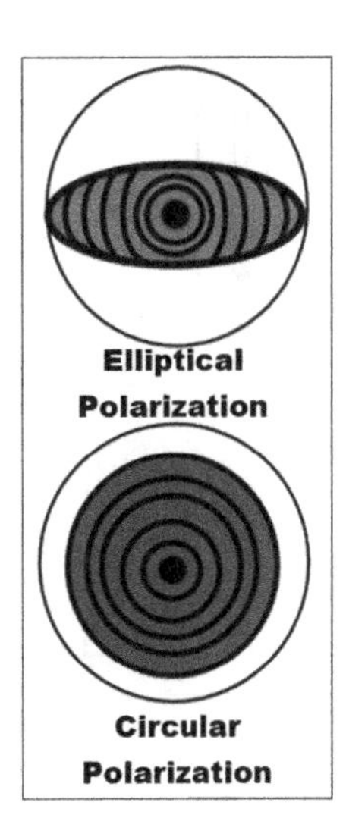

FIGURE 1.11
Visualization of various states of polarization of light.

1.12.1 Factors Affecting Polarized Emission of Radiation

There are a number of factors on which the emission of the linearly polarized laser light depends. For example, the gases that are involved in the active medium of the laser may be polarization dependent, like the Nd:YVO_4 or Nd:YLF lasers. Also, the losses may become polarization dependent in view of the presence of Brewster's plate or other optical components present inside the cavity. On the other hand, the polarization state of the laser output can be disturbed by random and temperature-dependent birefringence. Such a situation occurs in optical fibres if the fibres are not polarization maintaining or single-polarization fibres. It also may happen in laser crystal or glasses as a result of thermal effects that lead to depolarization loss. In the case of isotropic laser gain, small drifts of the birefringence may lead to large changes in the polarization state. Due to this, a significant variation in the polarization state also may occur across the beam profile.

1.12.2 Polarization Optics for Lasers

Many applications require polarized light in a specific state or direction, so it becomes mandatory to segregate the states of polarization of light. Many optical applications depend on good polarization control systems. These include laser materials processing, liquid crystal device characterization and manufacturing, fluorescence polarization assays and imaging, polarization diversity detection in communications and range-finding, polarized Raman spectroscopy, second harmonic-generation imaging and many other laser applications based on holography, interferometry and so forth. There are so many polarizers available, but not every problem is solved using one or the other. Thus, the number of important applications are lacking due to the lack of ideal polarization component. An ideal polarizer gives 100% transmission of the required polarization components and eliminates the unwanted components. The most significant parameter that governs an ideal polarizer is the 'contrast ratio'. The contrast ratio is defined as the ratio of the transmissions through a pair of polarizers when they are aligned parallel and when they are crossed. This ratio

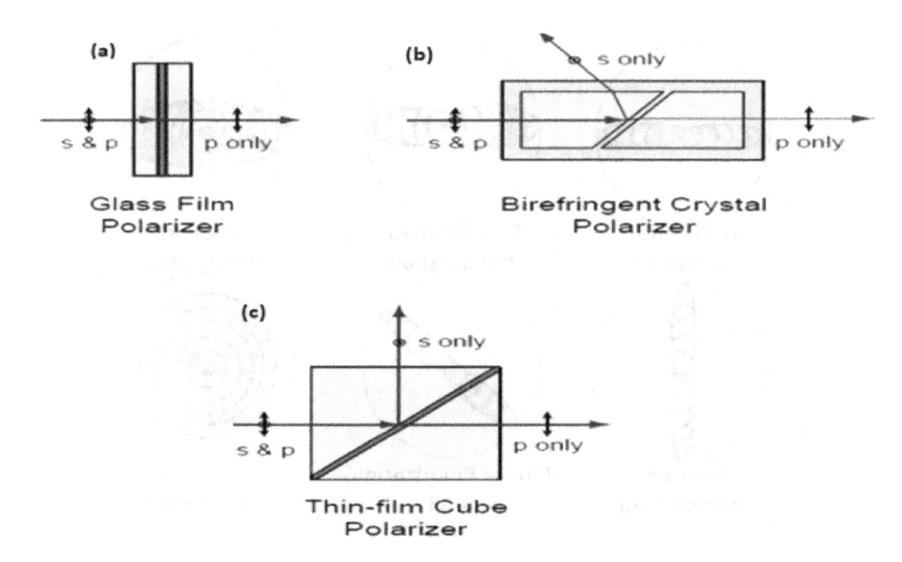

FIGURE 1.12
Beam splitters polarizers. (a) Glass film polarizer. (b) Birefringent crystal polarizer. (c) Thin-film cube polarizer.

varies from about 10^2:1 to as large as 10^4:1. Another parameter known as the 'extinction ratio' is the inverse of the contrast ratio.

Polarizers can be divided into the following two broad categories.

1.12.2.1 Absorptive Polarizers

The polarizer that absorbs or reflects one of the components and transmits the other component is known as the absorbing / absorptive polarizer. Wire grid, polymer sheet and glass film polarizers fall within this category. A glass film polarizer is shown in Figure 1.12(a), where it has been shown to absorb s-polarized light and allows the transmission of p-polarized field.

1.12.2.2 Beam Splitters Polarizers

With the use of this polarizer, the incident beam is divided into two parts having different polarizations. Since these polarizers do not absorb or reflect the energy of the unwanted polarization state, they are well suited for applications such as laser light and are very useful where the two polarization states are to be studied or simultaneously used. Birefringence crystal polarizer and thin-film plate polarizer are the examples of beam splitter polarizers (Figure 1.12b,c). These are shown to separate out the s- and p-polarized lights.

1.13 Fabry-Perot Cavity and Laser Oscillations

Optical resonators play an important role in amplifying the laser beam. Fabry-Perot resonators or etalons are optical cavities which are used for this purpose. These consist of two partially reflecting parallel mirrors with reflectance R_1 and R_2 separated by a length L

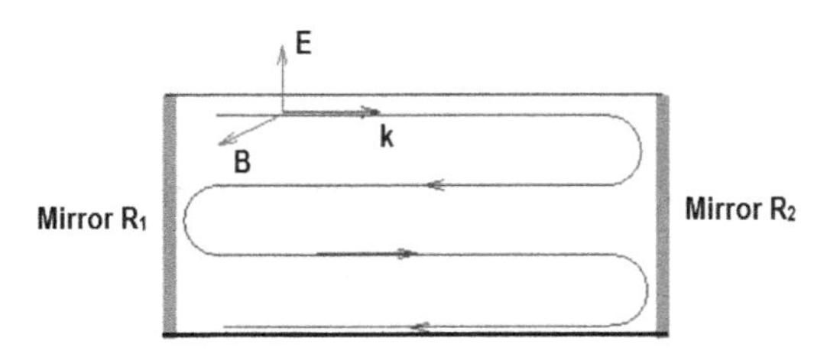

FIGURE 1.13
Ray suffering multiple reflections, bouncing to-and-fro between the mirrors in the Fabry-Perot cavity.

(cavity length). Consider an EM wave of amplitude E incident normally on one of the faces of the etalons. The ray suffers multiple reflections, bouncing to-and-fro between the mirrors (Figure 1.13). At each reflection, the field amplitude gets reduced by a factor $r_1 = \sqrt{R_1}$ or $r_2 = \sqrt{R_2}$. If we represent the wave vector of the field as $\vec{k}$, the wave suffers a phase change $2kL$ in addition to the phase change due to reflection.

The transmission intensity is maximum at the resonance, i.e. when $kL = m\pi$ (m are integers). Since the shape of intensity function repeats itself, the full-width at half-maximum (FWHM) of the intensity pattern can be obtained by looking at any of the peaks. Consider the peak near $kL = 0$. For large values of R (reflectivity), the peak is found to be very sharp and we may replace sin(kL) by kL itself. In view of this, FWHM is given by

$$\frac{1}{1+R(kL)^2} = \frac{1}{2}$$

A Fabry-Perot cavity can, therefore, be used to select wavelength by suitably adjusting the mirror separation L. The wavelengths at which resonances occur are given by

$$\lambda^{\max} = \frac{2L}{m}$$

This relation shows that the wavelength decreases as the value of m increases. For a medium whose refractive index is μ, the speed of light reads c/μ. Then the resonance frequencies are given as

$$\nu^{\max} = \frac{cm}{2\mu L}$$

which are equispaced. Each possible standing wave satisfying the above condition is called a cavity mode, with m being the mode number. If now, an active medium with a gain coefficient $g(\nu_0)$ is put inside the Fabry-Perot cavity, the optical power P in one round trip is given by $P = P_0 R_1 R_2 e^{2g(\nu_0)L}$. It means the optical amplification occurs if $P > P_0$, i.e.

$$g(\nu_s) \geq \frac{1}{2L} \ln\left(\frac{1}{R_1 R_2}\right)$$

The minimum value of the right-hand side of the above condition is termed as threshold gain $g_{th}(\nu_s)$, given by

$$g_{th}(\nu_s) = \frac{1}{2L} \ln\left(\frac{1}{R_1 R_2}\right)$$

1.14 Properties of Laser Beam

Laser beams are characterized by the following special properties.

1.14.1 Coherence

Laser beam is highly coherent which means different parts of the beam maintain a phase relationship for a long time. This results in an interference effect, which means when a laser beam reflects off a surface the reflected light can be seen to have bright regions separated by dark regions. Coherence can be of two types, namely, spatial coherence and temporal coherence. Temporal coherence with coherence time $\Delta\tau$ is defined as the time after which the phase correlation between two waves drops significantly; these waves were initially in phase or had a known phase difference between two points in the same wave. The reason for loss of coherence is that an optical source does not emit a CW for all the time to come. Thermal sources, for instance, have a typical lifetime of $\Delta t = 10^{-7}$ sec so that a wave which seems continuous, actually consists of a sequence of waves which are typically $c\Delta t = 0.3$ m long and have no phase correlation between parts of one wave train and the other.

The temporal coherence length can be defined as

$$l = c\Delta t, \text{ so that } l = \frac{c}{\Delta v} = \frac{\lambda^2}{\Delta\lambda}$$

Here λ is the mean wavelength and $\Delta\lambda$ is the spread of wavelengths about this mean. The spread of wavelength is a measure of the degree of monochromaticity of the wave.

On the other hand, spatial coherence describes the distance over which phase correlation exists between different points in the same wave in a direction perpendicular to the direction of observation. Spatial coherence arises for the reason that a source is never really a point source. We may arbitrarily define the spatial coherence length x to be the lateral distance at which the waves are out of phase by π. Thus $x\frac{d}{z} = \frac{\lambda}{2}$, where d is the size of the source and z is the distance away from it. Then the spatial coherence length is given by

$$x = \frac{\lambda}{2}\frac{z}{d}$$

Keeping in mind the above argument, the lateral coherence length is $l_c = \frac{\lambda z}{2D}$ for a source of circular shape of diameter D.

1.14.2 Directionality

Laser beam is highly collimated and can travel long distances without significant spread in the beam cross-section. However, in reality the beam spreads out as the collimated beam propagates. The full angle beam divergence $\Delta\phi = \frac{D}{z}$ is defined as the amount by which the beam diameter (D) increases over a distance after leaving the source. For a Gaussian beam profile, the intensity I varies as $I = I_0 e^{-\left(\frac{\tau}{w_0}\right)^2}$, where τ is the distance from the centre of the beam where the intensity is maximum (I_0) and w_0 is the beam width or waist radius. We define beam radius as the radial distance over which the intensity decreases by a factor $1/e^2$ from its maximum value at the centre. Thus nearly 94% of the energy is concentrated within the beam radius. The angular divergence of the beam is given by $\theta = \frac{\lambda}{\pi w_0}$. The radius of the beam varies with the distance as

$$w(z) = w_0 \sqrt{1 + \left(\frac{z}{z_0} \right)^2}$$

where $z_0 = \pi \omega_0^2 / \lambda$ is the Rayleigh length.

1.14.3 Laser Output Power and Threshold

The output power of the laser is the power coming out from the two end facets of the laser cavity. Since the power P in terms of output coupling efficiency η_0 is given by $P = \eta_o \hbar\omega \frac{N_P}{\tau_P} = \eta_o \hbar\omega \frac{n_P V_P}{\tau_P}$, the laser power above threshold in terms of η_i can be written as

$$P = \eta_o \hbar\omega \frac{N_P}{\tau_P} = \eta_i \hbar\omega \frac{n_P V_P}{\tau_P} (l - l_m)$$

The above expressions show that if $\eta_o = \eta_i$ were equal to unity then every electron injected into the laser per second above the threshold injection rate of l / q would end up producing a photon in the laser output. Above threshold, the laser is therefore a very efficient converter of electrical energy into optical energy. This property of the laser is commonly expressed in terms of the differential slope efficiency η_{dS} given by

$$\eta_{dS} = \frac{dP}{d\eta_o} = \eta_o \eta_i \frac{\hbar\omega}{q}$$

or the differential quantum efficiency η_{dQ}, given by

$$\eta_{dQ} = \frac{q}{\hbar\omega} \frac{dP}{dl} = \eta_o \eta_i$$

The effective mirror loss is given by

$$\alpha_m = \frac{1}{L} ln \left(\frac{1}{\sqrt{R_1 R_2}} \right)$$

The output coupling efficiency is

$$\eta_o = \frac{\alpha_m}{\alpha_m + \alpha}$$

1.15 Applications of Polarized Lasers

Polarized laser emission is important for a range of applications, and a few of them are listed below:

1. Polarized light is required in nonlinear frequency conversion, as the condition of phase matching in a nonlinear crystal is obtained only for one direction of polarization.

2. Polarization-dependent devices such as interferometers, semiconductor optical amplifiers and optical modulators require the processing of laser beams.
3. Implementation of polarization control is useful in imaging applications. Here it is possible to eliminate glare from reflective objects or explore surface defects by placing a linear polarizer over the light source or the lens or both.
4. Certain materials like glass and plastics become optically active on the application of stress (photoelastic effect). Here more stress produces pronounced effects. Hence, optical stress analysis of old sculptures and heritage buildings can be done by taking the small samples and analyzing them between the crossed filters, thereby calculating the strains in the buildings and sculptures. Similar technique can be applied to analyze complicated shapes by making plastic models of them. This effect depends on wavelength as well, and the wavelength dependence is exploited for artistic purposes.
5. Many important chemical compounds and oils are 'optically active', which rotate polarized light. Examples are active pharmaceutical ingredients, turpentine oil or sugar solution. Polarization technique checks for the presence of chirality of organic compounds. Here polarimetry technique is employed (which is based on the polarization phenomenon) to quantify many chemical compounds and food products in the chemical, pharmaceutical and food and beverage industries.
6. Astronomers make use of polarization phenomenon to get information on sources that account for the presence of radiations and scattering.
7. The 3D movie technology also relies on the polarization technique because the images for both eyes are projected from two different projectors having filters (polaroids) arranged in crossed positions.
8. All radio waves transmitting and receiving antennas are intrinsically polarized. Antennas are designed differently to transmit and receive a specific type of polarized light. Actually, for receiving different polarization components of the EM waves simultaneously, a specific radar antenna is designed.
9. Polarized light is used in the liquid crystal displays (LCDs) of television, watches and so forth. It involves the arrangement of liquid crystals between crossed polarizers sandwiched between two transparent electrodes. Here the light is blocked by a second polarizer and the display is made dark by applying appropriate potentials to the transparent electrodes. However, the potential difference is zero to make the display bright.

1.16 Conclusions

The EM waves travel with the speed of light (c) through the vacuum but in other media the speed is generally less than c due to the absorption and re-emission of the wave in the material medium. The amount of decrease in the speed depends on the optical density of the medium. When an EM wave propagates through the conductor, attenuation of its amplitude takes place, leading to the exponential decay in amplitude with distance. The distance through which the amplitude falls to the value $1/e$ times of its original value is known as skin depth. An incident EM wave having frequency less than the plasma frequency ($\omega < \omega_p$) will be unable to propagate through the plasma and will be totally reflected. This cutoff is modified when an external magnetic field is applied and that too depends on the direction of the applied magnetic field.

For a longitudinal wave $\vec{k} \parallel \vec{E}'$ and hence, $\vec{k} \times \vec{E}'$ vanishes, and the wave is also said to be electrostatic. If the wave is transverse, $\vec{k} \times \vec{E}'$ is finite, and the oscillating magnetic field $\vec{B}'$ is finite and the wave is electromagnetic in nature. In case of transverse waves with $\vec{k} \perp \vec{E}'$ the wave could be an ordinary wave, i.e. $\vec{E}' \parallel \vec{B}_0$ or extraordinary wave where $\vec{E}' \perp \vec{B}_0$. In the case of ordinary waves, the electric field causes the electrons to move parallel to $\vec{B}_0$, i.e. $\vec{v} \times \vec{B}_0$ vanishes and the wave shows the same behaviour as if it is propagating in the absence of magnetic field. In the presence of an external magnetic field, the waves concerning the low frequency ion oscillations are of two types, namely, the hydromagnetic wave (also known as the Alfven wave) propagating along $\vec{B}_0$, and the magnetosonic wave propagating across $\vec{B}_0$.

LASER, an acronym for light amplification by stimulated emission of radiation, is a highly intense, tunable and coherent source of light which may span almost the whole EM spectrum. The operation of laser involves two major phenomena namely amplification and stimulated emission of radiation / light. Lasers can be operated in different modes and are categorized on the basis of output of a laser. For example, CW operation and pulsed operation are the two basic modes of operations. In the CW mode of operation, laser emits an uninterrupted beam of light continuously with a stable output. The output wavelength emitted by the laser depends on the characteristics of the active medium. In some lasers, strong heating of the gain medium is caused in CW operation mode. Therefore, to minimize the heating of the active medium, another technique is used which is known as quasi-continuous wave operation. In this technique, the pump power is only switched on for short intervals of time. In the pulsed mode of operation, the output of the laser is in the form of pulses that varies with time like sine and cosine waveforms. Here, all the energy is dumped out in a single pulse which may last for nanoseconds to picoseconds. Different kinds of modes are also associated with pulse operation, which include Q-switching, mode locking and pulsed pumping.

The technique of producing pulses of extremely short duration typically of the order of picoseconds to femtoseconds is called mode locking. Here the laser is active for a very small time interval. Pulses so produced are separated by the time interval that a pulse takes to complete one to-and-fro movement inside the resonator cavity. Mode locking also can be either active or passive. Active mode locking produces modulated light as it depends on an external source used to stimulate oscillation in the resonant cavity, whereas an internal cavity source itself is used to instigate a change within the cavity in passive mode locking. Another technique to drive pulsed laser is pulsed pumping, where active material is pumped by a self-pumped source only. A self-pumped source could be a pulsed laser or flash pump producing flashes periodically.

For the laser-matter interaction, an important force, i.e. ponderomotive force, is involved. This force is a nonlinear electrorestrictive force experienced by a charged particle in an inhomogeneous oscillating EM field. In a non-relativistic regime, it is given as $\vec{F}_p = -\frac{e^2}{4m\omega^2}\vec{\nabla}|E|^2$ and in a relativistic regime, it is expressed as $\vec{F}_p = m\vec{\nabla}(\gamma - 1)$ where $m = \gamma m_0$.

The light in which the electric field is restricted to oscillate in a specific direction is called polarized light. Depending on the orientation of the electric field components, polarized light can have three types of polarizations. When the electric field vector components of light are restricted along one particular direction transverse to the direction of propagation, light is said to be linearly polarized. If the electric field vector has two mutually perpendicular components of equal amplitudes with a phase difference of $\pi/2$, then the resulting electric field so produced rotates in a circular manner about the direction of propagation of the wave. This type of light is called circularly polarized light. When the electric field vector oscillates with two linear components having different amplitudes and phase difference other than $\pi/2$, the resulting electric field traces an ellipse about the direction of wave propagation. This type of light is called elliptically polarized light. Circularly and linearly polarized lights can be regarded as special cases of elliptically polarized light.

Polarized light can be produced by a number of techniques. Light can be polarized on interaction with certain media. For example, an unpolarized light on entering a certain anisotropic medium splits up into two perpendicular components polarized in the perpendicular directions. Polarized light is also obtained by using polarizers, as they have the ability to block one of the components of the electric field more effectively and to transmit the other components. Such an ability is known as dichroism. Further, polarized light is also produced when an unpolarized light is reflected from the surface of a transparent dielectric medium, i.e. a non-metallic medium; the reflected light is partially polarized in TM mode. Here, the degree of polarization depends on the angle of incidence, and at a certain angle of incidence the reflected light is completely polarized. That angle of incidence is known as the angle of polarization or polarizing angle i_p. This angle is different for different media and depends on the nature of the reflecting surface.

There are a number of factors on which the emission of the linearly polarized laser light depends. For example, the gases that are involved in the active medium of the laser may be polarization dependent. Brewster's plate or other optical components present inside the cavity may disturb the polarization state of the polarized light. Thermal effects and small drifts of the birefringence may lead to large changes of the polarization state. The most significant parameter that governs an ideal polarizer is the 'contrast ratio', i.e. the ratio of the transmissions through a pair of polarizers when they are aligned parallel and when they are crossed. The polarizer that absorbs or reflects one of the components and transmits the other component is known as an absorbing or absorptive polarizer. Wire grid, polymer sheet and glass film polarizers fall within this category. The polarizer by which the incident beam is divided into two parts having different polarizations is known as a beam splitter polarizer. Since these polarizers do not absorb or reflect the energy of the unwanted polarization state, they are well suited to applications such as laser light and are very useful where the two polarization states are to be studied or simultaneously used. Polarized laser emission is important for a range of applications such as nonlinear frequency conversion, polarization-dependent devices such as interferometers and optical modulators, imaging applications, 3D movie technology and LCDs.

1.17 Selected Problems and Solutions

PROBLEM 1.1

Show that the equation of continuity is contained within the Ampère-Maxwell law and Gauss's law,

$$\text{i.e. } \vec{\nabla} \times \vec{B} = \mu_0 \vec{J} + \mu_0 \varepsilon_0 \frac{\partial \vec{E}}{\partial t}$$

SOLUTION

Divergence of the Ampère–Maxwell law gives

$$\vec{\nabla}\cdot\left(\vec{\nabla}\times\vec{B}\right)=\mu_0\vec{\nabla}\cdot\vec{J}+\vec{\nabla}\cdot\left(\mu_0\varepsilon_0\frac{\partial\vec{E}}{\partial t}\right)$$

Clearly under this treatment the left-hand side (LHS) of this equation vanishes. Then interchanging the divergence and time derivative on the RHS and cancelling the factor μ_0, gives

$$\vec{\nabla}\cdot\vec{J}+\frac{\partial}{\partial t}\left(\varepsilon_0\vec{\nabla}\cdot\vec{E}\right)=0$$

Using Gauss's law, $\vec{\nabla}\cdot\vec{E}=\frac{\rho}{\varepsilon_0}$ we finally obtain

$$\vec{\nabla}\cdot\vec{J}+\frac{\partial\rho}{\partial t}=0$$

This is the equation of continuity. Maxwell wrote down the equation of continuity alongside his other equations, but it is not counted as one of his fourth law of electromagnetism equations because it is a consequence of two of the other laws.

PROBLEM 1.2

Show that Maxwell's equations are invariant under the operation of time reversal, i.e. under the change $t\rightarrow -t$, $\vec{J}\rightarrow -\vec{J}$ and $\vec{B}\rightarrow -\vec{B}$, but leaving ρ and $\vec{E}$ unchanged.

SOLUTION

Applying the transformation rules for time reversal given in the question does not affect Gauss's law as $\vec{E}$ is not being asked to change. The remaining Maxwell's equations transform as follows

$$\vec{\nabla}\cdot(-\vec{B})=0$$

Or

$$\vec{\nabla}\cdot\vec{B}=0$$

$$\vec{\nabla}\times\vec{E}=-\frac{\partial(-\vec{B})}{\partial(-t)}$$

or

$$\vec{\nabla}\times\vec{E}=-\frac{\partial\vec{B}}{\partial t}$$

and $\vec{\nabla}\times(-\vec{B})=\mu_0(-\vec{J})+\mu_0\varepsilon_0\frac{\partial\vec{E}}{\partial(-t)}$.

$$\vec{\nabla}\times\vec{B}=\mu_0\vec{J}+\mu_0\varepsilon_0\frac{\partial\vec{E}}{\partial t}$$

It is clear from the above transformed equations that the original Maxwell's equations are recovered. Hence, the Maxwell's equations are unchanged under the operation of time reversal.

PROBLEM 1.3

How can we convert a linearly polarized beam to a circularly polarized beam?

SOLUTION

By inserting a $\lambda/4$ or $3\lambda/4$ plate with an axis at 45° to the input polarization, we can convert linearly polarized light into circularly polarized light.

PROBLEM 1.4

How can one isolate (eliminate or reroute) a reflected beam?

SOLUTION

It can be done by inserting a linear polarizer and $\lambda/4$ plate before the reflector. After the reflection, the second path of the $\lambda/4$ plate restores linear polarization, this time orthogonal to the linear polarizer.

PROBLEM 1.5

Refractive index of a polarizer is found to be 1.9128. Find the polarization angle and the angle of refraction.

SOLUTION

The refractive index of the polarizer is 1.9128. Brewster's law, i.e. $\mu = \tan i_p$ or $i_p = \tan^{-1} \mu$ gives the polarization angle $i_p = 62°24'$. The sum of the polarization angle and the angle of refraction is 90°, giving rise to the angle of refraction = $90° - 62°24' = 27°36'$.

PROBLEM 1.6

An unpolarized light is incident on a pair of polarizers with an intensity of $I_0 = 16$ W/m^2. The first polarizer is aligned to have its transmission axis at 50° from the vertical and the second polarizer has its transmission axis aligned at 20° from the vertical. What would be the intensity of light when it emerges from the first and second polarizers?

SOLUTION

According to Malus' law $I = I_0 \cos^2 \theta$, the beam is unpolarized before passing through the first polarizer. So $I_1 = 16 \times \frac{1}{2} = 8$ W/m^2. Now the intensity of light after passing through the second polarizer $I_2 = I_1 \cos^2 30° = 8 \times 3/4 = 6$ W/m^2.

PROBLEM 1.7

A He-Ne laser operating at 630 nm has an emission width of 10^{-6} nm. Calculate the temporal coherence length.

SOLUTION

The temporal coherence length is given by, $l = \frac{\lambda^2}{\Delta\lambda}$. So, $l = \frac{(630\times10^{-9})^2}{10^{-6}\times10^{-9}} = 396.9$ m

PROBLEM 1.8

An Argon laser operating in single mode has a linewidth of 7.5 MHz. What would be its coherence length?

SOLUTION

The coherence length is given by, $l = \frac{c}{\Delta\nu}$. So, $l = \frac{3\times10^8}{7.5\times10^6} = 40$ m

PROBLEM 1.9

A 10 mW He-Ne laser operating at 633 nm has a spot size of 10 mm. Assuming a Gaussian beam find the beam radius and intensity at a distance of 100 m from the source.

SOLUTION

The beam radius w_0 is half the spot size. The radius increases with distance z as per the following relation

$$w(z) = w_0\sqrt{1+\left(\frac{z}{z_0}\right)^2}$$

where $z_0 = \pi\omega_0^2/\lambda$ is the Rayleigh length.

$$w(z) = w_0\sqrt{1+\left(\frac{z}{\pi\omega_0^2/\lambda}\right)^2} = 5\times10^{-3}\times\sqrt{1+\left(\frac{100\times633\times10^{-9}}{3.14\times(5\times10^{-3})^2}\right)^2} = 6.423 \text{ mm}$$

$$\text{Area of spot} = \pi w^2 = 129.54\,\text{mm}^2$$

$$\text{Intensity, } I = \frac{\text{power}}{\text{area}} = \frac{10^{-2}}{129.54\times10^{-6}} = 77.19 \text{ W/m}^2$$

PROBLEM 1.10

A heterostructure diode has a cavity length of 500 microns. The peak wavelength of radiation is at 870 nm and the refractive index is 4. Find the index of the longitudinal mode and the separation $\Delta\lambda$.

SOLUTION

The index of the longitudinal mode is given by $m = \frac{2\mu L}{\lambda}$, where μ is the refractive index and L is the cavity length. So $m = \frac{2\times4\times500\times10^{-6}}{870\times10^{-9}} = 4598$.

The separation is given by $\Delta\lambda = \frac{2\mu L}{m} - \frac{2\mu L}{m+1} = \frac{2\mu L}{m^2} = \frac{\lambda^2}{2\mu L}$. So

$$\Delta\lambda = \frac{(870\times10^{-9})^2}{2\times4\times500\times10^{-6}} = 1.89\times10^{-10} = 0.189 \text{ nm}$$

PROBLEM 1.11

Find the threshold gain of the laser if the reflectivity of both the mirrors is $R_1 = R_2 = 0.099975$ and length of gain medium $l_g = 5$ cm.

SOLUTION

The threshold gain of laser is given by $\gamma_0 \geq \frac{1}{2l_g}\ln\left(\frac{1}{R_1R_2}\right)$. So the minimum value of RHS, i.e. the threshold gain is $\gamma_0 = \frac{1}{2l_g}\ln\left(\frac{1}{R_1R_2}\right)$

$$\gamma_0 = \frac{1}{2\times5}\ln\left(\frac{1}{0.0999\times0.0999}\right) = 0.46\%\ \text{cm}^{-1}$$

PROBLEM 1.12

Find the wavelength difference between two consecutive longitudinal modes of laser at an 800 nm wavelength in the cavity of a length of 300 μm. Given refractive index as 3.3.

SOLUTION

Difference in wavelength in two consecutive modes $\Delta\lambda$ is given by

$\Delta\lambda = \frac{\lambda^2}{2Ln_{\text{eff}}}$, where n_{eff} is the effective refractive index.

So,

$$\Delta\lambda = \frac{\left(8\times10^{-7}\right)^2}{2\times300\times10^{-6}\times3.3} = 3.23\times10^{-10}\text{m} = 0.32\ \text{nm}$$

PROBLEM 1.13

Considering an indium phosphide (InP) laser with laser length 400 μm and the reflectivity of both the mirrors as 0.45, find the effective mirror loss.

SOLUTION

The effective mirror loss α_m is given by

$\alpha_m = \frac{1}{L}\ln\left(\frac{1}{\sqrt{R_1R_2}}\right)$, where L is the laser length and R_1 and R_2 are the mirrors' reflectivity.

Hence

$$\alpha_m = \frac{1}{400\times10^{-6}} ln\left(\frac{1}{\sqrt{0.45\times0.45}}\right) = 19.96\ \text{cm}^{-1}$$

PROBLEM 1.14

Find the output coupling efficiency for an InGaAsP laser if the waveguide modal loss (α) is 13.7 cm^{-1} and effective mirror loss (α_m) is 26.2 cm^{-1}.

SOLUTION

The output coupling efficiency is given by $\eta_o = \frac{\alpha_m}{\alpha_m+\alpha}$. Hence,

$$\eta_o = \frac{26.2}{26.2+13.7} = 0.6566$$

PROBLEM 1.15

Find the photon lifetime τ_p in the cavity for which waveguide group velocity (v_g) is $c/3.1$, waveguide modal loss (α) is 14.5 cm^{-1}, laser length is 450 μm and reflectivity (R_1, R_2) of both the mirrors is 0.4.

SOLUTION

The photon lifetime τ_p can be obtained by the following relation

$$\frac{1}{\tau_p} = \frac{v_g}{L}\ln\left(\frac{1}{\sqrt{R_1R_2}}\right) + v_g\alpha$$

Hence, $\frac{1}{\tau_p} = \frac{3\times10^8}{3.1\times450\times10^{-6}}\ln\left(\frac{1}{\sqrt{0.4\times0.4}}\right) + \frac{3\times10^8}{3.1}\times14.5\times10^2$

or $\tau_p = 2.96\times10^{-12} = 2.96$ ps

PROBLEM 1.16

Calculate the threshold gain for a system with laser length (L) of 350 μm, waveguide group velocity (v_g) of $\frac{c}{4.2}$, waveguide model loss (α) of 18 cm^{-1}, facet reflectivities ($R_1 = R_2$) of 0.4 and active region mode confinement factor (Γ_a) of 0.09. What would be the threshold carrier density if the transparency carrier density (n_{tr}) is 2.24×10^{18} cm^{-3} and density constant (γ_0) is 1300 cm^{-1}?

SOLUTION

Threshold gain, $\gamma_{th} = \frac{1}{\tau_p\Gamma_a v_g}$, where photon lifetime in cavity is given by

$$\frac{1}{\tau_p} = \frac{v_g}{L}\ln\left(\frac{1}{\sqrt{R_1R_2}}\right) + v_g\alpha$$

$$\tau_p = 3.1688\times10^{-12} = 3.168 \text{ ps}$$

Threshold gain, $\gamma_{th} = \frac{1}{\tau_p\Gamma_a v_g} = 490.89$ cm^{-1}

The relation between threshold carrier density and threshold gain is given by

$$\gamma_{th} = \gamma_0 \ln\left(\frac{n_{th}}{n_{tr}}\right)$$

Hence

$$n_{th} = n_{tr}\exp\left(\frac{\gamma_{th}}{\gamma_0}\right)$$

Or, $n_{th} = 3.2676\times10^{18}$ cm^{-3}.

PROBLEM 1.17

Calculate the peak power of the pulses and the average output power of the Q-switched laser which emits pulses of 10 μJ of duration 1 ns. Given the repetition rate of the pulses as 10 kHz.

SOLUTION

The peak power P_{peak} of the pulse can be found simply by the ratio of the energy of the pulse and its duration. Hence,

$$P_{peak} = \frac{10 \times 10^{-6}}{10^{-9}} \frac{J}{\sec} = 10 \text{ kW}$$

The average output power P_{avg} is the pulse energy emitted in one period. Hence,

$$P_{avg} = 10 \times 10^{-6} \text{J} \times 10 \times 10^{3} \text{Hz} = 100 \text{ mW}$$

PROBLEM 1.18

In a special kind of surgery, each pulse produced by an argon-fluoride excimer laser lasts only for 10 ns and delivers 2.5 mJ of energy. Considering the laser to have a 0.85 mm diameter and if it emits 60 pulses in 1 second, calculate the power delivered in each pulse. Also find the average intensity of the beam during each pulse and the average power the laser generates.

SOLUTION

The power P_{pulse} in a pulse is given by the ratio of the energy delivered and the time in which the energy is delivered. Hence

$$P_{pulse} = \frac{2.50 \text{ mJ}}{10 \text{ ns}} = 2.5 \times 10^{5} \text{ W}$$

The average intensity is given by

$$I_{avg} = \frac{P_{pulse}}{\text{Area of spot}} = \frac{2.5 \times 10^{5}}{\pi \times \left(\frac{0.85 \times 10^{-3}}{2} \right)^{2}} = 4.4 \times 10^{11} \text{ W/m}^{2}$$

The average power of laser is given by the total energy delivered per unit time. Since the laser emits 60 pulses in 1 second, P_{avg} is obtained as

$$P_{avg} = \frac{U}{\Delta t} = \frac{60 \times 2.50 \times 10^{-3}}{1} = 0.15 \text{ W}$$

PROBLEM 1.19

CO_2 laser has a bandwidth of 50 MHz at 10.6 μm. Considering the spectral profile to be rectangular and the length of the cavity as 1 m, calculate the distance at which the cavity mirror must be placed to get at least one of the modes in the amplification band.

SOLUTION

The central frequency of CO_2 is situated exactly in the middle of the two longitudinal modes. Thus, the gap separating the end of the spectral band from the closest longitudinal mode would be $\frac{1}{2}\left(\frac{c}{2\mu L} - \Delta\nu\right)$.

The frequency of mode can be expressed as $\nu_m = \frac{mc}{2\mu L}$, where L is the length of the cavity. The distance that the cavity mirror must be moved would be $dL = \frac{L}{2}\left(\frac{\lambda}{c}\right)\left(\frac{c}{2L} - \Delta\nu\right)$.

Hence

$$dL = \frac{1}{2}\left(\frac{10.6\times10^{-6}}{3\times10^{8}}\right)\left(\frac{3\times10^{8}}{2\times1} - 50\times10^{6}\right) = 1.76\ \mu\text{m}$$

PROBLEM 1.20

Calculate the number of possible longitudinal modes and the separation between two modes in the He-Ne laser (wavelength 632.8 nm) having a cavity of length of 1 m.

SOLUTION

The number of possible modes can be estimated by using $2\mu L = N\lambda$, where L is the length of cavity, μ is the refractive index of the lasing medium, N is the number of all the possible modes and λ is the wavelength. Taking refractive index of the medium as 1, we can calculate the number of modes as

$$N = \frac{2L}{\lambda} = \frac{2\times1}{632.8\times10^{-9}} = 3.16\times10^{6}\,\text{modes}$$

Frequency separation is given by

$$\Delta f = c/2L = \frac{3\times10^{8}\,\text{m/sec}}{2\times1} = 1.5\times10^{8}\,Hz = 150\ \text{MHz}$$

Wavelength separation is evaluated using

$$\Delta\lambda = \frac{\left(632.8\times10^{-9}\right)^{2}}{2} = 2\times10^{-3}\,A^{\circ}$$

PROBLEM 1.21

Considering waist and laser in the same plane, find the radius of the beam at the waist for a He-Ne laser emitting at 633 nm and making a spot size with a radius 10 cm at $1/e^2$ times the distance of 5×10^2 m from the laser.

SOLUTION

The problem can be solved by using the formula that links the divergence of the beam and the waist size.

$\theta = \frac{\lambda}{\pi w_0}$. So $w_0 = \frac{\lambda}{\pi\theta}$, where θ is expressed in radians and is equal to $\frac{10\times10^{-2}}{5\times10^{2}}$.
Putting the values we get $w_0 = \frac{633\times10^{-9}}{\pi\times\theta} = 1$ mm

PROBLEM 1.22

Determine the skin depth for copper, where electrical conductivity and permeability at room temperature are 57×10^{6} $(\Omega\text{-m})^{-1}$ or S/m and $4\pi\times10^{-7}$ H, respectively. Frequency of incident EM wave is 60 Hz.

SOLUTION

Skin depth, $\delta = \sqrt{\frac{2}{\omega\mu\sigma}}$

$$\omega = 2\pi f = 2 \times 3.14 \times 60 \text{ Hz} = 376.8 \text{ rad/sec}$$

$$\mu = \mu_0 = 4\pi \times 10^{-7} \text{ H} = 12.56 \times 10^{-7} \text{ H}$$

$$\sigma = 57 \times 10^6 \text{ S/m}$$

So, $\delta = \sqrt{\frac{2}{376.8 \times 12.56 \times 10^{-7} \times 57 \times 10^6}} = 8.61 \text{ mm}$

PROBLEM 1.23

If frequency of an EM wave is increased from 60 to 10^6 Hz and then from 10^6 to 10^9 Hz, what would be the effect that one can see on the skin depth in the case of copper ($\sigma = 57 \times 10^6$ S/m; $\mu \sim \mu_0$)?

SOLUTION

Given: $f = 10^6$ Hz, $\omega = 2\pi f = 2 \times 3.14 \times 10^6 = 6.28 \times 10^6$ rad/sec, $\mu = \mu_0 = 4\pi \times 10^{-7}\text{H} = 12.56 \times 10^{-7}$ H and $\sigma = 57 \times 10^6$ S/m. Hence, skin depth would be

$$\delta = \sqrt{\frac{2}{\omega\mu\sigma}} = \sqrt{\frac{2}{6.28 \times 10^6 \times 12.56 \times 10^{-7} \times 57 \times 10^6}} = 0.067 \text{ mm}$$

At $f = 10^9$ Hz, Skin depth, $\delta = 2.11$ μm

This shows that high-frequency waves penetrate a shorter distance into a conductor than lower ones and this penetration goes down very fast.

PROBLEM 1.24

Calculate the relaxation time (τ) for mica, whose electrical conductivity and relative permeability at room temperature are 10^{-11} S/m and 5.8, respectively.

SOLUTION

Relaxation time τ is given by $\tau = \frac{\varepsilon}{\sigma}$. Given $\sigma = 10^{-11}$ S/m and $\varepsilon_r = \frac{\varepsilon}{\varepsilon_0} = 5.8$. Hence $\varepsilon = 5.8$ $\varepsilon_0 = 5.8 \times 8.85 \times 10^{-12} = 5.1 \times 10^{-11}$ F/m. So the relaxation time

$$\tau = \frac{\varepsilon}{\sigma} = \frac{5.1 \times 10^{-11}}{10^{-11}} = 5.1 \text{ second}$$

PROBLEM 1.25

In laser-plasma interaction, calculate the maximum energy gain and dephasing length for an electron if it is accelerated by the laser of frequency 10^{14} Hz in a plasma of density $10^{25}/\text{m}^3$.

SOLUTION

Maximum energy gain, $W_{\max} = \frac{4mc^2 n_c}{n_e}$, where n_e is the plasma density and n_c is the critical density. Now,

$$n_c = \frac{\epsilon_0 m\omega^2}{e^2} = \frac{8.85 \times 10^{-12} \times 9.1 \times 10^{-31} \times \left(2\pi \times 10^{14}\right)^2}{\left(1.6 \times 10^{-19}\right)^2} = 1.24 \times 10^{26}/\text{m}^3$$

Hence, $W_{max} = \frac{4mc^2}{n_e}\frac{n_c}{} = 406.22 \times 10^{-14}\text{J} = 25.38$ MeV.

Dephasing length, $L_d = \frac{n_c}{n_e}\lambda_p$ together with λ_p as the plasma wavelength.

Now, $\lambda_p = \frac{c}{v} = \frac{3\times10^8}{10^{14}} = 3\ \mu\text{m}$. Hence $L_d = \frac{1.24\times10^{26}\times3\times10^{-6}}{10^{25}} = 37.2\ \mu\text{m}$

PROBLEM 1.26

Consider a two-mirror cavity employing a high reflector ($R_1 = 99.8\%$) and an output coupler ($R_2 = 98.0\%$) and the gain medium filling the entire cavity length $m = L = 50$ cm. The gain medium is an electrical discharge in a 7:1 mixture of He:Ne with neon partial pressure $P_{\text{Ne}} = 0.3$ torr and temperature $T = 300$ K. Find threshold gain.

SOLUTION

The threshold gain is given by

$$\gamma_0 = \frac{1}{2l_g}\ln\left(\frac{1}{R_1R_2}\right)$$

Reflectivity of mirrors is given as

$$R_1 = 99.8\% = 0.998$$

$$R_2 = 99\% = 0.99$$

Now gain is calculated as

$$\gamma_0 = \frac{1}{2\times50}\ln\left(\frac{1}{0.998\times0.99}\right)\text{cm}^{-1} = 1.2\times10^{-4}\text{cm}^{-1}$$

Suggested Reading Material

Chen, F. F. (1984). *Introduction to plasma physics and controlled fusion*. New York: Plenum Press.

Khazanov, G. V., & Krivorutsky, E. N. (2013). Ponderomotive force in the presence of electric fields. *Physics of Plasmas, 20*(2), 022903.

Kupersztych, J., Raynaud, M., & Riconda, C. (2004). Electron acceleration by surface plasma waves in the interaction between femtosecond laser pulses and sharp-edged overdense plasmas. *Physics of Plasmas, 11*(4), 1669–1673.

Malik, H. K. (2013). Chapter 4 - Electromagnetic waves and their application to charged particle acceleration. *Wave Propagation/Book 2*, London: INTECH Open Science, Pages 73–112. http://dx.doi.org/10.5772/52246

Malik, H. K. (2014). Density bunch formation by microwave in a plasma-filled cylindrical wave-guide. *Europhysics Letters, 106*(5), 55002.

Malik, H. K., & Aria, A. K. (2010). Microwave and plasma interaction in a rectangular waveguide: Effect of ponderomotive force. *Journal of Applied Physics, 108*(1), 013109.

Malik, H. K., & Singh, A. K. (2018). 2nd Ed. *Engineering Physics. McGraw Hill Education.*

Tomar, S. K., & Malik, H. K. (2013). Density modification by two superposing TE_{10} modes in a plasma filled rectangular waveguide. *Physics of Plasmas, 20*(7), 072101.

2

Terahertz Radiation Generation Using Semiconducting Materials and Nanostructures

2.1 Introduction

Terahertz (THz) radiation spans a very significant part of the electromagnetic spectrum that remained unexplored for decades. Its frequency lies between about 300 GHz to 3 THz and it demarcates the important subjects, namely electronics and photonics, because its lower frequency is close to the microwave region and the upper frequency is close to the infrared (IR) region. The technological aspect of these types of radiations further increases because of their vital applications in numerous fields, which include materials characterization, medical imaging and diagnostics, non-destructive testing for security, tomography, and so forth. Only after the advent of mode-locked ultrafast pulsed lasers, it has become possible to explore the fresh techniques of THz generation and detection. THz radiation can be generated as continuous or pulsed waveforms through different techniques. For example, Grischkowsky (1993) has generated a continuous THz wave through difference frequency mechanism and THz pulses via optical rectification (OR) within electro-optic crystals.

Early researchers had made the use of laser-based photoconductive switches for the measurement of ultrashort electrical pulses that were guided by transmission lines. Later Mourou et al. (1981), Grischkowsky (1993), Kaiser (1993), and Robertson (1995) had realized that the photoconductive antennae (PCAs) could be used to couple radiation into free space or to collect radiation from the free space. The improvements of such devices were done from time to time using different materials and proposing novel geometries and antenna structure framework (Heidemann et al. 1983; DeFonzo and Lutz 1987) to optimize the performance of the photoconductive switches.

The production of the electromagnetic radiation by ultrafast laser pulses and its detection by time synchronized-gated technique has resulted in the development of the time-domain spectrometer, which proved to be an immensely useful tool while working in the far IR region to explore a range of THz frequencies (Grischkowsky 1993; Kaiser (1993). The spectrometer measurements provide complete information about the complex parameters, such as the reflection coefficient and transmission coefficient as a frequency function. Moreover, with the help of this technique, it has been possible to evaluate the complex dielectric function or complex conductivity of various samples (Darrow et al. 1990; Nuss et al. 1991; Robertson et al. 1991, 1992; Özbay et al. 1994; Ralph et al. 1994; Cheville and Grischkowsky 1995, 1999; Harde et al. 1997, 2001; Ronne et al. 1997; Jeon and Grischkowsky 1998; Gallot et al. 1999; Rønne et al. 1999; Schall et al. 1999; Corson et al. 2000; Dodge et al. 2000; Huggard et al. 2000; Jeon et al. 2000; Markelz et al. 2000; Kaindl et al. 2001).

The introduction of large-area photoconductive emitters led to the further development of THz frequencies with the use of ultrafast lasers; this type of work was performed by Zhang et al. (1990), Xu et al. (1991), and Leitenstorfer et al. (2000). Large gap PCAs are not suitable for generating the ultrafast electric field transients on transmission lines, but they are efficient for radiating THz pulses directly into free space. Kersting et al. (1998) have shown the major advantage that large gap PCAs provide over small gap PCAs, which is the high power THz output. Expanding the irradiated gap region eliminates the saturation effects, which may arise in photoconductive devices at high-power laser fluence. Such structures are easy to frame. The large emission area also produces a relatively collimated beam of THz radiation, which may be an important feature for various applications. Moreover, the direction of the emitted beam can be controlled by changing the pump laser's angle of incidence (Auston and Cheung 1985; Xu et al. 1991).

Large-area PCAs can be categorized based on the direction of the flow of the transient photocurrent. The devices in which electrical bias voltage is applied in the plane of surface with the help of pair of widely separated electrodes, the photocurrent transients are also directed along the surface. In another case, the bias field is applied perpendicular to the surface. This bias field is generated by intrinsic band bending, which is associated with trapped charges at the surface in a depletion field device. Such a field can also be applied externally in a p–i–n or Schottky-barrier structure. The lack of structure of antennae means that the spectral characteristics of the THz emission, which enables a broad spectral response, will be limited only by the dynamics of the inherent carrier and the properties of the pump laser (Han et al. 2000).

Although the details of the produced photocurrent depend on the geometrical structures, certain properties are independent of structures that can be analyzed within the same general framework of carrier transport and radiation properties. For example, an analysis of the resulting electric field transients can provide the details of the collective processes dynamics of the hot carrier and the material properties of semiconductors with sub-picosecond time resolution. Interest in exploiting these transients is increasing daily for the future innovative applications of coherent THz spectroscopy and imaging. A variety of semiconducting materials, such as InAs, InSb, GaSb, GaAs, InP, and novel geometry and framework ideas are being used (Auston and Cheung 1985; Zhang et al. 1990; Xu et al. 1991; Mittleman et al. 1997; Kersting et al. 1998; Han et al. 2000; Leitenstorfer et al. 2000; Krotkus et al. 2008) to enhance the THz emissions. Hence, the quest among the researchers is to produce high-power THz emission.

2.2 Pulsed THz Generation

The most common techniques for the generation of broadband pulsed THz radiation from semiconducting materials, using ultrafast excitation pulses, are the photoconductive emission, optical rectification, and the transient current effect.

2.2.1 Photoconductive Emission Mechanism

Devices used in the photoconductive emission mechanism as emitters are photoconductive switches or PCAs. The photoconductive approach is based on irradiating the gap of the PCA made up of special types of semiconducting materials (usually III-V group compounds) having high resistivity, low carrier lifetime, and high damage threshold. Suitable examples are LT-GaAs and Si-GaAs. The PCA is shown in Figure 2.1, which consists of two main components, i.e., highly resistive direct semiconductor thin film as a substrate and

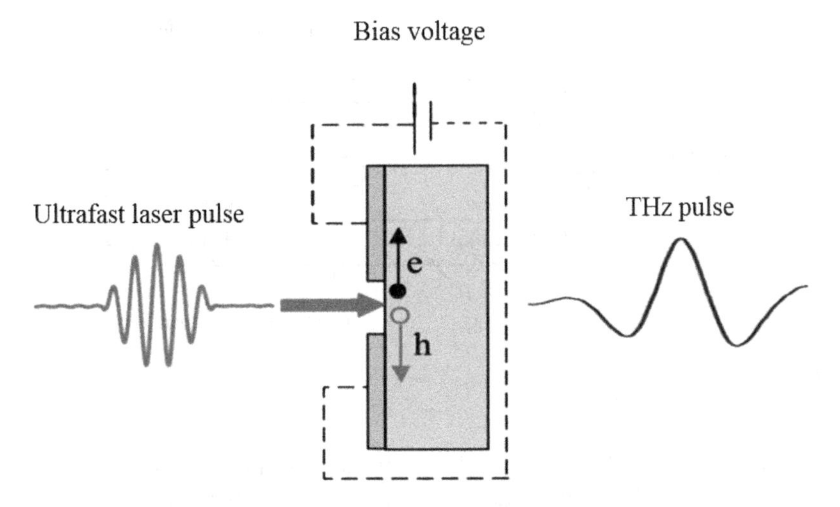

FIGURE 2.1
THz generation using a photoconductive antenna.

two electrical contact pads. A bias voltage is applied between the two contacts, which produces a negligible current because of the high resistivity of the semiconducting material. When the gap between the contacts is excited with the ultrafast laser pulses, the photocarriers are generated in the semiconductor, increasing its conductivity in a sub-picosecond timescale. The conductivity may decrease subsequently by recombination, trapping, and momentum scattering of the carriers in the sub-picosecond to nanosecond time domain depending on the material of the semiconductor. Under the influence of the optical pulse, these photocarriers are accelerated by the electric field, resulting in broadband electromagnetic pulse. The electric field of these pulses is time-dependent with frequencies in the THz region. Owing to the time-dependent THz pulse, a signal is produced in the outer circuit.

In the far-field approximation, the THz electric field $E_{THz}(r,t)$ emitted from a PCA at any point r and time t can be expressed as

$$E_{\text{THz}}(r,t) = -\frac{1}{4\pi\varepsilon c^2}\int\left[\frac{\partial J(r',t')}{\partial t'}\right]_{t'=t^r}\frac{\sin\theta}{|r-r'|}d^3x' \tag{2.1}$$

where ε and c are the dielectric constant of the medium and the speed of light, respectively. The time derivative is taken at the retarded time $t^r = t - |r - r'|/c$, where θ is the angle between the generated photocurrent and the direction of detection, and $J(r',t')$ is the current density on the PCA; the coordinating variables with prime are of the source. Hu et al. (1991) have given the relationship between the current density and the applied bias voltage E_{bias}. Hence, the equation can be written as

$$J(t) = \frac{\sigma(t)E_{\text{bias}}}{\frac{\sigma(t)Z_0}{1+n_d}+1} \tag{2.2}$$

where $\sigma(t)$ is the conductivity, Z_0 is the characteristic impedance of the vacuum, and n_d is the refractive index of the substrate.

LT-GaAs is the most commonly used substrate due to its ultrashort carrier lifetime, high carrier mobility, and high damage field. Many other materials, such as ion-implemented GaAs,

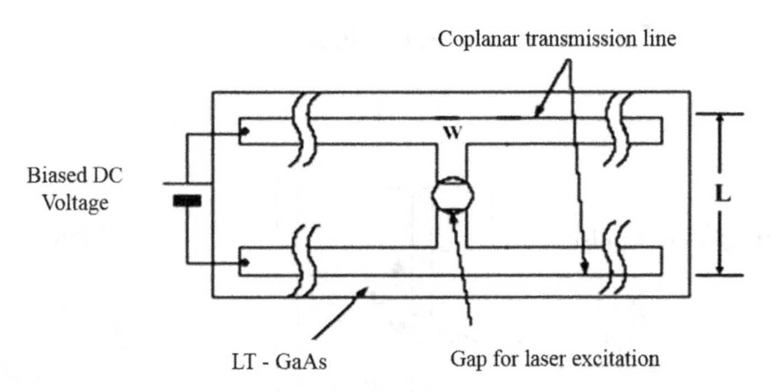

FIGURE 2.2
Photoconductive antenna structure.

InGaAs, GaAsN, and GaAsBi, also have been studied as photoconductive emitters (Castro-Camus and Alfaro 2016). The schematic diagram of the PCA structure is shown in Figure 2.2. The LT-GaAs substrate is mainly used, but it can be replaced with other semiconductors such as GaSe. Biased DC voltage is applied in the gap, which is encircled in the diagram. This is also irradiated by laser pulses from the pump laser. The gap is usually of the order of 5–10 μm. The overall size of the antenna is generally 10–20 μm (width) × 30–50 μm (length).

The PCA can also be used for the detection of THz waves. When the THz waves strike the antenna, they cause a slight change in the transient current (TC). This change in the current $J(t)$ describes the incident electromagnetic field as

$$E_{\text{THz}} \propto \frac{dJ(t)}{dt} \tag{2.3}$$

The maximum field of THz wave that can be measured by the PCA depends on the mobility (μ) of the carriers, laser energy ($h\upsilon$), biasing voltage (V_b), reflectance of the substrate (R), and average power of the laser (P_{in}), and is given by

$$E_{\max} = e\mu\left(\frac{1-R}{h\upsilon}\right)\frac{P_{in}V_b}{D} \tag{2.4}$$

Here the quantity D is the gap that is excited by the laser. Moreover, the quality of emitted THz radiation depends on various factors, but largely, antenna geometry decides this in addition to the material of the photoconductive (PC) substrate such as bandgap, carrier lifetime, and carrier mobility. The structure of the antenna influences the efficiency of THz generation. Bow-tie and log-periodic tooth antennas are found to be more efficient (Castro-Camus and Alfaro 2016). The basic geometry of the four antennae is shown in Figure 2.3.

The range of THz frequencies emitted by antenna also depends on the factors such as duration of optical pulse, applied bias voltage, and carrier scattering time. Nanostructures are included to reduce the lifetime of carriers without affecting their mobility. Lai et al. (2017) reported that PCA coated with manganese ferrite nanoparticles (NPs) enhances the THz efficiency to a greater extent.

2.2.2 Optical Rectification Mechanism

Optical rectification is a nonlinear second-order optical process that does not depend on external bias voltage for producing THz radiation. In this process, two frequencies

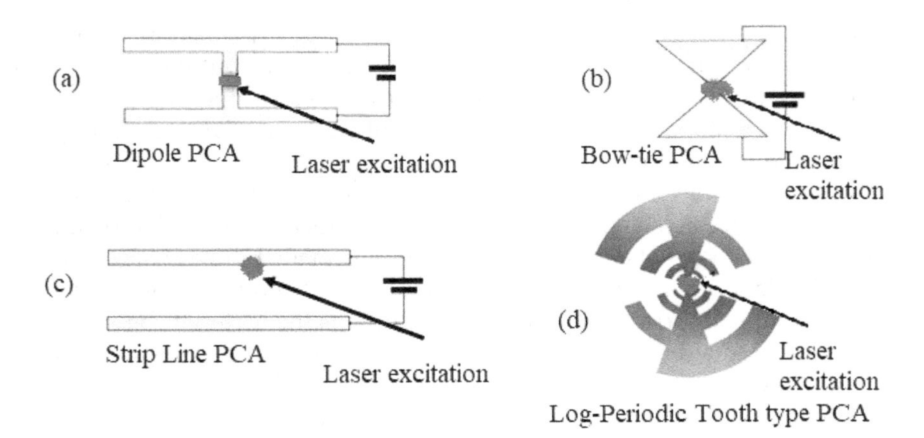

FIGURE 2.3
Various structures of photoconductive antennae (PCA). (a) Dipole. (b) Bow-tie. (c) Strip line. (d) Log-periodic tooth.

either from a near-IR (NIR) source or laser fields mix through the second-order susceptibility tensor to generate a THz radiation field. A schematic of this mechanism is shown in Figure 2.4.

When laser or NIR radiation falls on the crystal, the nonlinearity of the material is exploited by optical excitation. It alters the state of polarization of the crystal to produce two components, namely, sum frequency and difference frequency components. For the generation of THz radiation, the second-order nonlinear polarization component [e.g., $P^{(2)}(\omega)$] is important. According to Hu et al. (1991) and Castro-Camus and Alfaro (2016), the THz electric field in terms of this can be written as

$$E_i^{\mathrm{THz}}(\omega) \propto \frac{\partial^2 P_i^{(2)}(\omega)}{\partial t^2} = \frac{\partial^2 \chi_{ijk}^{(2)}}{\partial t^2}(\omega = \omega_1 - \omega_2; \omega_1, \omega_2) E_j^{\mathrm{NIR}}(\omega_1) E_k^{\mathrm{NIR}}(\omega_2) \tag{2.5}$$

where $\chi_{ijk}^{(2)}$ is the second-order nonlinear susceptibility tensor; E_1 and E_2 are the incident excitation electric field components at the frequencies ω_1 and ω_2, respectively; and $P_i^{(2)}(\omega)$ and $E_i^{\mathrm{THz}}(\omega)$ are the ith component of induced second-order polarization at frequency $\omega = \omega_1 - \omega_2$ and the corresponding THz field, respectively (Khurgin 1994).

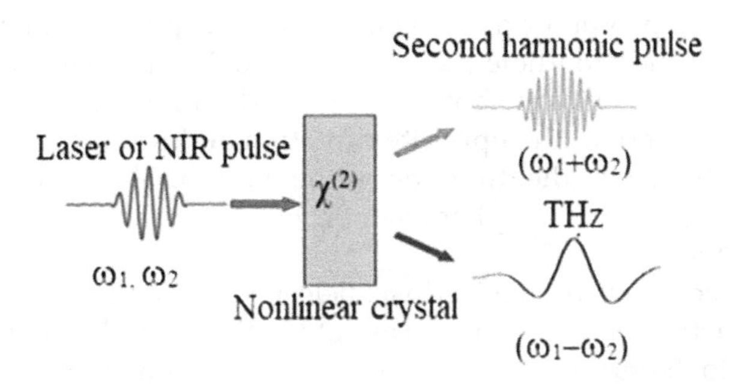

FIGURE 2.4
THz generation using optical rectification effect.

The underlying phenomenon in OR mechanism can be explained as follows. When a nonlinear medium is irradiated by an ultrafast femtosecond pulsed laser, the medium gets polarized and the molecular dipoles are formed within the medium. Because the spectral bandwidth associated with such short pulses is very large, a beating polarization is realized due to the mixing of different frequency components. These oscillating dipoles then result in the emission of electromagnetic radiation in the THz regime. The OR effect is analogous to the classical electrodynamic emission of radiation by an accelerating or decelerating charge. However, the difference is made by the charges that are now in a bound dipole form, and the THz generation depends on the second-order susceptibility of the nonlinear optical medium. Zinc telluride is the material commonly used for the generation of radiation in the 0.5–3 THz range. Because THz generation through OR mechanism relies on the nonlinear property of the material and not on the generation of electron-hole pairs, the THz radiation can be observed even with below-band-gap excitation via this mechanism. Enhanced nonlinear coefficients of a material can be realized by above-band-gap excitations, which effectively improve the THz output power. The OR emitters are highly efficient and are mainly used because of a simpler execution of the mechanism. Broadband THz radiation is generated by this method. The efficiency of the OR mechanism depends on the incident pulse characteristics, nonlinear coefficients viz. nonlinear absorption and refraction coefficients of the material, the orientation of the crystal, THz absorption in the material, and the phase-matching conditions. Till date, numbers of nonlinear materials have been investigated for THz emission through the OR mechanism, which include semiconductors, such as ZnTe, GaP, GaAs, InAs, InP, InSb, GaSe, and CdTe; dielectric crystals such as lithium niobate ($LiNbO_3$); organic materials such as DAST; and the metals such as gold.

2.2.3 Transient Current Effect

The transient current or current surge mechanism is a linear effect. TCs may arise either due to the drift of the carrier because of the surface field (SF) effect or due to diffusion current because of the difference in the rates of diffusion of the charge carriers. In view of this, the TC effect can be classified as a SF effect or photo-Dember effect.

2.2.3.1 Surface Field (SF) Effect

Schematic of the mechanism for two different types of semiconductors is shown in Figure 2.5. For many semiconductors, a surface depletion field is formed due to the presence of the Fermi level and bending of the conduction band and valence band at the surface. When a p-type or an n-type semiconductor is irradiated with ultrafast photoexcitation, the electron-hole pairs are liberated in the semiconductor (electrons are shown by 'e' and holes are shown by 'h' in the figures). The intrinsic SF drives the two kinds of carriers in the opposite direction and produces a photocurrent with formation of the dipoles in the direction of the normal. These transient dipoles oscillate and emit THz pulses. This phenomenon has been shown by Dember (1931) and Dekorsy et al. (1993).

Thus, the SF effect relies on the intrinsic field instead of the external bias in the case of the photoconductive emission. In the case of the SF effect, the flow of current is in the direction normal to the surface. In this process, the direction of the surface depletion field is decided by the types of doping and the position of the surface states with respect to the Fermi level. The surface depletion field $E_d(x)$ as a function of the distance x perpendicular

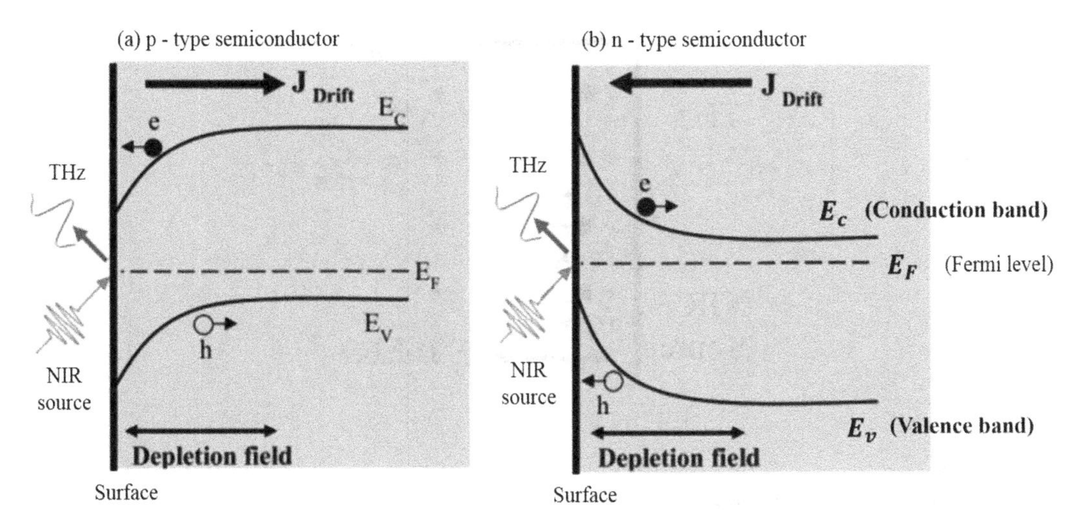

FIGURE 2.5
THz generation using the surface field effect.

to the surface can be written in accordance with the work carried out by Dember (1931) and Zhang and Auston (1992) as

$$E_d(x) = (eN / k)(W - x) \tag{2.6}$$

where N is the impurity concentration and W is the depletion width. The depletion width can be given as

$$W = \sqrt{(2k / eN)\left[V - (kT / e)\right]} \tag{2.7}$$

together with V as the potential barrier and kT/e as the thermal energy. The generated drift current of the photoexcited carriers, due to this SF effect, and the corresponding THz radiation field can be represented as

$$J_{\text{drift}} = eE\left(n\mu_n + p\mu_p\right) \tag{2.8}$$

$$E_{\text{THz}} \propto \frac{\partial J_{\text{drift}}}{\partial t} \tag{2.9}$$

2.2.3.2 Photo-Dember Effect

In the photo-Dember effect (Figure 2.6), a strong spatial gradient is formed near the semiconductor surface because of the difference in the diffusion rates of electrons and holes. Electrons, that are more mobile, tend to diffuse rapidly away from the surface, whereas holes remain in the vicinity of the surface owing to their comparatively lower mobility. Hence, transient dipoles are formed near the surface, along the normal direction, which finally emit THz pulses due to their oscillations in view of the field of NIR source.

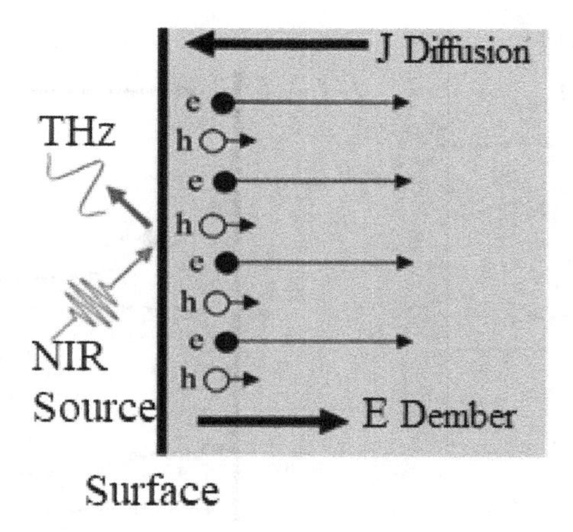

FIGURE 2.6
Schematic for THz generation using the photo-Dember effect.

The diffusive currents of the electrons (J_e) and the holes (J_h) can be represented as

$$J_e \propto eD_e \frac{d\Delta n}{dx} \tag{2.10a}$$

$$J_h \propto eD_h \frac{d\Delta p}{dx} \tag{2.10b}$$

Here e is the electronic charge; Δn and Δp are the densities of photoexcited electrons and holes, respectively; and D_e and D_h are the diffusion coefficients of the electrons and the holes, respectively. Diffusion coefficients can be evaluated in terms of the temperature T and mobility μ of the respective carriers, as

$$D = kT\mu / e \tag{2.11}$$

here k is the Boltzmann constant. The diffusion current is obtained as $J_{\text{diffusion}} = J_e + J_h$. Hence, it is proportional to the carrier mobility. Further, the diffusion current of the electrons, due to large mobility and kinetic energy, is much larger than that of the holes. Hence, the amplitude of the THz radiation is approximately proportional to the electron mobility, i.e.

$$E_{\text{THz}} \propto \frac{\partial J_n}{\partial t} \propto \mu_e(T) \tag{2.12}$$

According to Zhang and Auston (1992), the photo-Dember effect is more pronounced in narrow bandgap semiconductors, such as InAs and InSb, due to the higher electron mobility, weak depletion field, and the short absorption depth.

2.3 THz Radiation Generation by Metallic Nanoparticles

Nanomaterials are superior due to their unique size-dependent properties. These are classified into 0D, 1D, 2D, and 3D based on the dimensions. Nanoparticles (NPs) being 0D (all dimensions <100 nm) structures have enormous applications in different areas, mainly in materials science, pharmaceuticals, optics, and photonics. The NPs, either metallic or semiconducting, behave differently when they are irradiated with light. In metallic NPs, the relatively free electrons oscillate collectively with a frequency, called a plasmon frequency. The high value of third-order susceptibility near plasma frequency in NPs makes them potential candidates for nanoelectronic, nano-optic, nanophotonics, and biomedical applications (Sharma et al. 2016).

The easy and low-cost synthesis techniques of NPs like hydrothermal synthesis, sol-gel, and coprecipitation techniques make them even more useful in the field of science and technology. The investigations of their physical, chemical, and structural properties have been largely explored to understand the dynamics of the NPs. Many researchers such as Gao et al. (2009), Welsh and Wynne (2009), Zhang et al. (2015), Suvorov et al. (2012), and Javan and Erdi (2017) have studied the THz wave radiation from nanostructured metallic surfaces. The reason behind this phenomenon could be explained by the electron accelerations because of the ponderomotive potential linked with the surface plasmon (SP) excitation.

Some metal films, for example, gold or silver, with particular nanostructured surfaces have been found to exhibit strong field enhancement on the illumination by the femtosecond laser pulse. It is due to the surface plasmons (SPs), which are resonantly driven oscillations of collective electrons. When the laser pulse is incident, a coupling of the incident photons with an SP wave (SPW) occurs. It further leads the electromagnetic energy to be highly confined to the sub-wavelength spatial region, leading to increment near the surfaces and, hence, introducing the multiphoton photoelectric emission effect. A large amount of work has been done using femtosecond lasers for the generation of THz radiation from nanostructured metal surfaces. It includes work done by Suvorov et al. (2012) and Ramakrishnan and Planken (2011) on films of metal, Garwe et al. (2011) and Schmidt et al. (2012) on shallow metallic gratings, and Polyushkin et al. (2011) and Ramakrishnan et al. (2012) on metallic nanoparticle (NP) ordered arrays.

However, not all the metallic NPs have a frequency range in the THz regime; some of them have frequency out of the THz radiation. The optical properties of NPs can be tailored by changing the number of parameters, including their orientations concerning the field of the incident laser. Also, many other semiconducting NPs can be explored for the THz radiation generation with significantly low plasmon frequency. However, the THz radiations generated from the above mentioned techniques have several flaws and drawbacks which limit their many applications.

2.3.1 Analytical Treatment for Laser and Nanoparticle Interaction

Initially, a bulk medium is considered which consists of NPs, and this medium is subjected to an external electric field $\vec{E}$ of the incident laser radiation. The NPs are considered to have basal planes (plane perpendicular to the principal axis in the crystal systems is called basal plane) aligned in two different directions, i.e., parallel to the electric field ($\vec{E} \parallel \hat{s}$) and normal to the electric field ($\vec{E} \perp \hat{s}$), where $\hat{s}$ is the symmetry axis normal to the plane inside the NP.

The overall macroscopic density of NPs is written as

$$n_0 = \sum_s \frac{4\pi}{3} l_s n_{0s} \tag{2.13}$$

where $s = \|, \perp$ gives components parallel and perpendicular to the field, respectively. The factor $\frac{4\pi}{3} l_s$ represents the proportion of volume occupied by the NPs for both the orientations and volume of the voids of the unit cell. The parameter l_s is given as

$$l_s = \left(\frac{r_s}{d_s}\right)^3 \tag{2.14}$$

where r_s and d_s are the average radius of NPs and distance between any two consecutive NPs, respectively. Here, NPs are assumed to be spherical in shape, and the charge to be uniformly distributed throughout them. Hence, the macroscopic density can be written as

$$n_0 = \sum_s \frac{\frac{4\pi}{3}(r_s)^3}{(d_s)^3} n_{0s} \tag{2.15}$$

We also consider the density ripples such that the ripples density is varied in the following manner

$$n' = n_q e^{iqz} \tag{2.16}$$

together with q as the wavenumber of the density ripples, representing the periodicity. In order to create such ripples, the stream of NPs can be made to flow through a nozzle with maintained density modulation with a background host noble gas (Ar).

Let the two Gaussian laser beams are propagating simultaneously through the medium in the z-direction with their electric fields as

$$\vec{E}_j = \frac{1}{2} E_0 \hat{y} e^{-\frac{y^2}{a_0^2}} e^{i(k_j z - \omega_j t)} + c.c. \quad j = 1, 2 \tag{2.17}$$

where E_0 and a_0 are the laser beam parameters such as amplitude and beam width, respectively, and k_j and ω_j represent, respectively, their wave number and frequency.

The first-order equation of motion of electrons due to nonrelativistic interaction of the electric field of the lasers and the electrons in the electron cloud of NP is given by

$$\frac{d\vec{v}_j}{dt} + \Gamma \vec{v}_j + \frac{\omega_p^2}{3} \vec{r}_j = \frac{-e}{m} \vec{E}_j \tag{2.18}$$

where Γ is the damping factor, which is mainly due to the electron scattering and other related loss mechanisms; e is the magnitude of the electronic charge; m is the mass of an electron; and $\omega_p = \sqrt{\frac{4\pi e^2 n_0}{m}}$ is the plasma frequency in centimetre-gram-second (CGS) units. The term $\frac{\omega_p^2}{3} \vec{r}_j$ represents the restoring force that tends to bring the electronic cloud back to the equilibrium position, due to the immobile ions in spherical NPs (SNPs). Eq. (2.18) is solved to calculate the electron's velocity due to the interaction of the NPs with the laser beams. Owing to the nonlinear interaction with the laser beams, the conduction electrons of the NPs experience a ponderomotive force at the beat frequency. It can be calculated through the ponderomotive potential, as given by Malik et al. (2012),

$$\phi_p = -m\vec{v}_1 \cdot \vec{v}_2^* / 2e \tag{2.19}$$

Hence, the ponderomotive force $\vec{F}_P^{NL}$ can be evaluated. By considering the ponderomotive force, the nonlinear equation of motion can be expressed as

$$\frac{\partial^2 \vec{X}^{NL}}{\partial t^2} + \Gamma \frac{\partial \vec{X}^{NL}}{\partial t} + \frac{\omega_p^2}{3} \vec{X}^{NL} = \frac{\vec{F}_P^{NL}}{m} \tag{2.20}$$

where $\vec{X}^{NL}$ is the electrons' cloud displacement due to the ponderomotive force. Solving the equation (2.20), we can obtain the nonlinear velocity $\vec{v}^{NL}\left(\equiv \frac{d\vec{X}^{NL}}{dt}\right)$ of the electrons, and, therefore, by taking the contribution of all the particles, the nonlinear macroscopic current density $\vec{J}^{NL}$ is evaluated.

The wave equation obtained from Maxwell's equations that indicate the THz radiation generation due to the nonlinear current $\vec{J}^{NL}$ is given by

$$-\nabla^2 \vec{E}_{\text{THz}} + \vec{\nabla}\left(\vec{\nabla} \cdot \vec{E}_{\text{THz}}\right) = \frac{4\pi i\omega}{c^2} \vec{J}^{NL} + \frac{\omega^2}{c^2} \varepsilon_{\text{eff}} \vec{E}_{\text{THz}} \tag{2.21}$$

In the above Eq. (2.21), the effective electric permittivity of the medium has been introduced via ε_{eff}. This can be calculated by the Maxwell-Garnett approach given by Ruppin (2000), assuming that the medium contains two types of graphite NPs, one having their basal planes perpendicular to the field of incident radiation and the other having their basal planes parallel to the field. The divergence of Eq. (2.21) gives rise to the electric field of the THz radiation, as is shown by Malik et al. (2012),

$$\vec{E} = -\frac{4\pi i}{\omega \varepsilon_{\text{eff}}(\omega)} \vec{J}^{NL} \tag{2.22}$$

Hence, by knowing the value of $\vec{J}^{NL}$, one gets the expression for the THz field from Eq. (2.21).

2.3.2 Special Cases

NPs are not necessarily spherically shaped and can be cylindrical or disc-shaped too; and the effective permittivity for these cases is different (Sihvola 1999). For example

(1) when the NPs are cylindrical (c):

$$\varepsilon_{\text{eff}}(c) = \varepsilon_h + \sum_{k=\perp,\parallel} f_{ck} \frac{(\varepsilon_k - \varepsilon_h)(\varepsilon_k + 5\varepsilon_h)}{(3 - 2f_{ck})\varepsilon_k + (3 + 2f_{ck})\varepsilon_h} \tag{2.23}$$

(2) when the NPs are disc-shaped (ds):

$$\varepsilon_{\text{eff}}(ds) = \varepsilon_h + \sum_{k=\perp,\parallel} f_{dk}(\varepsilon_k - \varepsilon_h) \frac{2\varepsilon_k + \varepsilon_h}{(3 - f_{dk})\varepsilon_k + f_{dk}\varepsilon_h} \tag{2.24}$$

where ε_h is the permittivity of the host medium and ε_k is the permittivity of two different types of NPs, and f_{ck} and f_{dk} are the volume filling factors for cylindrical and disc-shaped NPs, respectively.

Let us consider a case of a medium that contains conducting spherical and cylindrical NPs, and shape-dependent THz radiation generation takes place by the beating of two super-Gaussian laser beams in such a medium. Spherical and cylindrical geometries of graphite NPs in bulk form are considered and put through to external electric field $\overrightarrow{E_j}$ of the incident lasers, having their basal planes aligned in two dissimilar directions, i.e., parallel to the electric field ($\overrightarrow{E_j} \perp k_{i(=y)}$) and perpendicular to the electric field ($\overrightarrow{E_j} \parallel k_{i(=z)}$), where k_i is the symmetry axis normal to the basal plane inside the NP with dissimilar configurations.

The overall macroscopic density of NPs is written as

$$n_0 = \sum_{k_i}(n_{0s} + n_{0c}) = \sum_{k_i}(f_{sk_i} + f_{ck_i})n_{0k_i} \tag{2.25}$$

where n_{0s} and n_{0c} represent the densities of spherical nanoparticles (SNPs) and cylindrical NPs (CNPs), respectively. Both the factors $f_{sk_i} = \frac{4\pi r_{sk_i}^3}{3a_{sk_i}^3}$ and $f_{ck_i} = \frac{\pi r_{ck_i}^2 h_{ck_i}}{a_{ck_i}^3}$ signify the volume fraction, i.e. proportion of volume occupied by SNPs and CNPs to the volume of the unit cell for the configurations. Here r_{sk_i}, r_{ck_i} and a_{sk_i}, a_{ck_i} are, respectively, the average radius and distances between the two consecutive SNPs and CNPs, respectively, in a particular configuration. It is assumed that the NPs have a uniform distribution of charges through them. We also assume ripples in the density, varying as

$$n' = n_\beta e^{i\beta z} \tag{2.26}$$

where β is the density ripple wavenumber.

Let us consider the two super-Gaussian laser beams propagating simultaneously through the medium; the electric field of such beams is expressed as

$$\vec{E}_j = \frac{1}{2}E_0 e^{-\left(\frac{y^2}{b^2}\right)^p} e^{i(k_j z - \omega_j t)}\hat{y} + c.c.\ j = 1,2 \tag{2.27}$$

where E_0, p, and b are the beam parameters such as amplitude, index, and beam width of each of the lasers, respectively.

When the laser is injected, cloud of electrons in NPs shows collective behavior; these can be treated as a single fluid. Therefore, the equation of motion (Eq. 2.18) of the first order for nonrelativistic interaction of electrons with the lasers electric field is modified as

$$\frac{d\overrightarrow{v_j}}{dt} + \Gamma\overrightarrow{v_j} + S_s\frac{\omega_p^2}{3}\overrightarrow{r_j} + S_c\frac{\omega_p^2}{2}\overrightarrow{r_j} = \frac{-e}{m}\overrightarrow{E_j}$$

where S_s and S_c are the coefficients for SNPs and CNPs, respectively. The equation of motion leads to the following oscillatory velocity of electrons

$$\overrightarrow{v_j} = \frac{-i\omega_j}{m}\frac{eE_0 e^{i(k_j z - \omega_j t)} e^{-\left(\frac{y^2}{b^2}\right)^p}}{\left(\omega_j^2 + i\Gamma\omega_j - S_s\omega_p^2/3 - S_c\omega_p^2/2\right)} \tag{2.28}$$

A ponderomotive force is experienced by the electrons due to the nonlinear interaction of laser beams at the beat frequency $\omega = \omega_1 - \omega_2$. This force is obtained with the help of ponderomotive potential ϕ_p as,

$$\vec{F}_P^{NL} = -e\vec{\nabla}\phi_p$$

$$= \frac{-\omega_1\omega_2 e^2 E_0^2}{2m\left(\omega_1^2 + i\Gamma\omega_1 - S_s\omega_p^2/3 - S_c\omega_p^2/2\right)\left(\omega_2^2 - i\Gamma\omega_2 - S_s\omega_p^2/3 - S_c\omega_p^2/2\right)}$$

$$\left\{\left(-\frac{4py^{2p-1}}{b^{2p}}\right)\hat{y} + (ik)\hat{z}\right\} e^{-2\left(\frac{y^2}{b^2}\right)^p} e^{i(kz-\omega t)} \tag{2.29}$$

This ponderomotive force, which is nonlinear in nature, oscillates at frequency ω and wave number $k(=k_1 - k_2)$. One can study the motion of electrons under the effect of such a nonlinear force by the following equation (modified form of Eq. 2.20)

$$\frac{\partial^2 \vec{X}^{NL}}{\partial t^2} + \Gamma\frac{\partial \vec{X}^{NL}}{\partial t} + S_s\frac{\omega_p^2}{3}\vec{X}^{NL} + S_c\frac{\omega_p^2}{2}\vec{X}^{NL} = \frac{\vec{F}_P^{NL}}{m} \tag{2.30}$$

where $\vec{X}^{NL}$ is the overall displacement of the electron cloud. To calculate the nonlinear velocity $\vec{v}^{NL}$of the electrons, Eq. (2.30) is solved, whose y-component comes out to be

$$v_y^{NL} =$$

$$\frac{2ie^2\omega\omega_1\omega_2E_0^2py^{2p-1}e^{-\frac{2y^{2p}}{b^{2p}}}e^{i[kz-\omega t]}}{m^2b^{2p}\left(\omega_1^2 + i\Gamma\omega_1 - S_s\omega_p^2/3 - S_c\omega_p^2/2\right)\left(\omega_2^2 - i\Gamma\omega_2 - S_s\omega_p^2/3 - S_c\omega_p^2/2\right)\left(\omega^2 + i\Gamma\omega - S_s\omega_p^2/3 - S_c\omega_p^2/2\right)} \tag{2.31}$$

Therefore, to calculate the total nonlinear macroscopic current density, one must take the contribution of both types of the configured NPs, such as

$$J_{NL} = \sum_{k_i}(f_{sk_i} + f_{ck_i})n'ev_y^{NL}$$

$$J_{NL} = \sum_{k_i}\left(-\frac{4\pi r_{sk_i}^3}{3a_{sk_i}^3} - \frac{\pi r_{ck_i}^2 h_{ck_i}}{a_{ck_i}^3}\right)$$

$$\frac{2n_{\beta k_i}ie^3\omega\omega_1\omega_2E_0^2py^{2p-1}e^{-\frac{2y^{2p}}{b^{2p}}}e^{i[(k+\beta)z-\omega t]}}{m^2b^{2p}\left(\omega_1^2 + i\Gamma\omega_1 - S_s\omega_p^2/3 - S_c\omega_p^2/2\right)\left(\omega_2^2 - i\Gamma\omega_2 - S_s\omega_p^2/3 - S_c\omega_p^2/2\right)\left(\omega^2 + i\Gamma\omega - S_s\omega_p^2/3 - S_c\omega_p^2/2\right)} \tag{2.32}$$

2.3.3 THz Generation and Efficiency of the Scheme

The wave equation that explains the THz radiation generation is reproduced as

$$-\nabla^2\vec{E}_{\text{THz}} + \vec{\nabla}\left(\vec{\nabla}\cdot\vec{E}_{\text{THz}}\right) = \frac{4\pi i\omega}{c^2}\vec{J}^{NL} + \frac{\omega^2}{c^2}\varepsilon_{\text{eff}}\vec{E}_{\text{THz}} \tag{2.33}$$

where ε_{eff} is the effective permittivity of the medium, which can further be obtained by electromagnetic mixing approach given by Sihvola (1999). The divergence of Eq. (2.33), by following the technique discussed previously, gives the expression for the THz radiation field as

$$\left|\frac{E_{0\text{THz}}}{E_0}\right| = \frac{2}{\varepsilon_{\text{eff}}}\sum_{k_i} - \left(\frac{4\pi r_{sk_i}^3}{3a_{sk_i}^3} + \frac{\pi r_{ck_i}^2 h_{ck_i}}{a_{ck_i}^3}\right)$$

$$\frac{n_{\beta k_i}}{n_{0k_i}} \frac{\omega_1\omega_2\omega_{pk_i}^2 E_0 p y^{2p-1} e^{-\frac{2y^{2p}}{b^{2p}}} e^{i[(k+\beta)z-\omega t]}}{mb^{2p}\left(\omega_1^2 + i\Gamma\omega_1 - S_s\omega_{pki}^2/3 - S_c\omega_{pki}^2/2\right)\left(\omega_2^2 - i\Gamma\omega_2 - S_s\omega_{pki}^2/3 - S_c\omega_{pki}^2/2\right)\left(\omega^2 + i\Gamma\omega - S_s\omega_{pki}^2/3 - S_c\omega_{pki}^2/2\right)} \tag{2.34}$$

In the case of clustered plasma, free electrons of plasma contribute towards the nonlinearity, whereas in the case of NPs, the nonlinearity occurs mainly due to the conduction electrons of the graphite NPs. While calculating the current density, the contribution of conduction electrons has already been taken into consideration. So, to calculate the effective permittivity of the medium, the contribution of non-conducting bound electrons is now considered.

The overall effective permittivity of the host medium gets changed as now it consists of graphite NPs with two distinct shapes and orientations. Because of the addition of SNPs and CNPs in equal numbers, the effective permittivity is written as $\varepsilon_{\text{eff}} = \frac{\varepsilon(s)+\varepsilon(c)}{2}$, where

$$\varepsilon(s) = \varepsilon_h + 3\varepsilon_h \frac{\sum_{k_i} f_{sk_i} \frac{\varepsilon_{k_i} - \varepsilon_h}{\varepsilon_{k_i} + 2\varepsilon_h}}{1 - \sum_{k_i} f_{sk_i} \frac{\varepsilon_{k_i} - \varepsilon_h}{\varepsilon_{k_i} + 2\varepsilon_h}} \tag{2.35}$$

$$\varepsilon(c) = \varepsilon_h + \sum_{k_i} f_{ck_i} \frac{\left(\varepsilon_{k_i} - \varepsilon_h\right)\left(\varepsilon_{k_i} + 5\varepsilon_h\right)}{\left(3 - 2f_{ck_i}\right)\varepsilon_{k_i} + \left(3 + 2f_{ck_i}\right)\varepsilon_h} \tag{2.36}$$

where ε_h is the permittivity of the host medium and ε_{k_i} is the permittivity of bound electrons (only) for distinct configurations of NPs. According to the case, ε_{eff} is considered to get the amplitude of THz field from Eq. (2.34) by taking $\varepsilon_{\text{eff}} = \varepsilon(s)$ in the case of only SNPs and $\varepsilon_{\text{eff}} = \varepsilon(c)$ in the case of only CNPs. For the combinations of both SNPs and CNPs, ε_{eff} has already been considered.

Based on the ratio of the average energy densities of the emitted THz radiation $(< W_{\text{THz}} >)$ and the lasers $(< W_{\text{pump}} >)$, efficiency of the scheme is also obtained. Average electromagnetic energy stored per unit volume in the electric field is evaluated, as calculated by Rothwell and Cloud (2018) based on the relations $W_{Ei} = \frac{1}{8\pi}\varepsilon_0 \frac{\partial}{\partial\omega_i}\left[\omega_i\left(1 - \frac{\omega_p{}^2}{\omega_i{}^2}\right)\right] < |E_i|^2 >$. Hence,

$$< W_{\text{pump}} > = \frac{\varepsilon_0 a_0}{p2^{1/2p}} |E_0|^2 \Gamma(1/2p) \tag{2.37}$$

$$< W_{\text{THz}} > = \frac{2\varepsilon_0}{\varepsilon_{\text{eff}}^2} \sum_{k_i} \left(f_{sk_i} + f_{ck_i}\right)^2 \times \left(\frac{n_{\beta k_i}}{n_{0k_i}}\right)^2$$

$$\frac{\omega_1{}^2\omega_2{}^2\omega_{pk_i}^4 E_0^4 p e^2}{m^2\, b\left(\omega_1^2 + i\Gamma\omega_1 - S_s\omega_{pk_i}^2/3 - S_c\omega_{pk_i}^2/2\right)^2\left(\omega_2^2 - i\Gamma\omega_2 - S_s\omega_{pk_i}^2/3 - S_c\omega_{pk_i}^2/2\right)^2\left(\omega^2 + i\Gamma\omega - S_s\omega_{pk_i}^2/3 - S_c\omega_{pk_i}^2/2\right)^2}$$

$$\frac{\Gamma(2 - 1/2p)}{2^{(3-1/p)}} \tag{2.38}$$

The ratio $\frac{<W_{THz}>}{<W_{pump}>}$ finally yields the efficiency η as

$$\eta = \frac{2}{\varepsilon_{eff}^2}\sum_{k_i}\left(f_{sk_i}+f_{ck_i}\right)^2\left(\frac{n_{\beta k_i}}{n_{0k_i}}\right)^2$$

$$\frac{\omega_1{}^2\omega_2{}^2\omega_{pk_i}^4 E_0^2 p^2 e^2}{m^2\, b^2\left(\omega_1^2+i\Gamma\omega_1-S_s\omega_{pk_i}^2/3-S_c\omega_{pk_i}^2/2\right)^2\left(\omega_2^2-i\Gamma\omega_2-S_s\omega_{pk_i}^2/3-S_c\omega_{pk_i}^2/2\right)^2\left(\omega^2+i\Gamma\omega-S_s\omega_{pk_i}^2/3-S_c\omega_{pk_i}^2/2\right)^2}$$

$$\frac{\Gamma(2-1/2p)}{2^{(3-3/2p)}\,\Gamma(1/2p)} \tag{2.39}$$

2.3.3.1 Results and Discussion

A CO_2 gas laser with $\lambda = 9.9\ \mu$m and $v = 1.9 \times 10^{14}$ s^{-1} is taken as the incident laser source. Reference parameters from Javan and Erdi (2017) are considered, and the different physical parameters supporting the mechanism of emission of THz amplitude are analyzed. The dependence of normalized THz field amplitude with normalized laser beat wave frequency for two different index values of lasers, $p = 2$ and $p = 4$ for CNPs ($S_s = 0$, $S_c = 1$) and SNPs ($S_c = 0$, $S_s = 1$) separately is shown in Figure 2.7. The large magnitude of the THz field is obtained for the higher index super-Gaussian laser beam, when a comparative study of the laser index for $p = 2$ and 4 is done; this is due to the stronger ponderomotive force because of the laser field profile (the envelope of the laser). Actually, the sharpness in the gradient of the laser field or intensity envelope is greater for large index ($p = 4$) super-Gaussian laser beam than the low index ($p = 2$) laser. Because of this, a strong ponderomotive force is produced that further leads to the higher nonlinear current in the first case.

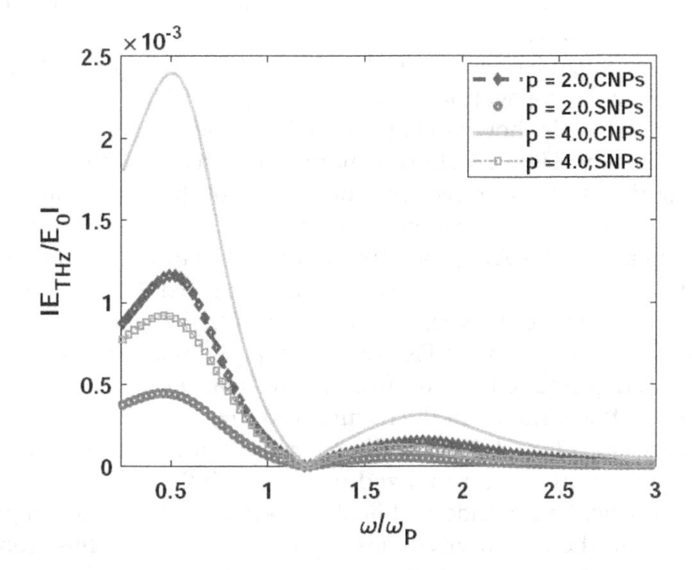

FIGURE 2.7
Normalized THz field versus normalized frequency for different values of the laser index, when $I = 10^{19}$ W/m^2, $\left(\frac{n_{\beta k_i}}{n_{0k_i}}\right) = 0.3$, $y = 0.9b$, $a_{sk_i} = a_{ck_i} = 5r_{sk_i}$, $r_{sk_i} = r_{ck_i}$ and $h_{ck_i} = 4.0\, r_{ck_i}$. Here CNPs stands for cylindrical nanoparticles and SNPs for spherical nanoparticles.

The laser interacts with the medium, which is an assembly of two types of density-modulated NPs i.e. SNPs and CNPs with distinct configurations laying their basal plane normal parallel and perpendicular to the laser field, leads to two resonances, i.e. primary resonance and secondary resonance, respectively. So, the two maxima for the generated THz amplitude are observed in the two cases of distinct laser index and different NPs. For both the values of the laser index p = 2 and p = 4, the first maximum appears near the frequency 12×10^{12} Hz, and the second with much less THz amplitude is displaced relatively towards higher frequencies. It is also observed that even for the same inter-particle separation and radii as that of SNPs, CNPs contribute more towards the THz generation. The first maximum in SNPs corresponds to the contribution of plasmon oscillations on the surface of SNPs when laser beat wave frequency approaches $\pm\omega_{pk_i} / \sqrt{3}$, which further leads to the excitation of large amplitude nonlinear currents. In contrast, CNPs resonate when the laser beat wave frequency approaches $\pm\omega_{pk_i} / \sqrt{2}$ leading to the excitation of nonlinear currents with a much higher value than in the case of SNPs.

In accordance with the Gans theory, the plasmon resonance wavelength depends on the shape and size of the particles because the polarizability of the material is highly dependent on the above mentioned factors. Normally, a particle has two surface plasmon (SP) resonances such as longitudinal SP resonance and transverse SP resonance when it gains additional modes of plasmon resonance due to the perturbation in its symmetry. Therefore, the secondary maximum might correspond to the transverse plasmon resonance in this case.

In the case of SNPs, the contribution is mainly due to longitudinal resonance because the SNPs have symmetric geometric structure, so they have unity aspect ratio (ratio of height to width), whereas in CNPs, due to deviation of shape from spherical symmetry (non-unity aspect ratio), transverse resonance along with longitudinal resonance also contributes. Due to this, at a slightly higher frequency, a secondary maximum appears but its contribution is quite a bit less compared with the primary maximum. Therefore, THz emission through CNPs is higher due to the greater third dimension in CNPs compared with SNPs because CNPs interact more effectively with the incident laser radiation, bringing out secondary resonance.

The profile of normalized THz amplitude, which is a function of the transverse distance from the z-axis of laser propagation for different types of NPs, i.e. spherical, cylindrical, and a mixture of the two, is shown in Figure 2.8. The above mentioned reason also justifies the observation that the THz radiation is more in the case of CNPs than only the SNPs. The highest magnitude of the transverse component of the ponderomotive force, which produces the largest oscillating current resonantly, is due to the higher intensity of emitted THz radiation at $y = 0.9b$. Also, the mixture of two shaped particles (SNPs and CNPs) together contribute more effectively towards the THz generation than the single-shaped NPs because of the higher nonlinearity in the system. On the other hand, the lower electric field amplitude of THz radiation in Figure 2.8 compared with that of Figure 2.7 is due to the selection of beating frequency ω far from the resonances.

Figure 2.9 depicts the variation of the normalized THz field with normalized frequency for various aspect ratios for CNPs as well as for SNPs. It is observed that increasing the height to radius ratio with the normalized height of CNPs, $h = 6.0$ to $h = 8.0$, leads to an increase in the THz field amplitude that is of the order of 10^{-3} due to a higher aspect ratio in CNPs. Whereas in the case of SNPs, just by increasing the radius from $r = 6$ to $r = 8$, THz field amplitude increases by a factor of 10 compared with CNPs because the distance between any two consecutive NPs diminishes due to the increment in the radius and the situation becomes closer to a cluster of particles. As a result, a very large number of conduction electrons takes part in plasmon excitation per unit volume effectively. Therefore,

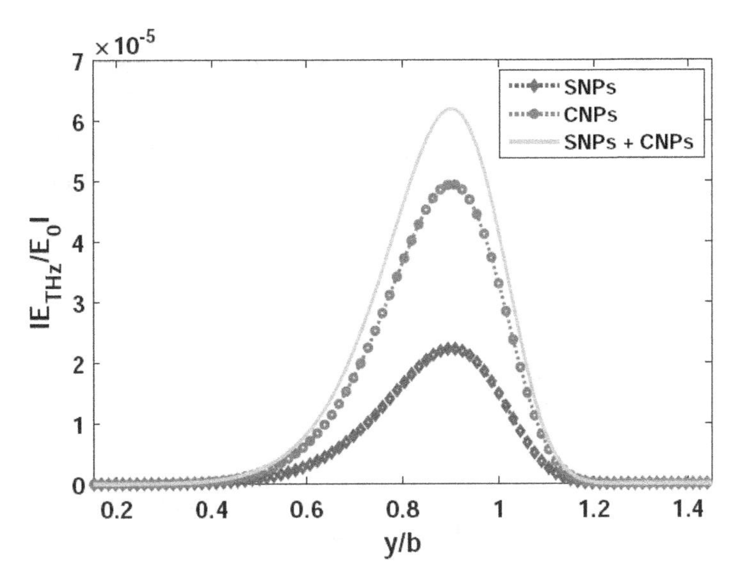

FIGURE 2.8
Normalized THz field versus transverse distance (y/b) from the z-axis for different shapes of NPs and a combination of both, when $b = 5 \times 10^{-5}$ m, $p = 4.0$, $a_{sk_i} = a_{ck_i} = 5r_{sk_i}$, $r_{sk_i} = r_{ck_i}$, $h_{ck_i} = 4.0\ r_{ck_i}$, $I = 10^{19}$ W/m^2, $\left(\frac{n_{\beta ki}}{n_{0ki}}\right) = 0.3$, and and $\omega = 1.8 \times 10^{14}$ Hz. Here SNPs (CNPs) stands for spherical (cylindrical) nanoparticles and SNPs + CNPs stands for the combination of the two.

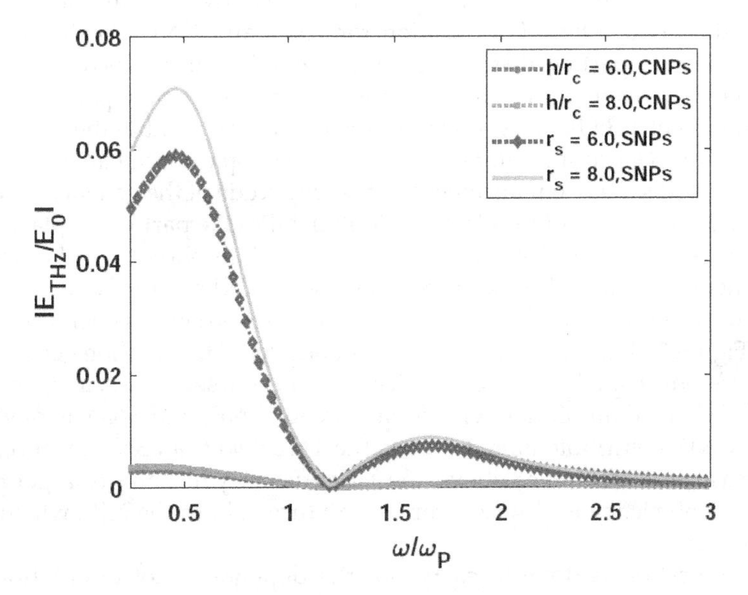

FIGURE 2.9
Normalized THz field versus normalized frequency for varying normalized heights and radius of CNPs and SNPs, respectively, when $b = 5 \times 10^{-5}$ m, $p = 4.0$, $I = 10^{19}$ W/m^2, $\left(\frac{n_{\beta ki}}{n_{0ki}}\right) = 0.3$, and $a_{sk_i} = a_{ck_i} = 5r_{sk_i}$, $r_{sk_i} = r_{ck_i}$ and $y = 0.9b$. In the figure, $r_s = 6.0$ means $r_s = 6.0$ nm and $r_s = 8.0$ means $r_s = 6.0$ nm.

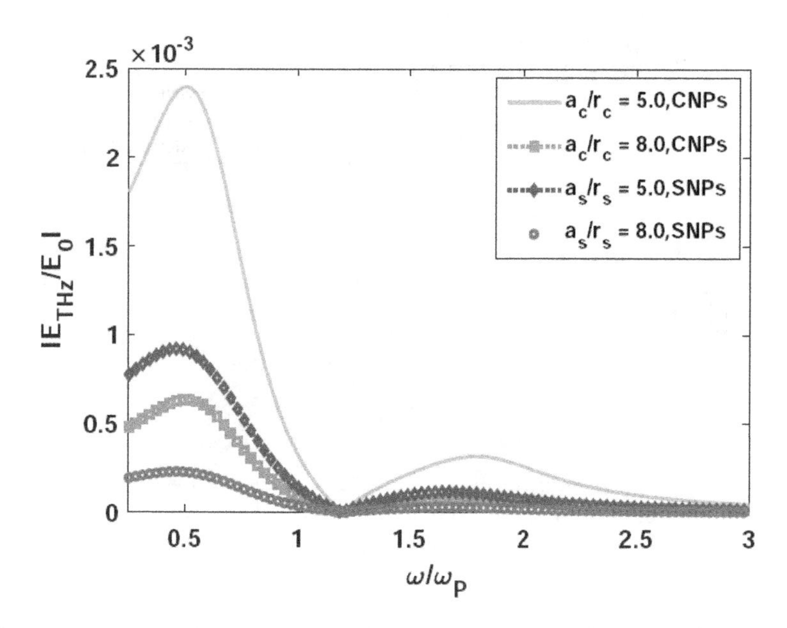

FIGURE 2.10
Normalized THz field versus normalized frequency for CNPs and SNPs with different inter-NP separation, when $b = 5 \times 10^{-5}$ m, $p = 4.0$, $I = 10^{19}$ W/m^2, $\left(\frac{n_{\beta k_i}}{n_{0k_i}}\right) = 0.3$, and $h_{ck_i} = 4.0\, r_{ck_i}$.

a higher amount of macroscopic nonlinear current is produced, which finally results in a very sharp increment in the THz field amplitude. Therefore, THz field amplitude generated is of the order of 10^{-2} in the case of SNPs.

Figure 2.10 shows the nature of the precise comparative magnitudes of emitted THz radiation with varied inter-NP separation for SNPs and CNPs. With the radius of the respective NPs, inter-particle distance is normalized. It is also considered that both SNPs and CNPs have the same radii. With decrement in inter-particle separation, a sudden rise in the magnitude of THz fields is observed due to the increment in the density of NPs in a particular volume. Hence, a greater amount of macroscopic nonlinear current is produced, which finally results in the enhancement in the amplitude of the generated THz field since more conduction electrons of the electron cloud of NPs take part in the plasmon excitation per unit volume effectively. The rise in the THz field amplitude for the decrease in the same amount of inter-particle separation is greater in CNPs than SNPs.

Analysis of the efficiency of the mechanism of THz radiation generation is made with the help of Figure 2.11. In this figure, the dependence of THz radiation generation on beam width for different shaped NPs is shown. The trend is the same as that of Figure 2.7, where normalized THz field amplitude with frequency for CNPs and SNPs is plotted. It is concluded that CNPs contribute more towards the THz field than SNPs. Hence, on the basis of similar reasons as mentioned above, the THz efficiency in CNPs is larger than in SNPs, and maximum efficiency is observed for the mixture of both the NPs, which is consistent with Figure 2.7.

Finally, exploration is done to carry out the dependence of orientation of the NPs basal planes with respect to the incident laser radiation towards emitted THz radiation efficiency, as depicted by Figure 2.12. The efficiency of the mechanism of THz field amplitude gets slightly affected by the orientation of NPs; however, perpendicularly oriented NPs contribute less towards better efficiency than parallel oriented NPs in

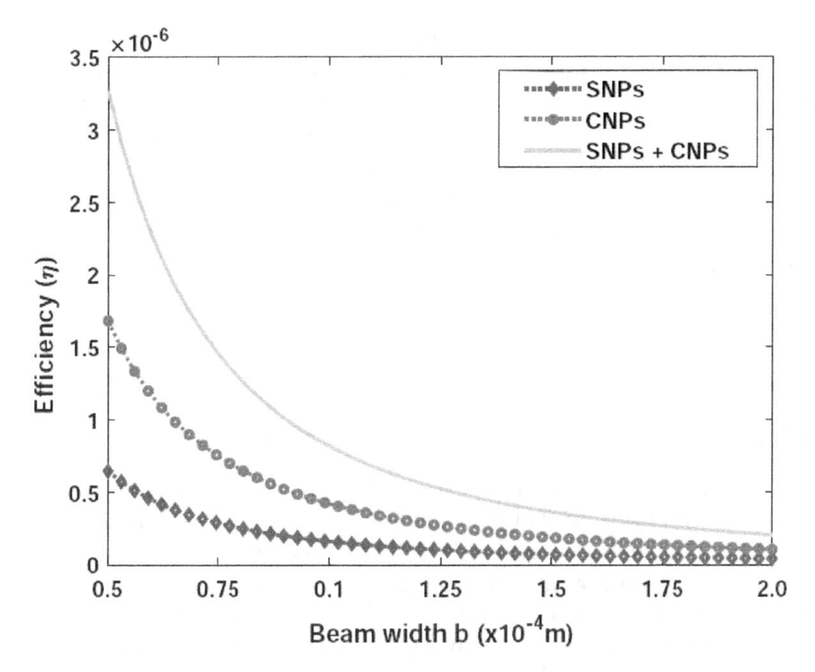

FIGURE 2.11
Efficiency versus beam width for different shapes of NPs and a combination of both the NPs, when $p = 4.0$, $I = 10^{19}$ W/m^2, $\left(\frac{n_{\beta ki}}{n_{0ki}}\right) = 0.3$, $a_{sk_i} = a_{ck_i} = 5r_{sk_i}$, $r_{sk_i} = r_{ck_i}$, $h_{ck_i} = 4.0\ r_{ck_i}$ $y = 0.5b$ and $\omega = 1.8 \times 10^{14}$ rad/sec. Here SNPs (CNPs) stands for spherical (cylindrical) nanoparticles and SNPs + CNPs stands for the combination of the two.

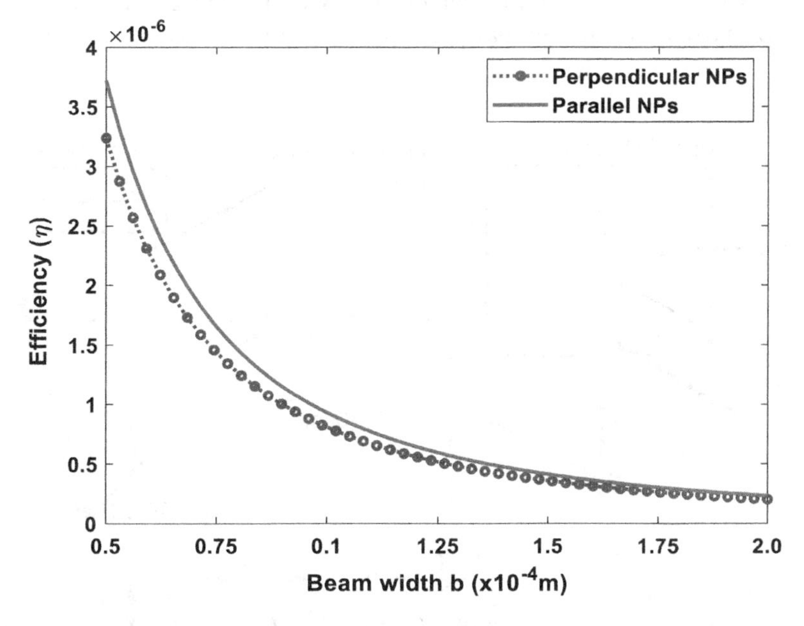

FIGURE 2.12
Efficiency versus beam width for perpendicular and parallel orientation for the combination of both CNPs and SNPs, when $\omega = 1.8 \times 10^{14}$ rad/sec, $p = 4.0$, $I = 10^{19}$ W/m^2, $\left(\frac{n_{\beta ki}}{n_{0ki}}\right) = 0.3$, $a_{sk_i} = a_{ck_i} = 5r_{sk_i}$, $r_{sk_i} = r_{ck_i}$ and $h_{ck_i} = 4.0\ r_{ck_i}$.

comparison. This might be due to the difference in the optical properties of graphite NPs for different configurations of basal planes with respect to the incident electric field polarization. Parallel NPs are affected more and lead to greater THz generation and hence greater efficiency because the damping factor and plasmon frequency of the parallel oriented NPs are of slightly lesser order of magnitude than those of the perpendicularly oriented NPs. Moreover, in parallel oriented NPs, the excited plasmon wave and the phase alignment between the incident laser polarization also match, which results in larger efficiency.

2.4 THz Radiation Generation by Metal Gratings

Nanodimension structures also interact strongly with the laser light and concentrate fields, and exhibit enhanced nonlinear optical phenomena due to very high field strength. Here the surface plasma wave is localized to take the form of SP. The SP can be excited in metal films deposited on a shallow grating. For gold as the metal, SPs –assisted multiphoton excitation of electrons in this metal can occur by irradiating the surface by femtosecond laser pulses. It leads to the emission of photoelectrons even in the vacuum conditions. An evanescent field produced inside the surface exerts a ponderomotive force, emitting the photoelectrons away from the surface. This way a femtosecond current surge is created because of multiphoton ionization and ponderomotive acceleration of electrons, resulting in the emission of THz pulses.

Figure 2.13 shows a typical experimental setup for the generation of THz radiation based on a metal grating with a 500 mm grating element and 40 nm etch depth. The incident 100 fs laser beam with a wavelength of 800 nm, energy of 1 mJ, and repetition rate of 1 kHz is divided into two parts, i.e. pump beam and weak probe beam, into the

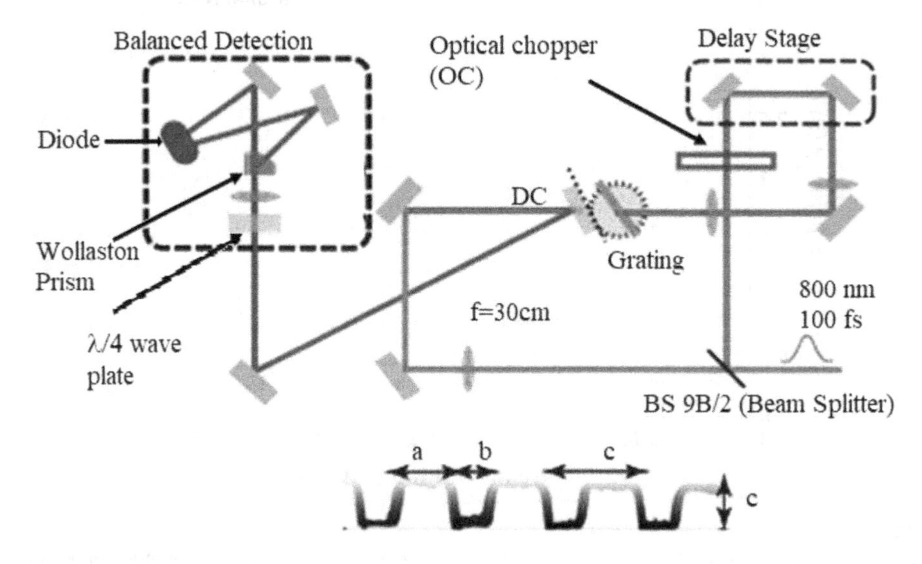

FIGURE 2.13
Typical experimental set up for THz radiation generation by metal gratings.

ratio of 98:2 using a beam splitter. A well-collimated pump beam, 5 nm in diameter, is incident on the sample (grating) after passing through the delay stage and the telescope (Jeon and Grischkowsky 1998). The power density in this experiment can be varied between 0.25 and 4 mJ/cm^2. The sample is oriented differently for different polarization states, i.e. for s-polarized or p-polarized light. Robertson (1995) has shown that the electro-optic sampling method is used to measure and detect the THz radiation emitted by the samples. Customized shallow 10 × 10 mm^2 grating is produced in UV-grade fused silica. A 30 nm gold coating is made on the grating surface using a standard vacuum evaporator. The thickness of the film is generally chosen to have the maximum absorption of the pump laser, which is calculated by standard Fresnel equations for the multilayered grating surface. The grating is held on a rotation stage in air and is irradiated from the back, i.e. facing away from the incident laser beam, as shown in Figure 2.13. To vary the power continuously, a half-wave plate and polarizing beam cube are used while keeping the pump-probe time delay constant.

When light is incident on the corrugated surfaces such as grating, the SP effect comes into the picture because surface waves (electromagnetic) are produced on the boundary between the metal and air. The electric fields of these waves decay exponentially away from the boundary over a distance of the order of one wavelength. Hence, by appropriate phase-matching, SPs can be excited. The phase-matching can be expressed as

$$\sin\theta + \frac{N\lambda}{\Lambda} = n_{sp} \tag{2.40}$$

where N is the order of diffraction, θ is the angle of incidence on the grating surface, λ is the wavelength, and Λ is the grating period.

The refractive index in terms of dielectric function ε of the metal corresponding to SP is given by

$$n_{sp} = \varepsilon^{1/2} / (\varepsilon + 1)^{1/2} \tag{2.41}$$

In the said experiment, it is observed that the energy of the THz pulse is maximum for the angle when SPs are excited for p-polarized light. The energy is found to decrease when SPs are excited for s-polarized light.

The underlying mechanism behind the generation of THz radiation by metal grating can be understood as follows. Actually an evanescent wave is generated inside the surface of the metal (grating). This wave exerts a ponderomotive force on the electrons directed away from the surface. So high energy electrons (~ 1 keV) are produced, which are called photoelectrons (Cheville and D. Grischkowsky 1999; Gallot et al. 1999; Jeon et al. 2000). These photoelectrons are accelerated in the ponderomotive potential. Since accelerating charges emit radiation, one gets THz radiation with the electric field given by

$$\vec{E}(t) \propto \frac{d\vec{j}(t)}{dt} \tag{2.42}$$

where $\vec{j}(t)$ is the current density (photoinduced). In the experiment performed by Welsh et al. (2007), a third- to fourth-order power dependence is observed. So, it cannot be considered as a second-order OR process; rather, it involves the photoelectron production and the evanescent wave acceleration.

2.5 THz Radiation Generation by Quantum Dot Materials

Quantum dot (QD) materials exhibit unique properties owing to their extremely small size and quantum confinements in all the three dimensions. The exceptionally different properties, such as high thermal and optoelectronics efficiency and short carrier lifetime are also very significant for the production of THz radiation through PCA-based THz devices. Estacio et al. (2009) and Kruczek et al. (2012) have embedded QD layers in a superconducting structure that serves as a highly efficient, ultrafast device for photoconductive-based emissions.

The implanted QDs act as an active medium in the host crystal, contributing to the efficient THz signal conversion. Figure 2.14 shows the structure of the QD PCA. The upper region carries multilayers of InAs QDs incorporated into a GaAs matrix while the middle region is a distributed Bragg reflector. Here all semiconductor structures are grown by the molecular beam epitaxy (MBE) method. On the top, a layer is made up of LT-GaAs as it is more appropriate for Ti/Au metal contact deposition. It is used to significantly reduce the device dark current.

The PCA with embedded QDs is excited by a pump laser (ultrafast laser pulses) along with a bias voltage between the two contacts. It induces a photocurrent which is responsible for the generation of THz radiation. The induced photocurrent generated by pumping such structures depends on a number of factors. It depends on which layer or material is optically excited and which layer / material charge carriers relax to. Porte et al. (2009) have shown that the current also depends on the extent of optical excitation applied to the assembly. In the case of THz antennae, an external electric field is also applied across the active region, which contributes to carrier drift velocities and their subsequent capture times (Lagatsky et al. 2005).

The mechanism can be understood as follows. Actually, the incident laser radiation is focussed on the active region between the biased electrodes of antennae and is absorbed by the semiconductor. External bias voltage further adds to the number of processes like ultrafast carrier generation, carrier capturing, carrier recombination, and so forth. These result in the large nonlinear current that leads to the THz generation. This finding is similar to that of Leyman et al. (2016).

Such a PCA embedded with QDs is referred to as a QD PCA. As an example, for the optical to THz signal conversion, the QD PCD is excited by a Ti:sapphire laser, which operates at a short wavelength. The bulk GaAs barrier layers generate the THz radiation on the excitation. The emitted THz radiation depends on the electric field applied to the

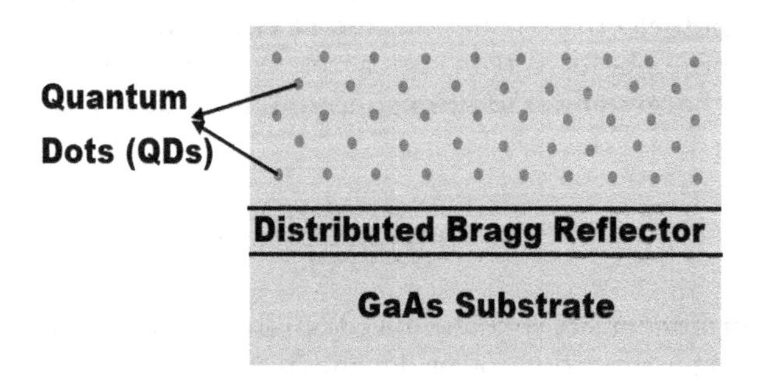

FIGURE 2.14
Side view of the assembly of a QD photoconductive antenna.

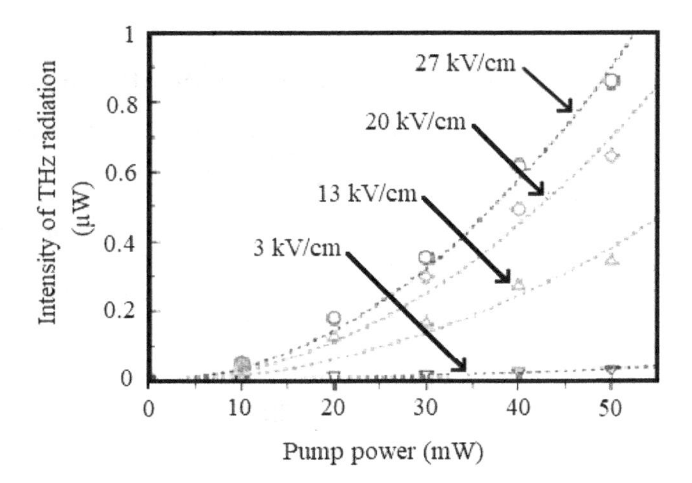

FIGURE 2.15
Dependence of intensity of THz radiation on laser pump power and antenna field.

PCA, i.e. the antenna field (a few tens of kV/cm) and the laser pump power (several tens of mW). Figure 2.15 shows the dependence of the intensity of the THz radiation on the pump power for different values of the antenna field. It is clear from the figure that the THz intensity increases from 0.05 to 0.9μW for 50 mW power of the laser pump when the antenna field is increased from 3 to 27 kV/cm. However, there is not always an enhancement in the intensity of the THz radiation with the increasing antenna field, which is clear from Figure 2.16. Actually, the intensity of THz radiation saturates in almost all the cases of laser pump power. The interesting point is that this saturation depends on the power of the laser pump. In the present mechanism of QD PCA, the enhancement and saturation of THz intensity are due to the fast electron trapping by QDs, which is the consequence of increased electron velocity.

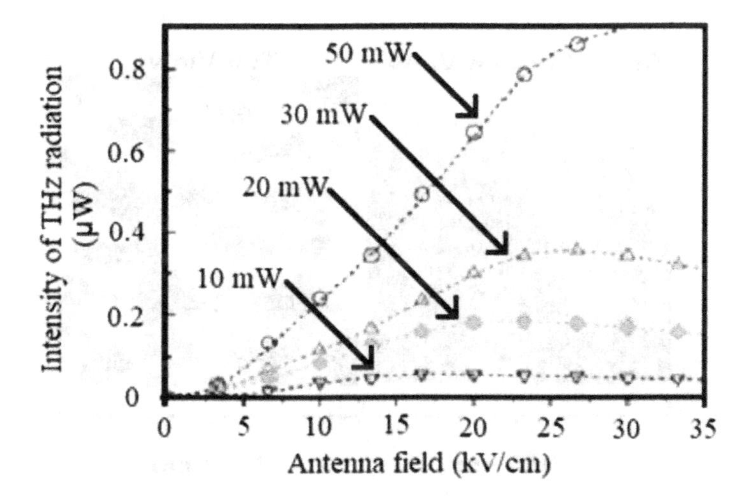

FIGURE 2.16
Saturation of THz intensity with antenna field for different values of pump power.

2.6 THz Radiation Generation by Random 2D Metallic Structures

High-power THz emissions have been reported by Zhang et al. (2015) by irradiating random 2D metallic structures, using femtosecond laser radiations. They deposited metal film as a substrate with randomly arranged nanoscale pores (Figure 2.17) due to which the surface acquires a nanoscale and contributes to enhancing the coupling of light in the film, giving rise to several absorption mechanisms too. The most popular method to couple photons and SPs is by forming a grating on the surface. In this case, the coupling condition is given in terms of the grating period Λ. Actually, the difference in the wave numbers of p-polarized light and SPs is written as

$$K_P - K_{SP} = \frac{2\pi}{\Lambda} \tag{2.43}$$

It means the effective coupling depends on matching of the momenta of the photon and plasmons. Actually, a randomly roughened surface enables the momenta matching condition over a wide range of wave vectors. Also, it assists in localizing and enhancing the SPs. On exciting the metallic surface by femtosecond laser radiation, a part of laser energy absorbed is trapped inside the surface layer of the sample and then dissipates into a bulk sample through conduction as a residual thermal energy. This way, an appreciable amount of thermal energy can be trapped in a sample through the surface roughness effect. It also increases the temperature of the bulk sample that later acts as a thermal radiation source. In the experiment, a 100 fs (pulse duration) Ti:sapphire laser with a wavelength of 800 nm and pulse energy of 1.3 mJ is employed as an excitation source with a repetition rate of 1 kHz. The radiation spectra emitted by the sample can be measured and characterized based on a Fourier transform Michelson interferometer (Figure 2.18).

The power of the emitted THz radiation is different for different metallic films, which means it shows the dependence on refractive index n. For example, the THz radiation with the highest power is obtained for the case of Ru (n = 1.82), after which the film prepared with Pt (n = 1.81) produced a significant amount of THz radiation. A weak THz signal is obtained for the case of Au (n = 1.50) film (Figure 2.19).

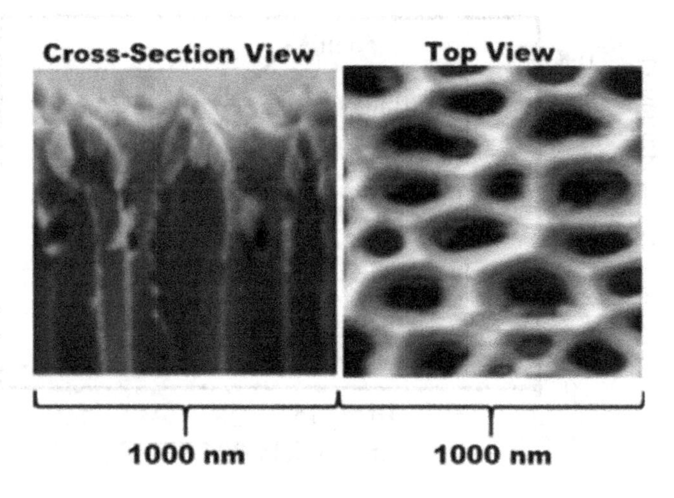

FIGURE 2.17
Cross-sectional and top views (SEM images) of 100-nm-thick Ru metal film with nanopores.

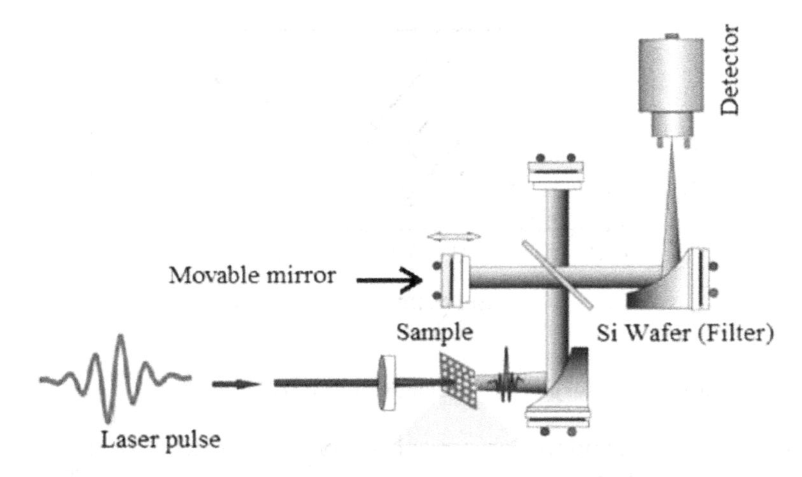

FIGURE 2.18
Schematic of the FT Michelson interferometer.

The energy conversion efficiency of an optical input laser to THz to IR radiation is found to linearly increase with the pump fluence, as shown in Figure 2.20. Consistent with Figure 2.19, the maximum energy conversion efficiency (2.45%) is observed for the Ru film ($n = 1.82$) resulted from a maximum emitted radiation power of 24.5 mW.

The variation of the peak-to-peak amplitude of the THz electric field with the energy of a laser pulse shows a linear behavior until the thermal effects on THz emission are negligible (for pump energies <15 μJ). It confirms the OR process. However, the variation of THz amplitude deviates from linear dependence due to the increased surface temperature and super-linear law is observed for the optical to THz conversion efficiency.

2.7 THz Radiation Generation by Carbon Nanotube Array

The THz radiation can be generated using a carbon nanotube (CNT) array mounted on a dielectric surface when a laser radiation is incident on the structure (Figure 2.21). The ponderomotive force of the laser imparts oscillatory velocity to the free electrons of the

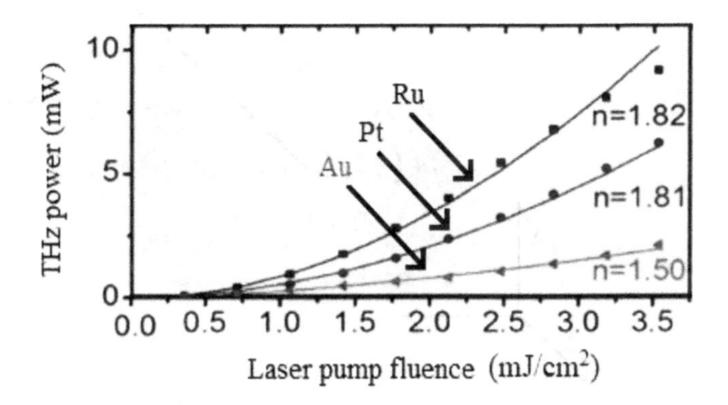

FIGURE 2.19
Dependence of THz power on laser pump fluence for different metallic films.

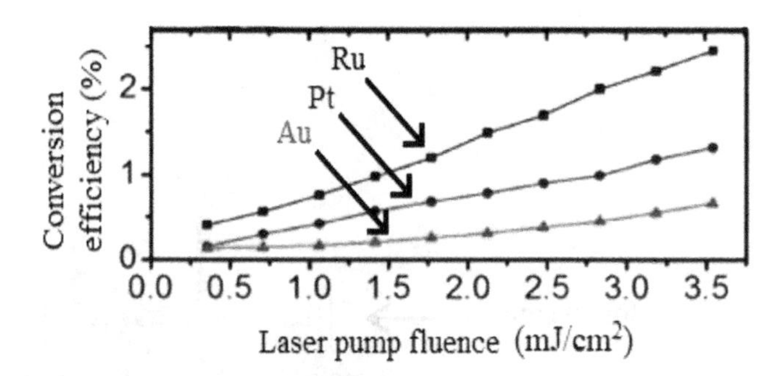

FIGURE 2.20
Energy conversion efficiency of laser input to THz for different metallic films.

nanotubes, generating the nonlinear current responsible for the THz generation. In other words, it leads the nanotubes into oscillating dipoles antennae emitting THz radiation. In such a mechanism, THz power has been observed when SP resonance takes place between the free electrons of the nanotubes and the incident laser field, then the beat frequency becomes $\omega_p / \sqrt{2}$, where ω_p is the SP frequency. It has been realized that the THz power depends on the direction of the angle of incidence of beating lasers' radiation on the CNT array. Specifically, the THz power initially increases with the incident angle, reaches a maximum value, and after that, decreases slowly for higher values of the angle. Similar work has been done by Malik and Uma (2018), which shows that the emitted THz power increases in a direction where the phase difference between the fields due to successive nanotubes is an integral multiple of 2π.

2.7.1 Mathematical Treatment

The electric field of the lasers incident on the CNT array can be written in the following form, considering the reflected waves as well

$$\vec{E}_j = \left(\hat{z} + \cot\theta_i \hat{x}\right)\sin\theta_0 A_j e^{-i\left(\omega_j t - \frac{\omega_j}{c}\cos\theta_i z + \frac{\omega_j}{c}\sin\theta_i x\right)} + \left(\hat{z} - \cot\theta_i \hat{x}\right)\sin\theta_i R_A A_j e^{-i\left(\omega_j t - \frac{\omega_j}{c}\cos\theta_i z - \frac{\omega_j}{c}\sin\theta_i x\right)} \quad (2.44)$$

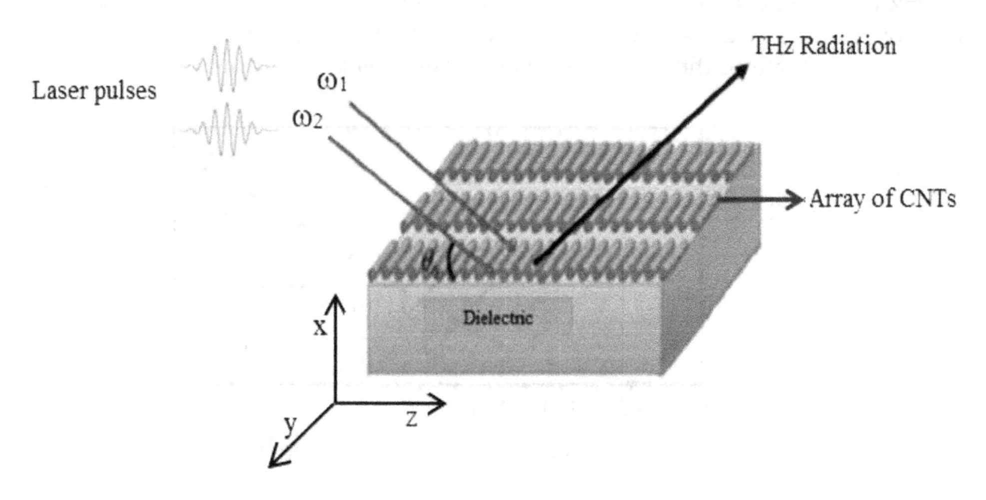

FIGURE 2.21
Schematic of laser-CNT interaction for THz generation.

Since laser beating is considered, j = 1 and 2 stand for laser 1 and 2, respectively. ω_j is the frequency of the corresponding laser, θ_i is the angle of incidence of the laser, and R_A is the amplitude of reflection coefficient of the waves reflected from the dielectric surface.

Under the effect of lasers' fields, the electrons of the CNTs are displaced by a certain distance. It leads to the space charge field. Then the electric field of lasers and the space charge field impart drift velocity of the electrons, which can be calculated by using the equation of motion. The corresponding ponderomotive force on the electrons of the CNTs can be calculated using the following formula

$$\vec{F}_P = -\frac{e}{2}\vec{v}\times\vec{B}^* \tag{2.45}$$

Here $\vec{B}$ is the magnetic field corresponding to the field $\vec{E}$ of the lasers (* refers to the complex conjugate). The oscillatory velocity of the electrons because of ponderomotive force can be expressed as

$$\vec{v}_\omega^{NL} = \frac{\vec{F}_P\ \omega}{mi\left(\omega^2 - \frac{\omega_p^2}{2}\right)} \tag{2.46}$$

The nonlinear current density inside a single nanotube at frequency ω can be deduced using the formula

$$\vec{j}_\omega^{NL} = -n_{oe}e\vec{v}_\omega^{NL} = \frac{n_{oe}e\vec{F}_{pz}\ \omega}{mi\left(\omega^2 - \frac{\omega_p^2}{2}\right)} \tag{2.47}$$

One can talk about the vector potential corresponding to the field of an oscillating dipole in a CNT. The total vector potential due to the entire CNT array at a far point can be calculated using the relation

$$\vec{A}(\vec{r},t) = \frac{\mu_0}{4\pi r}\pi r_c^2 l_c\left(\vec{j}_N^{NL}\right)_{t-\frac{r_N}{c}} \tag{2.48}$$

Here r_c is the radius and l_c is the length of CNTs, and the concept of retarded time has been taken into account. Corresponding to this total vector potential $\vec{A}(\vec{r},t)$, the magnetic field of the emitted radiation at a distance r can be calculated as

$$\vec{B} = \vec{\nabla}\times\vec{A}(\vec{r},t) \tag{2.49}$$

The time average of the pointing vector $\vec{S}$ provides the power density of the emitted radiation. It is obtained as

$$\vec{S}_{av} = \hat{r}\frac{|B|^2}{2\mu_0}c \tag{2.50}$$

where c is the speed of light.

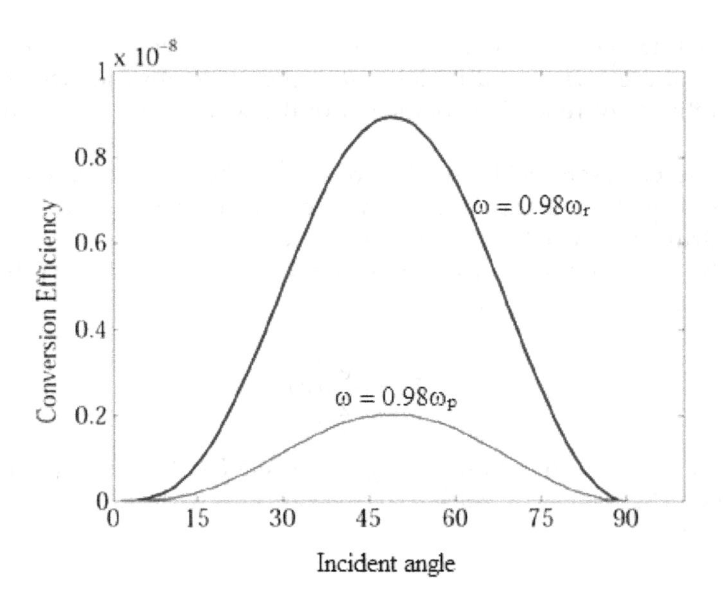

FIGURE 2.22
Dependence of THz power conversion efficiency on the angle (degrees) of laser incidence.

The efficiency of generated THz radiation can be evaluated as the ratio of the power of the emitted THz radiation to the power of incident laser radiation. This is given by

$$\eta = \frac{P_{\text{THz}}}{P_{\text{Laser}}} = \frac{\vec{S}_{av}\ 4\pi r^2}{\left|\vec{E}_1\right|^2 / 2\mu_0 c\pi r_0^2} \tag{2.51}$$

where r_o is the beam width. Hence, by putting in the value of $\vec{S}_{av}$, efficiency can be calculated. The calculations show that the emitted THz power increases with the angle and attains a maximum for $\theta_0 = 48.75°$, after that it decreases slowly for the higher values of incident angle θ_i, as shown in Figure 2.22. The typical parameters for the figure are $\omega_2 = 2 \times 10^{14}$ rad/sec, $\frac{r_c\,\omega_p}{c} = 0.003$, $N_1 = 1000$, $N_2 = 60$, and $a_2 = 0.03$.

It is seen that the THz field or power attains a maximum in the directions given by

$$\psi_1 = \frac{\omega d}{c}(\cos\theta - \cos\theta_0) = 2m_1\pi$$

and

$$\psi_2 = \frac{\omega b}{c}\sin\theta \sin\phi = 2m_2\pi$$

where b and d are the separations between the CNTs in the y- and z-directions, ϕ is the azimuthal angle, and m_1 and m_2 are 1, 2, 3, …. A sharp increase in the THz power is attained due to plasmon resonance when $\omega \sim \frac{\omega_p}{\sqrt{2}}$.

The phenomenon of resonance in the present mechanism is shown through Figure 2.23, when the variation of THz power conversion efficiency is plotted with the beating frequency ω (in the multiple of $\omega_p/\sqrt{2}$). Clearly, the efficiency increases abruptly when ω reaches the value $\omega_p/\sqrt{2}$; otherwise, it attains lower values. A sharp increase in the THz power is attained due to plasmon resonance when $\omega \sim \omega_p/\sqrt{2}$.

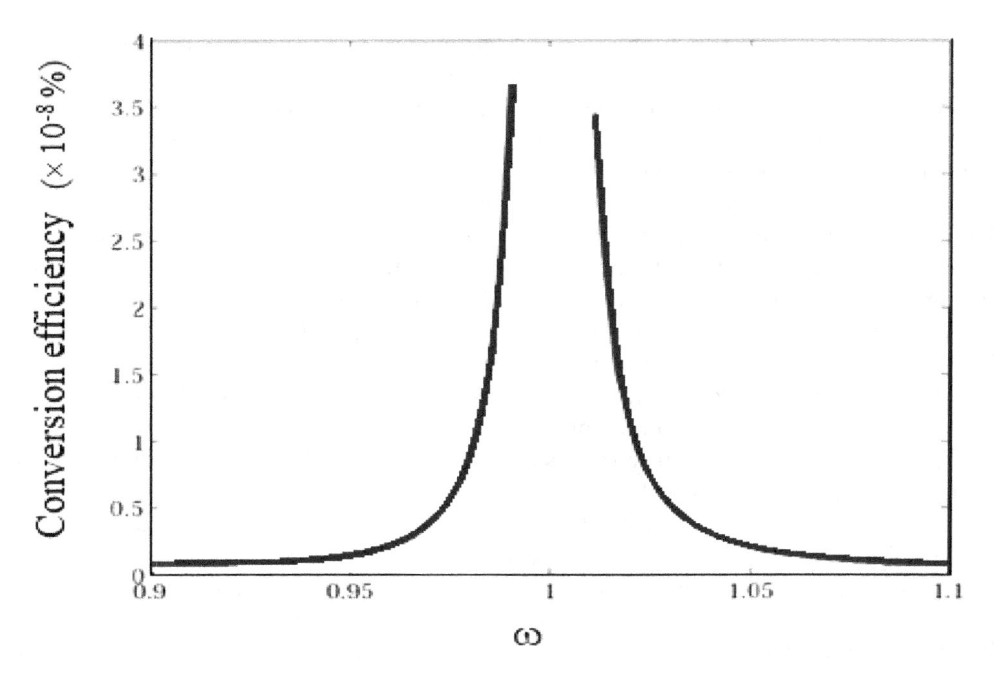

FIGURE 2.23
Variation of THz power conversion efficiency with beating frequency ω (in the multiple of $\omega_p/\sqrt{2}$) for $\theta = 48.75°$ and other parameters the same as in Figure 2.22.

2.8 Conclusions

The most common techniques for the generation of broadband pulsed THz radiations from semiconducting materials using ultrafast excitation pulses are photoconductive emission, OR effect, and the TC effect. The photoconductive approach is based on irradiating the gap of PCA made up of semiconducting materials, usually from III-V group compounds. The THz field here depends on the mobility (μ) of the carriers, laser energy ($h\upsilon$), biasing voltage (V_b), the reflectance of the substrate (R), and average power of the laser (P_{in}). This approach saturates at high excitation densities due to screening of the applied direct current (DC) bias and high bias voltage increases the dark current. The OR emitters, nevertheless, can be used at high excitation power. Further, to observe high emission efficiency, the photoconductive approach often requires proper focusing due to the infinite size of the antenna gap. On the other hand, the efficiency of the OR mechanism depends on the incident pulse characteristics, nonlinear coefficients viz. nonlinear absorption and nonlinear refraction coefficients of the material, the orientation of the crystal, THz absorption in the material, and the phase-matching conditions.

In SF effect, the intrinsic SF drives two kinds of carriers in the opposite direction and produces a photocurrent with the formation of dipoles in the direction of the normal. These transient dipoles oscillate and emit THz pulses. In the photo-Dember effect, a strong spatial gradient is formed near the semiconductor surface because of a difference in the diffusion rates of electrons and holes, leading to a transient dipole emitting THz pulses. This effect is more pronounced in narrow bandgap semiconductors, such as InAs and InSb, due to the higher electron mobility, weak depletion field, and short absorption depth.

The field of THz radiation in the case of NPs depends on their orientation with respect to the incident electric field of the lasers, the shape of the NPs (SNPs or CNPs), and separation between the NPs and parameters of the laser such as beam width, index of laser, and so forth. The NPs oriented parallel to the incident electric field contribute more towards the THz emission. Moreover, CNPs contribute more effectively towards THz generation than SNPs. On the other hand, QD layers in the superconducting structure have resulted in a highly efficient, ultrafast device for photoconductive-based emissions (QD PCA), where external bias voltage adds on to the processes like ultrafast carrier generation, carrier capturing, carrier recombination, and so forth. In such systems, the THz intensity increases when the antenna field is increased, with the fixed value of the power of the laser pump. However, there is not always an enhancement in the intensity of the THz radiation with the increasing antenna field; rather, it saturates in almost all the cases of laser pump power, and this saturation depends on the power of the laser pump.

The irradiation of a randomly roughed 2D metallic structure using femtosecond laser radiations enables the momenta matching condition over a wide range of wave vectors. Here a part of laser energy absorbed is trapped inside the surface layer of the sample and then dissipates into a bulk sample through conduction. In such a case, the energy conversion efficiency of optical input laser to THz to IR radiation is found to linearly increase with the pump fluence. However, this linear dependence is found to deviate in the case of increased surface temperature. In the case of CNT array mounted on a dielectric surface, free electrons of the nanotubes generate nonlinear current and hence the THz radiation. Here the THz power depends on the direction of angle of incidence of beating lasers and the power initially increases with the incident angle, reaches a maximum value, and then decreases slowly for higher values of the angle. The THz power also increases in a direction where the phase difference between the fields due to successive nanotubes is an integral multiple of 2π.

2.9 Selected Problems and Solutions

PROBLEM 2.1

Calculate the THz electric field output amplitude from the ZnTe photodetector. Given the electro-optic coefficient of the ZnTe detector = 3.9 pm/V, length of detector = 1 mm, the NIR refractive index for the ZnTe detector = 2.8. Consider the value of $\Delta P/P_{\text{probe}}$ around 3×10^{-3}.

SOLUTION

The electric field of the THz radiation at the output of the detector is given by

$$\frac{\Delta P}{P_{\text{probe}}} = \frac{\omega n^3 E_{THz} r_{41} L}{c}$$

where $r_{41} = 3.9$ pm/V, $L = 1$ mm, $n = 2.8$, $\omega = 2 \times 10^{15}$ rad/sec, and $c = 3 \times 10^8$ m/sec.

Hence, $3 \times 10^{-3} = \frac{2\times10^{15}\times(2.8)^3\times E_{THz}\times3.9\times10^{-12}\times10^{-3}}{3\times10^8} = \frac{171\ E_{THz}}{3\times10^8}$

or $E_{THz} = \frac{3\times10^{-3}\times3\times10^8}{171} = 5.26\times10^3$ V/m

PROBLEM 2.2

The diagram below (Figure 2.24) shows an emitter of pulsed THz radiation and the beam path THz radiation takes before arriving at a detector (collimation and focusing optics not shown). The incident IR pulse has an electric field E_{IR} in the y-direction. In what directions are the electric field and magnetic field of the radiated THz pulse when it reaches the detector?

SOLUTION

The THz pulse is generated when:

An IR laser pulse hits the (intrinsic) semiconductor between the metal contacts (shown in Figure 2.24). This photoexcites electrons and holes. Electrons and holes separate in opposite directions under the applied bias of the emitter, forming a dipole. In the far field, the dominant radiation from an electric dipole is parallel to the dipole, i.e. in the x-direction (vertical on this diagram). In view of the propagation of the THz pulse in the z-direction, the magnetic field of the THz would be in the y-direction (electromagnetic waves are transverse plane waves in the far field). Note that this is *not* necessarily the direction of the polarization of the incident beam. The electric field of the THz radiation will be in the x-direction.

PROBLEM 2.3

How could the electric field direction of the THz pulse be altered in the problem 2.2?

SOLUTION

There are various options:

1. Rotate the emitter about the z-axis.
2. Place some polarization modifying component in the THz beam path, such as a wire-grid polarizer, a waveplate (e.g. quarter / half waveplates for circular / rotated linear polarization).

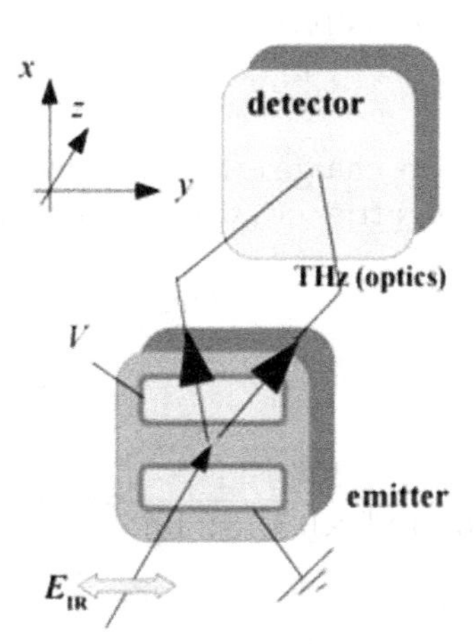

FIGURE 2.24
Schematic for pulsed THz radiation emitter.

PROBLEM 2.4

Name two ways in which the amplitude E_{THz} of the THz radiation pulse could be increased, and explain why with reference to simple mathematical formula(e) by seeing the above figure (Figure 2.24).

SOLUTION

The emitted amplitude E_{THz} is proportional to dJ/dt. It means increasing the current density will increase E_{THz}. This can be done either by increasing the IR laser's power (increasing carrier density), or choosing a semiconductor with a higher mobility. Alternatively, use a laser with shorter pulse duration (to get a faster rise in current density), i.e. to reduce dt.

PROBLEM 2.5

Find the maximum THz field that is capable of being measured by the PCA. Given $e = 1.6 \times 10^{-19}$C, $\mu = 8000$ cm^{2}V^{-1}s^{-1}, $R = 0.29$, $D = 2$ μm, $P_{in} = 10$ mW, $V_b = 48$ kV, and $h\nu = 3.2 \times 10^{-19}$J.

SOLUTION

The maximum electric field is given by

$$E_{\max} = e\mu\left(\frac{1-R}{h\upsilon}\right)\frac{P_{in}V_b}{D}$$

where μ is the mobility of the carriers, $h\nu$ is the laser photon energy, R is the reflectance of the substrate, P_{in} is the average laser power, and V_b is the biasing voltage.

So, $E_{\max} = 1.6\times10^{-19}\times8000\times10^{-4}\left(\frac{1-0.29}{3.2\times10^{-19}}\right)\frac{10\times10^{-3}\times48\times10^{3}}{2\times10^{-6}}$

$$E_{\max} = 6.8\times10^{7}\ \text{V/m}$$

PROBLEM 2.6

A visible 68-nm wavelength light is used for the photoconductive emission of THz pulses. Calculate the carrier density of electrons moving with a velocity 0.3c and producing nonlinear current of an amount 1×10^8 Amp/m^2.

SOLUTION

If the electron velocity is v and its charge e, then the photoconductive current is given by $J = Nev$. From this the density of carrier electrons can be written as

$$N = J/ev$$

Putting in the values we find

$$N = \frac{10^8}{1.6\times10^{-19}\times0.3\times3\times10^8} = 6.94\times10^{18}\ \text{m}^{-3}$$

PROBLEM 2.7

During the OR process, the lasers hit the semiconductor and the semiconductor gets polarized. Find out the amount of polarization when the lasers of electric field 4.4×10^8 V/m are used.

SOLUTION

The equation governing the OR process is given by

$$P(\omega_{\text{THz}}) = \varepsilon_0 \chi^{(2)} E^2$$

In terms of the atomic electric field E_{at} (= $\frac{e}{4\pi\varepsilon_0 a_0^2}$, where $a_0$is the Bohr radius) the susceptibility is given by $\chi^{(2)} = \frac{\chi^{(1)}}{E_{at}} \sim \frac{1}{E_{at}}$. This can also be written as

$$\chi^{(2)} = \frac{(4\pi\varepsilon_0)^3 \hbar^4}{m^2 e^5} \cong 1.94 \times 10^{-12} \text{m}/\text{V}$$

Hence, the amount of polarization can be calculated as

$$P = 8.85 \times 10^{-12} \times 1.94 \times 10^{-12} \times 4.4 \times 10^8$$
$$= 75.54 \times 10^{-16} \text{C} - \text{m}$$

PROBLEM 2.8

An ultrashort excitation pulse is incident on the photoconductive medium at an angle θ, as shown in Figure 2.25. The dielectric constant of the medium is ε. Calculate the angle at which the transmitted THz emits through the medium.

SOLUTION

If ω and Ω denote the optical and THz frequencies, then Snell's law can be written as

$$n_1(\omega)\sin\theta = n_1(\Omega)\sin\theta_1 = n_2\sin\theta_2$$

Or $n_1 \sin\theta = n_2 \sin\theta_2$

$$\Rightarrow \sin\theta = \sin\theta_2$$

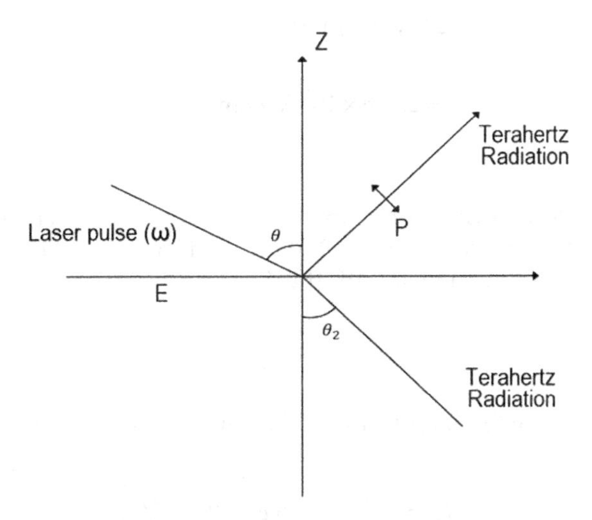

FIGURE 2.25
Schematic for THz radiation when an ultrashort pulse interacts with photoconductive medium.

Or $\sin\theta_2 = \frac{\sin\theta}{\sqrt{\varepsilon}}$
Hence,

$$\theta_2 = \sin^{-1}\left(\frac{\sin\theta}{\sqrt{\varepsilon}}\right)$$

PROBLEM 2.9

A THz wave is emitted from TC in large-area emitters excited at the normal incidence. The dielectric constant of the photoconductive medium is 13.1 and mobility of electrons is 8500 cm²/Vs, and the number of charge carrier excited by the laser of electric field 5×10^8 V/cm is 4.7×10^{17} cm^{-3}. What is the electric field of emitted THz wave?

SOLUTION

For normal incidence, the electric field of THz wave in terms of photocurrent J is given by

$$E = \frac{-4\pi}{c}\frac{1}{1+\sqrt{\varepsilon}}J$$

The current J in terms of electron mobility μ, density of charge carriers N, and field E is calculated based on the following expression

$$J = -Ne\mu E$$
$$J = -4.7\times10^{17}\times1.6\times10^{-19}\times8500\times5\times10^{8} = -3.196\times10^{11}\,\text{Amp}/\text{cm}^2 = -3.196\times10^{15}\,\text{Amp/m}^2$$

Hence,

$$E = \frac{-4\pi}{c}\frac{1}{1+\sqrt{13.1}}\times3.19\times10^{11}$$

$$= 2.898\times10^{7}\,\text{V/m}$$

$$= 2.898\times10^{5}\ \text{V}/\text{cm}$$

PROBLEM 2.10

Calculate the ponderomotive force experienced by the electrons when Gaussian laser beams of electric field $E = 5 \times 10^8$ V/cm and beam width $b_w = 0.01$ cm interact with the plasma. The frequencies of the beating lasers are 2×10^{14} rad/sec and 2.3×10^{14} rad/sec.

SOLUTION

The ponderomotive force (in SI units) exerted by an oscillating electric field is given by

$$F_p = \frac{e^2E^2}{2\,m\omega_1\omega_2}\times e^{-2y^2/b_w^2}\left(\frac{-4y}{b_w^2}\right)$$

Take $y/b_w = 1$,

$$F_p = \frac{\left(1.6\times10^{-19}\right)^2\times\left(5\times10^{10}\right)^2\times0.13533\times(-4)}{2\times9.1\times10^{-31}\times2\times2.3\times10^{28}\times0.01\times10^{-2}}$$

$$= 4.13829\times10^{-13}\ N$$

Suggested Reading Material

Auston, D. H., & Cheung, K. P. (1985). Coherent time-domain far-infrared spectroscopy. *JOSA B, 2*(4), 606–612.

Castro-Camus, E., & Alfaro, M. (2016). Photoconductive devices for terahertz pulsed spectroscopy: A review. *Photonics Research, 4*(3), A36–A42.

Cheville, R. A., & Grischkowsky, D. (1995). Far-infrared terahertz time-domain spectroscopy of flames. *Optics Letters, 20*(15), 1646–1648.

Cheville, R. A., & Grischkowsky, D. (1999). Far-infrared foreign and self-broadened rotational linewidths of high-temperature water vapor. *JOSA B, 16*(2), 317–322.

Corson, J., Orenstein, J., Oh, S., O'Donnell, J., & Eckstein, J. N. (2000). Nodal quasiparticle lifetime in the superconducting state of Bi 2 Sr 2 CaCu 2 O 8+ δ. *Physical Review Letters, 85*(12), 2569.

Darrow, J. T., Hu, B. B., Zhang, X.-C., & Auston, D. H. (1990). Subpicosecond electromagnetic pulses from large-aperture photoconducting antennas. *Optics Letters, 15*(6), 323–325.

DeFonzo, A. P., & Lutz, C. R. (1987). Optoelectronic transmission and reception of ultrashort electrical pulses. *Applied Physics Letters, 51*(4), 212–214.

Dekorsy, T., Pfeifer, T., Kütt, W., & Kurz, H. (1993). Subpicosecond carrier transport in GaAs surface-space-charge fields. *Physical Review B, 47*(7), 3842.

Dember, H. (1931). Photoelectromotive force in cuprous oxide crystals. *Phys. Z, 32*(1), 554–556.

Dodge, J. S., Weber, C. P., Corson, J., & … Beasley, M. R. (2000). Low-frequency crossover of the fractional power-law conductivity in SrRuO 3. *Physical Review Letters, 85*(23), 4932.

Estacio, E., Pham, M. H., Takatori, S., … Awitan, F. C. B. (2009). Strong enhancement of terahertz emission from GaAs in InAs/GaAs quantum dot structures. *Applied Physics Letters, 94*(23), 232104.

Gallot, G., Zhang, J., McGowan, R. W., Jeon, T.-I., & Grischkowsky, D. (1999). Measurements of the THz absorption and dispersion of ZnTe and their relevance to the electro-optic detection of THz radiation. *Applied Physics Letters, 74*(23), 3450–3452.

Gao, Y., Chen, M.-K., Yang, C.-E., & … Luo, C. (2009). Analysis of terahertz generation via nanostructure enhanced plasmonic excitations. *Journal of Applied Physics, 106*(7), 74302.

Garwe, F., Schmidt, A., Zieger, G., … Meyer, H.-G. (2011). Bi-directional terahertz emission from gold-coated nanogratings by excitation via femtosecond laser pulses. *Applied Physics B, 102*(3), 551–554.

Grischkowsky, D. R. (1993). Nonlinear generation of sub-psec pulses of THz electromagnetic radiation by optoelectronics—Applications to time-domain spectroscopy, *Frontiers in Nonlinear Optics*. CRC Press, Pages 196–227.

Han, P. Y., Cho, G. C., & Zhang, X.-C. (2000). Time-domain transillumination of biological tissues with terahertz pulses. *Optics Letters, 25*(4), 242–244.

Harde, H., Cheville, R. A., & Grischkowsky, D. (1997). Terahertz studies of collision-broadened rotational lines. *The Journal of Physical Chemistry A, 101*(20), 3646–3660.

Harde, H., Zhao, J., Wolff, M., Cheville, R. A., & Grischkowsky, D. (2001). THz time-domain spectroscopy on ammonia. *The Journal of Physical Chemistry A, 105*(25), 6038–6047.

Heidemann, R., Pfeiffer, T., & Jäger, D. (1983). Optoelectronically pulsed slot-line antennas. *Electronics Letters, 19*(9), 316–317.
Hu, B. B., Zhang, X.-C., & Auston, D. H. (1991). Terahertz radiation induced by subband-gap femtosecond optical excitation of GaAs. *Physical Review Letters, 67*(19), 2709.
Huggard, P. G., Cluff, J. A., Moore, G. P., … Ritchie, D. A. (2000). Drude conductivity of highly doped GaAs at terahertz frequencies. *Journal of Applied Physics, 87*(5), 2382–2385.
Javan, N. S., & Erdi, F. R. (2017). Theoretical study of the generation of terahertz radiation by the interaction of two laser beams with graphite nanoparticles. *Journal of Applied Physics, 122*(22), 223103.
Javan, N. S., & Erdi, F. R. (2019). Magnetic field effect on Fresnel coefficients of the thin slab of graphite nanocomposite. *Plasmonics, 14*(1), 219–230.
Jeon, T.-I., & Grischkowsky, D. (1998). Characterization of optically dense, doped semiconductors by reflection THz time domain spectroscopy. *Applied Physics Letters, 72*(23), 3032–3034.
Jeon, T.-I., Grischkowsky, D., Mukherjee, A. K., & Menon, R. (2000). Electrical characterization of conducting polypyrrole by THz time-domain spectroscopy. *Applied Physics Letters, 77*(16), 2452–2454.
Kaindl, R. A., Carnahan, M. A., Orenstein, J., … Lowndes, D. H. (2001). Far-infrared optical conductivity gap in superconducting MgB 2 films. *Physical Review Letters, 88*(2), 27003.
Kaiser, W. (1993). *Ultrashort laser pulses: generation and applications.* Springer-Verlag Berlin Heidelberg.
Kersting, R., Heyman, J. N., Strasser, G., & Unterrainer, K. (1998). Coherent plasmons in n-doped GaAs. *Physical Review B, 58*(8), 4553.
Khurgin, J. B. (1994). Optical rectification and terahertz emission in semiconductors excited above the band gap. *JOSA B, 11*(12), 2492–2501.
Krotkus, A., Adomavičius, R., & Pačebutas, V. (2008). Characterizing semiconductor materials with terahertz radiation pulses. In Sixth International Conference on Advanced Optical Materials and Devices (AOMD-6), Vol. 7142, p. 714205. International Society for Optics and Photonics.
Kruczek, T., Leyman, R., Carnegie, D., … Rafailov, E. U. (2012). Continuous wave terahertz radiation from an InAs/GaAs quantum-dot photomixer device. *Applied Physics Letters, 101*(8), 81114.
Lagatsky, A. A., Rafailov, E. U., Sibbett, W., Livshits, D. A., Zhukov, A. E., & Ustinov, V. M. (2005). Quantum-dot-based saturable absorber with p-n junction for mode-locking of solid-state lasers. *IEEE Photonics Technology Letters, 17*(2), 294–296.
Lai, W., Abdulmunem, O. M., Del Pino, P., … Zhang, H. (2017). Enhanced terahertz radiation generation of photoconductive antennas based on manganese ferrite nanoparticles. *Scientific Reports, 7*, 46261.
Leitenstorfer, A., Hunsche, S., Shah, J., Nuss, M. C., & Knox, W. H. (2000). Femtosecond high-field transport in compound semiconductors. *Physical Review B, 61*(24), 16642.
Leyman, R. R., Gorodetsky, A., Bazieva, N., … Rafailov, E. U. (2016). Quantum dot materials for terahertz generation applications. *Laser & Photonics Reviews, 10*(5), 772–779.
Malik, A. K., Malik, H. K., & Stroth, U. (2012). Terahertz radiation generation by beating of two spatial-gaussian lasers in the presence of a static magnetic field. *Physical Review E, 85*(1), 16401.
Malik, R., & Uma, R. (2018). THz generation by laser coupling to carbon nanotube array. *Physics of Plasmas, 25*(1), 13106.
Markelz, A. G., Roitberg, A., & Heilweil, E. J. (2000). Pulsed terahertz spectroscopy of DNA, bovine serum albumin and collagen between 0.1 and 2.0 THz. *Chemical Physics Letters, 320*(1–2), 42–48.
Mittleman, D. M., Hunsche, S., Boivin, L., & Nuss, M. C. (1997). T-ray tomography. In *Ultrafast Electronics and Optoelectronics, Optical Society of America*, p. UF5.
Mourou, G., Stancampiano, C. V., Antonetti, A., & Orszag, A. (1981). Picosecond microwave pulses generated with a subpicosecond laser-driven semiconductor switch. *Applied Physics Letters, 39*(4), 295–296.
Nuss, M. C., Goossen, K. W., Gordon, J. P., Mankiewich, P. M., O'Malley, M. L., & Bhushan, M. (1991). Terahertz time-domain measurement of the conductivity and superconducting band gap in niobium. *Journal of Applied Physics, 70*(4), 2238–2241.
Özbay, E., Michel, E., Tuttle, G., … Bloom, D. M. (1994). Terahertz spectroscopy of three-dimensional photonic band-gap crystals. *Optics Letters, 19*(15), 1155–1157.

Polyushkin, D. K., Hendry, E., Stone, E. K., & Barnes, W. L. (2011). THz generation from plasmonic nanoparticle arrays. *Nano Letters, 11*(11), 4718–4724.

Porte, H. P., Uhd Jepsen, P., Daghestani, N., Rafailov, E. U., & Turchinovich, D. (2009). Ultrafast release and capture of carriers in InGaAs/GaAs quantum dots observed by time-resolved terahertz spectroscopy. *Applied Physics Letters, 94*(26), 262104.

Ralph, S. E., Perkowitz, S., Katzenellenbogen, N., & Grischkowsky, D. (1994). Terahertz spectroscopy of optically thick multilayered semiconductor structures. *JOSA B, 11*(12), 2528–2532.

Ramakrishnan, G., Kumar, N., Planken, P. C. M., Tanaka, D., & Kajikawa, K. (2012). Surface plasmon-enhanced terahertz emission from a hemicyanine self-assembled monolayer. *Optics Express, 20*(4), 4067–4073.

Ramakrishnan, G., & Planken, P. C. M. (2011). Percolation-enhanced generation of terahertz pulses by optical rectification on ultrathin gold films. *Optics Letters, 36*(13), 2572–2574.

Robertson, W. M. (1995). *Optoelectronic techniques for microwave and millimeter-wave engineering*. Artech House.

Robertson, W. M., Arjavalingam, G., Meade, R. D., Brommer, K. D., Rappe, A. M., & Joannopoulos, J. D. (1992). Measurement of photonic band structure in a two-dimensional periodic dielectric array. *Physical Review Letters, 68*(13), 2023.

Robertson, W. M., Arjavalingam, G., & Shinde, S. L. (1991). Microwave dielectric measurements of zirconia-alumina ceramic composites: A test of the Clausius–Mossotti mixture equations. *Journal of Applied Physics, 70*(12), 7648–7650.

Rønne, C., Åstrand, P.-O., & Keiding, S. R. (1999). THz spectroscopy of liquid H_2O and D_2O. *Physical Review Letters, 82*(14), 2888.

Ronne, C., Thrane, L., Åstrand, P.-O., Wallqvist, A., Mikkelsen, K. V, & Keiding, S. R. (1997). Investigation of the temperature dependence of dielectric relaxation in liquid water by THz reflection spectroscopy and molecular dynamics simulation. *The Journal of Chemical Physics, 107*(14), 5319–5331.

Rothwell, E. J., & Cloud, M. J. (2018). *Electromagnetics*. CRC Press.

Ruppin, R. (2000). Evaluation of extended Maxwell-Garnett theories. *Optics Communications, 182*(4–6), 273–279.

Schall, M., Helm, H., & Keiding, S. R. (1999). Far infrared properties of electro-optic crystals measured by THz time-domain spectroscopy. *International Journal of Infrared and Millimeter Waves, 20*(4), 595–604.

Schmidt, A., Garwe, F., Hübner, U., … Stafast, H. (2012). Experimental characterization of bi-directional terahertz emission from gold-coated nanogratings. *Applied Physics B, 109*(4), 631–642.

Sharma, D., Malik, B. P., & Gaur, A. (2016). Sensitive measurement of nonlinear absorption and optical limiting in undoped and Fe-doped ZnO quantum dots using pulsed laser. *Indian Journal of Physics, 90*(11), 1293–1298.

Sihvola, A. H. (1999). *Electromagnetic mixing formulas and applications*. Institution of Electrical Engineers, IEE electromagnetic waves series 47. Institution of Electrical Engineers.

Suvorov, E. V., Akhmedzhanov, R. A., Fadeev, D. A., Ilyakov, I. E., Mironov, V. A., & Shishkin, B. V. (2012). Terahertz emission from a metallic surface induced by a femtosecond optic pulse. *Optics Letters, 37*(13), 2520–2522.

Welsh, G. H., Hunt, N. T., & Wynne, K. (2007). Terahertz-pulse emission through laser excitation of surface plasmons in a metal grating. *Physical Review Letters, 98*(2), 26803.

Welsh, G. H., & Wynne, K. (2009). Generation of ultrafast terahertz radiation pulses on metallic nanostructured surfaces. *Optics Express, 17*(4), 2470–2480.

Xu, L., Zhang, X., Auston, D. H., & Jalali, B. (1991). Terahertz radiation from large aperture Si p-i-n diodes. *Applied Physics Letters, 59*(26), 3357–3359.

Zhang, L., Mu, K., Zhou, Y., Wang, H., Zhang, C., & Zhang, X.-C. (2015). High-power THz to IR emission by femtosecond laser irradiation of random 2D metallic nanostructures. *Scientific Reports, 5*, 12536.

Zhang, X., & Auston, D. H. (1992). Optoelectronic measurement of semiconductor surfaces and interfaces with femtosecond optics. *Journal of Applied Physics, 71*(1), 326–338.

Zhang, X., Hu, B. B., Darrow, J. T., & Auston, D. H. (1990). Generation of femtosecond electromagnetic pulses from semiconductor surfaces. *Applied Physics Letters, 56*(11), 1011–1013.

3

Surface Plasmon Resonance and THz Radiation

3.1 Introduction

Wood (1902) observed that when polarized light is allowed to fall on a system of mirror and gratings engraved on its surface, reflected light consists of an anomalous series of alternate light and dark bands. Later, it was physically interpreted by Rayleigh (1907), and further explored by Fano (1941). Only after the phenomenon of excitation of surface plasmons was reported by Otto (1968), was a clear and complete explanation of the phenomenon given by Kretschmann and Raether (1968).

In the case of light incident at a particular angle on an interface between negative and positive permittivity materials, resonant oscillation of conduction electrons is seen. These oscillations take place due to absorption of light, and the resonant excitation of the electrons is called surface plasmon resonance (SPR). These resonating electrons or surface plasmons are sensitive to the surrounding environment, and there is a loss of intensity in the reflected beam that exhibited a dark band, which is analyzed as a dip in the reflection intensity curve of the phenomenon.

3.2 What Is Surface Plasmon Resonance and How Does It Work?

Now we can explain the SPR as a phenomenon in which light is allowed to fall on an interface through the prism at varying angle of incidence (Figure 3.1). Under this situation, there will be an interaction between the incident light's photons and free electrons of the metal layer through the evanescent waves. At certain angle of incidence, the free electrons are excited by absorbing the incident light intensity, causing a decrease in the intensity of the reflected light and set to resonate. The angle at which the reflected light intensity shows a sudden decrease is known as the angle of resonance or SPR-dip. At this angle, the wave vector of the incident light matches with that of surface plasmons, giving rise to oscillations of the free electrons known as surface plasmons.

The angle of resonance depends on many factors of the optical system, but most important are the refractive indices of both the media involved in the process. SPR is an effective technique to observe and record changes in the index of refraction in the nearby region of the metal surface. As the index of refraction changes, the angle of resonance and the position of intensity minimum changes; hence, the angle at which it is observed also changes (dips A and B in Figure 3.1).

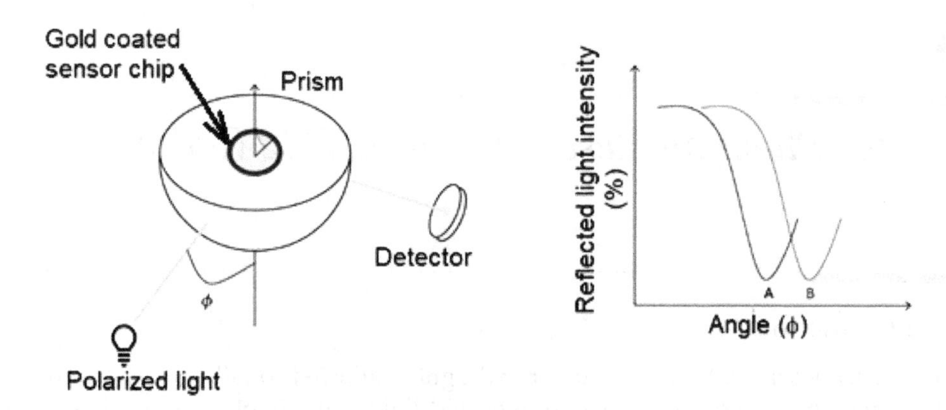

FIGURE 3.1
Schematic of excitation of SPR (left) and the intensity curve of reflected light (right).

3.3 Necessary Condition for Excitation of Surface Plasmons

Let us consider the interface separating the two media along the x-direction. For surface plasmon all the components of the TM (Transverse Magnetic) mode exists and those of the TE (Transverse Electric) mode are zero. We can write the electric fields in the two media as

$$\vec{E}_1 = \left(E_{x1}\hat{x} + E_{z1}\hat{z}\right)e^{i(k_{x1}x + k_{z1}z - \omega t)} \tag{3.1a}$$

$$\vec{E}_2 = \left(E_{x2}\hat{x} + E_{z2}\hat{z}\right)e^{i(k_{x2}x + k_{z2}z - \omega t)} \tag{3.1b}$$

Magnetic field vector in medium 1 can be expressed as

$$\vec{H}_1 = \left(H_{yl}\hat{y}\right)e^{i(k_{x1}x + k_{z1}z - \omega t)} \tag{3.2}$$

From Maxwell's equation $\vec{\nabla} \cdot \vec{D} = 0$, we write

$$\vec{\nabla} \cdot \left(\vec{E}_1 \varepsilon_1\right) = 0 \tag{3.3}$$

$$=> k_{x1}E_{x1} + k_{z1}E_{z1} = 0 \tag{3.4}$$

Now

$$\vec{\nabla} \times \vec{H}_1 = \partial\vec{D}/\partial t \text{ (assuming } J = 0) \tag{3.5}$$

Putting in the value of H_1, we can get

$$H_{y1} = \frac{\varepsilon_1 \omega}{k_{z1}} E_{x1} \tag{3.6}$$

Using boundary condition $E_1^{\parallel} = E_2^{\parallel}$ we get

$$E_{x1} = E_{x2} \tag{3.7}$$

Using boundary condition $D_{z1}^{\perp} = D_{z2}^{\perp}$ along with the use of Eqs. (3.1a) and (3.1b), we get

$$\varepsilon_1\left(-\frac{k_{x1}}{k_{z1}}.E_{x1}\right) = \varepsilon_2\left(-\frac{k_{x2}}{k_{z2}}.E_{x2}\right) \tag{3.8}$$

Since boundary condition must be satisfied for every x and t, putting $z = 0$ and $t = 0$ in the above equations, we get

$$k_{x1} = k_{x2} \tag{3.9}$$

From above Eqs. (3.6) and (3.7) we get a very important relation for surface plasmons:

$$\frac{\varepsilon_1}{k_{z1}} = \frac{\varepsilon_2}{k_{z2}} \tag{3.10}$$

Now what are k_{z1} and k_{z2}? To understand this, let us examine the decay of fields

$$\vec{E}_1 = \left(E_{x1}\hat{x} + E_{z1}\hat{z}\right)e^{i(k_{x1}x + k_{z1}z - \omega t)} \tag{3.11}$$

$$\vec{E}_2 = \left(E_{x2}\hat{x} + E_{z2}\hat{z}\right)e^{i(k_{x2}x + k_{z2}z - \omega t)} \tag{3.12}$$

Suppose the field is present on both the sides of the interface, then the z-component is purely imaginary, i.e. can be expressed as $e^{i\alpha_1 z}$, $e^{i\alpha_2 z}$, where, α_1 and α_2 and are real quantities. Therefore, $k_{z1}$and k_{z2} must be imaginary.

For the first medium z_1 is positive, therefore α_1 must be positive. However, for the second medium z_2 is negative, which means α_2 must be negative.

$$\alpha_1 > 0 => k_{z1} = i\alpha_1 > 0 \tag{3.13}$$

$$\alpha_2 < 0 => k_{z2} = i\alpha_2 < 0 \tag{3.14}$$

In view of these and the observation that $\frac{\varepsilon_1}{k_{z1}} = \frac{\varepsilon_2}{k_{z2}}$, the quantities ε_1 and ε_2 must have an opposite sign; this is the most important condition for the surface plasmon to get excited. This also conveys the meaning that both the media should have opposite permittivity, which means if one has positive permittivity then the other has negative permittivity. Now, how do we achieve the negative permittivity?

To answer this question, we recall that the permittivity relation for metals is given by

$$\varepsilon_r = \frac{\varepsilon_{\text{metal}}}{\varepsilon_0} = 1 - \frac{\omega_p^2}{\omega^2} \tag{3.15}$$

In terms of plasma frequency ω_p of the metal, the relative permittivity ε_r is hence written as

$$\varepsilon_r = 1 - \frac{\omega_p^2}{\omega^2} \tag{3.16}$$

Here ω is the frequency of the wave defined in Eq. (3.1). If $\omega_p > \omega$, then the relative permittivity of the metal becomes negative, $\varepsilon_r < 0$. Hence, it is essential that we choose a second medium as the metal in view of the fact that $\omega_p > \omega$ can be achieved.

3.4 Dispersion Relation for Surface Plasmons

We can write the relation between ω and k for the two media as:

$$\frac{\omega_1^2}{k_1^2} = \frac{1}{\mu\varepsilon_1} \tag{3.17}$$

$$\frac{\omega_2^2}{k_2^2} = \frac{1}{\mu\varepsilon_2} \tag{3.18}$$

In our case, the wave is traveling in the (x, z)-plane, which means

$$k_1^2 = k_{x1}^2 + k_{z1}^2 \text{ and } k_2^2 = k_{x2}^2 + k_{z2}^2 \tag{3.19}$$

Since $k_{x1} = k_{x2} = k_x$, the above equation is written as

$$k_{z1}^2 = \omega^2\varepsilon_1\mu - k_x^2 \tag{3.20a}$$

$$k_{z2}^2 = \omega^2\varepsilon_2\mu - k_x^2 \tag{3.20b}$$

Dividing Eqs. (3.20a) and (3.20b) and using Eq. (3.10) we get

$$\frac{k_{z1}^2}{k_{z2}^2} = \frac{\omega^2\varepsilon_1\mu - k_x^2}{\omega^2\varepsilon_2\mu - k_x^2} = \frac{\varepsilon_1^2}{\varepsilon_2^2}$$

On solving this we get

$$k_x = \omega\sqrt{\frac{\mu\varepsilon_1\varepsilon_2}{\varepsilon_1 + \varepsilon_2}} = \frac{\omega}{c}\sqrt{\frac{\varepsilon_{r1}\varepsilon_{r2}}{\varepsilon_{r1} + \varepsilon_{r2}}} \tag{3.21}$$

Since ε_{r1} and ε_{r2} are of opposite signs, $\varepsilon_{r1} + \varepsilon_{r2}$ should be negative to keep the propagation constant k_x a real quantity. It means

$$\varepsilon_{r1} + \varepsilon_{r2} < 0 \tag{3.22}$$

But the permittivity of metals (second medium) is expressed as $\varepsilon_{r2} = 1 - \frac{\omega_p^2}{\omega^2}$. Hence, we get

$$\omega < \frac{\omega}{\sqrt{1 + \varepsilon_{r1}}} = \omega_{sp} \tag{3.23}$$

where ω_{sp} is the surface plasmon frequency. It means, for surface plasmon to exist, the frequency of the incident wave must be less than the frequency of surface plasmon. It is not advisable to excite the surface plasmons on both sides of the interface for practical purposes. Therefore, to confine the energy in a smaller width, the decay length should be small on both the sides (decay length < wavelength of incident light).

Further we know that $\varepsilon_{r1} + \varepsilon_{r2} < 0$ and $\varepsilon_{r2} < 0$

$$\Rightarrow g\left|\varepsilon_{r2}\right| > \left|\varepsilon_{r1}\right|$$

$$\Rightarrow g\left|\frac{k_{z1}}{k_{z2}}\right| = \sqrt{\frac{-\varepsilon_{r1}}{\varepsilon_{r2}}} < 1 \because k_{z1} = \frac{\omega}{c}\frac{\varepsilon_{r1}}{\sqrt{\varepsilon_{r1}+\varepsilon_{r2}}};\ k_{z2} = \frac{\omega}{c}\frac{\varepsilon_{r2}}{\sqrt{\varepsilon_{r1}+\varepsilon_{r2}}}$$

$$\Rightarrow g\left|\frac{k_{z2}}{k_{z1}}\right| = \sqrt{\frac{-\varepsilon_{r2}}{\varepsilon_{r1}}} > 1 \tag{3.24}$$

Let us suppose $\varepsilon_{r1} \ggg \varepsilon_{r2}$, then on one side the decay length will be very small, but on the other side it will be very large, which is undesirable. So the best possible way is to have $\varepsilon_{r1} \approx \varepsilon_{r2}$.

3.5 How Do We Excite Surface Plasmons?

Let a wave be incident on an interface and make an angle θ_1 with the interface normal, as shown in Figure 3.2. For exciting the surface plasmon, k_x of the incident wave along the interface must match with the wave vector $(k_x)_{sp}$ of surface plasmons.

Therefore, $(k_x)_{sp} = \frac{\omega}{c}\sqrt{\frac{\varepsilon_{r1}\varepsilon_{r2}}{\varepsilon_{r1}+\varepsilon_{r2}}} = \frac{\omega\sqrt{\varepsilon_{r1}}}{c}\sqrt{\frac{1}{1+\frac{\varepsilon_{r1}}{\varepsilon_{r2}}}}$ (note: $\frac{\varepsilon_{r1}}{\varepsilon_{r2}}$ is a negative term)

$$k_x \sin\theta_1 = \frac{\omega\sqrt{\varepsilon_{r1}}}{c}\sin\theta_1 < \frac{\omega\sqrt{\varepsilon_{r1}}}{c} \tag{3.25}$$

But

$$(k_x)_{sp} > \frac{\omega\sqrt{\varepsilon_{r1}}}{c} \tag{3.26}$$

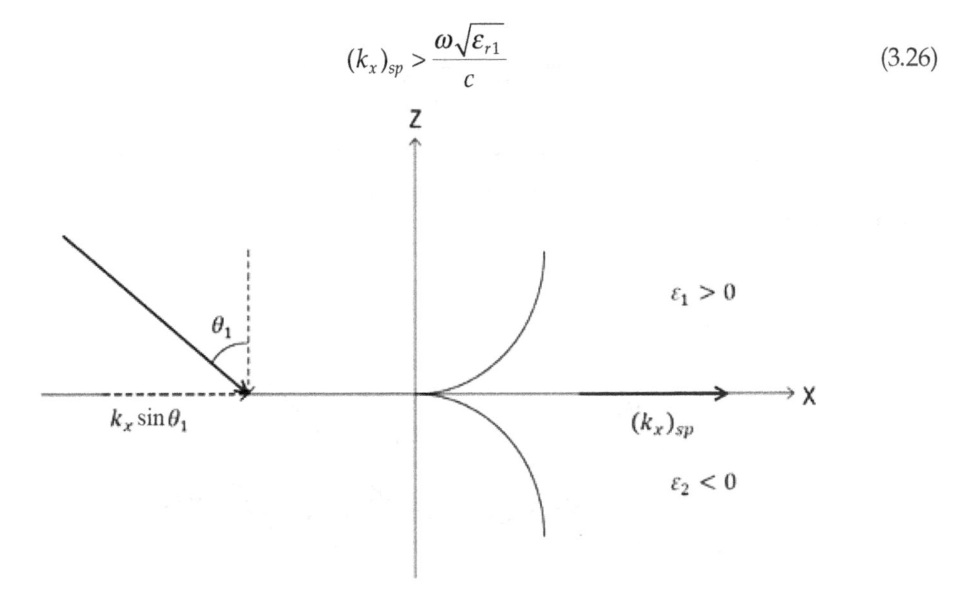

FIGURE 3.2
Showing x-components of wave vectors of incident wave and surface plasmons for momenta matching of the two and exciting the SPR.

This cannot be true for plane waves for any angle of incidence. Therefore, plane wave cannot be used directly to excite the surface plasmons. Hence, the surface plasmons are excited through the evanescent waves generated. Thus, a prism is kept near the metal interface and a wave is allowed to incident on it so that the total internal reflection takes place at the bottom surface of the prism due to which evanescent wave is produced parallel to the metal surface. As the angle of the incidence is varied, the wave vector of the evanescent wave changes. At a certain angle of incidence, it matches with that of the surface plasmon, and significant energy transfer takes place and the dip is observed in the reflected light. This is how the surface plasmons are excited.

3.6 Configurations of SPR

To measure the dip and the shift in the dip of reflected light, SPR instruments can be configured in different ways. There are mainly three different optical configurations by which excitation of plasmons are achieved.

3.6.1 Grating Configuration

In this arrangement, surface plasmons are excited by an incident light via a grating coupler, as shown in Figure 3.3. The wave vector of the incident light is increased by a factor of $2\pi/a$ for the grating constant a. Then, by selecting a particular grating, we will be able to match a component of the incident wave vector parallel to the interface to satisfy the dispersion relation; hence, the condition of resonance for θ_i is in accordance with the following relation

$$\frac{\omega}{c}\sin\theta_i + \frac{2\pi}{a} = \frac{\omega}{c}\sqrt{\frac{\varepsilon_2}{\varepsilon_2 + \varepsilon_1}} \tag{3.27}$$

where ε_1 and ε_2 are the permittivity of air and metal, respectively. Angle θ_i corresponds to the resonance angle at which an intensity minimum is observed in the reflected light due to the excitation of SPR.

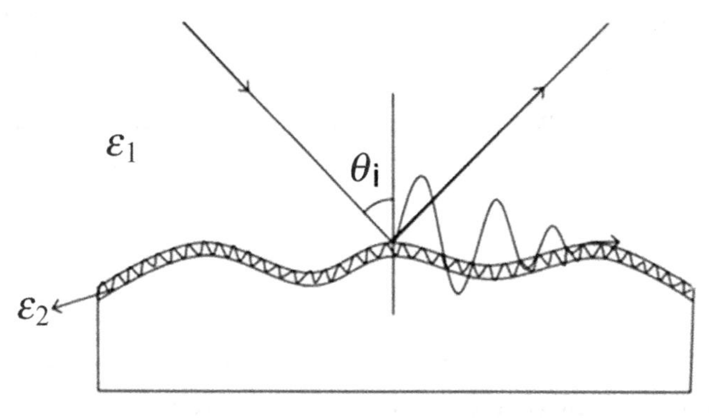

FIGURE 3.3
Schematic diagram of grating configuration.

The main drawback of the grating configuration is that the incident light radiation is passed through the sample solution in the absence of a prism, which may contribute to absorption of light depending on the nature of the sample, and the results may not be so accurate (Wood 1912).

3.6.2 Otto Configuration

In this configuration, the metal surface (ε_2) and the prism surface or surface of glass block (ε_p) are separated by an air gap (ε_1), as shown in Figure 3.4. An air gap acts as a dielectric medium having the dielectric constant $\varepsilon_1 < \varepsilon_p$.

The thickness of the air gap is very crucial in this configuration, which is approximately 1 μm wide. The resonance takes place at this metal-dielectric interface only. The presence of this gap reduces the SPR efficiency; therefore, it is useful for very specific applications, for example, the studies of surface quality.

Resonance condition is achieved by exact matching of the components of the wave vectors of the light photons and plasmons, given by

$$\frac{\omega}{c}\sqrt{\varepsilon_p}\sin\theta_1 = \frac{\omega}{c}\sqrt{\frac{\varepsilon_1\varepsilon_2}{\varepsilon_1+\varepsilon_2}} \tag{3.28}$$

The right-hand side of this equation is a complex quantity. The real part gives the wavelength of surface plasma waves (plasmons), while the imaginary part gives the propagation length of surface plasmon waves along the interface, which is also responsible for the evanescent field. This field decays exponentially into both the media, but mainly the field resides in the dielectric medium as there occurs more damping in the metal.

3.6.3 Kretschmann Configuration

In this configuration, the metal layer (ε_2) is directly deposited over the prism or surface of glass block (ε_p). Photons are made to fall on the surface of the prism at an angle greater than the critical angle, so that they can undergo total internal reflection and tunnel through the metal layer to excite surface plasmon waves. This enables more efficient plasmon generation. This technique is comparatively easy to handle. Therefore, this is the most frequently

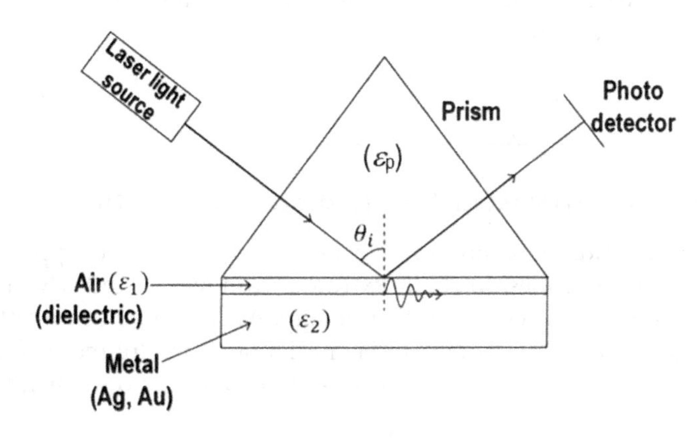

FIGURE 3.4
Schematic diagram of Otto configuration.

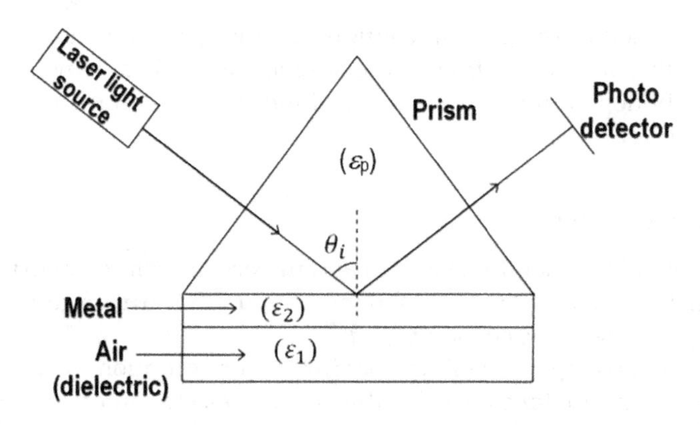

FIGURE 3.5
Schematic diagram of the Kretschmann configuration.

employed configuration (Figure 3.5). In this chapter, our discussion will be confined to the Kretschmann configuration only.

In this configuration, the thickness of the film is crucial. It should be optimum enough so that the evanescent wave is produced and coupled to the surface plasmon on the opposite surface of the metal. The necessary condition for the excitement of SPR is filled by varying the incident angle so that the totally reflected wave reaches an appropriate value inside the glass block or the prism.

3.7 Light Wave Coupling

There is a coupling of the light beam with the surface plasmon present in the sites at the interface of the metal and dielectric medium if the component of its wave vector parallel to the interface matches with the wave vector of the surface plasmon. Since $k_{sp} \gg k_{\text{incident}}$, surface plasmon cannot be excited directly by the incident light. Therefore, the wave number of the incident wave has to be increased by attenuated total reflection or diffraction using different coupling devices. The most frequently used coupler is the prism coupler along with the grating and waveguide couplers.

3.8 Surface Plasmon-Assisted THz Radiation Generation

Terahertz (THz) radiation generation has been the most interesting part of research for the last more than two decades due to its extremely innovative and advantageous applications in different fields of science and technology. Although a number of techniques are being used for the efficient generation of THz radiation, thirst for more efficient and handy techniques never ends. Surface plasmon-assisted THz radiation generation has its own advantages.

Groups all over the world have worked for the emission of coherent THz pulses by illuminating plain and nanostructured gold surfaces by ultrafast lasers (Kadlec et al. 2004,

2005; Gladun et al. 2007; Welsh et al. 2007; Gao et al. 2009; Welsh and Wynne 2009; Garwe et al. 2011; Polyushkin et al. 2011; Ramakrishnan and Planken 2011). Kadlec et al. (2004) considered a few hundred nanometer thick plain gold films and illuminated them with an amplified Ti:Sapphire laser. They established a relation between fluence of the emitted THz radiation and the fluence of the pump laser such that the THz fluence scales as the square of the incident laser fluence. In contrast, in the case of nanostructured Au surfaces illuminated by amplified lasers, Polyushkin et al. (2011) and Welsh et al. (2007) observed a third-, fourth-, or higher-order dependence of the emitted THz fluence because of the multiphoton ionization and ponderomotive acceleration of electrons in the evanescent field of surface plasmons (Zawadzka et al. 2001). There has been no established defined power rule of the dependence of the THz fluence on the incident laser fluence. For example, when the nanostructured gold surface with a lower thickness of 8 nm is illuminated with laser pulses from the Ti:Sapphire laser oscillator, a second-order dependence of the emitted THz amplitude on the incident laser power is noted (Ramakrishnan and Planken 2011). Still, plain flat gold surfaces could not prove to be useful for THz emission due to zero output.

We focus on the work of Ramakrishnan et al. (2012) who have investigated the THz emission from a hemicyanine-terminated alkanethiol self-assembled monolayer (SAM) in the Kretschmann geometry. They observed the enhanced radiation due to the generation of surface plasmons on a thin film of gold (Figure 3.6). A Ti:Sapphire oscillator was

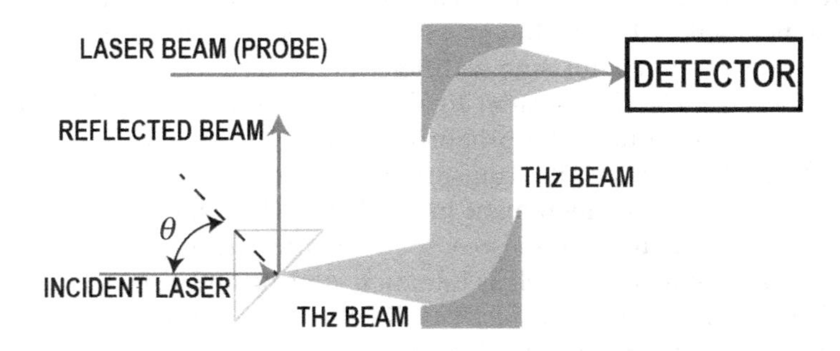

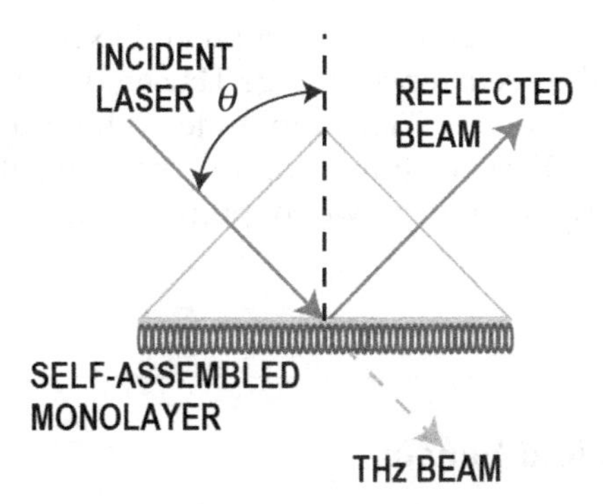

FIGURE 3.6
Schematic diagram of the experimental setup (top) and Kretschmann geometry for attenuated total reflection (bottom).

taken as the laser source, which could generate plane polarized 50 fs laser pulses with an 800 nm wavelength (at the centre) and 800 mW average power. Using vacuum evaporation, 44 nm-thick Au films were prepared on the hypotenuse side of the right-angled glass prisms. Laser pulses were incident on the other side of the prism. THz radiation produced on the other side was collected with the help of parabolic mirrors and fed to the electro-optic (EO) detection crystal, which is generally thick ZnTe.

In this experiment, surface plasmon–mediated THz radiation was obtained when poorly focussed laser pulses were allowed to incident on the Au film. However, no noticeable quantity of THz emission was detected when the side of the film, which is exposed to air, was directly excited. This strongly suggests the enhancement in THz output by surface plasmon excitation. They observed almost three factors of enhancement in the amplitude of the emitted THz electric field when the Au surface was shielded with a SAM of hemicyanine-terminated alkanethiol molecules. In this mechanism, the excitation of surface plasmons plays a key role in the enhancement of THz radiation and even single-layer hemicyanine-terminated alkanethiol molecules contributed significantly to enhance the THz emission compared with a bare Au surface (Naraoka et al. 2002).

Using this technique, the pump light is eliminated by reflection at 90° and only the THz beam so produced get transmitted through the prism. Hence, it is an efficacious technique to keep the pump beam from going along the direction of the detection crystal, which eases the whole process further.

Another sampling pulse is also synchronized to fall on the detection crystal. Elliptically polarization of the probe beam by THz electric field is proportional to its own instantaneous electric field value. Hence, the probe beam, step by step, is passed through a differential detection setup, comprising a differential detector such as a Wollaston prism and a quarter-wave plate. The system finally evaluates the ellipticity produced in the probe beam. This was found to be proportional to the instantaneous magnitude of the THz electric field. At the SAM, the transmission of localized high intensity is achieved. Consequently, a considerable increment in the THz polarization, which is grown in the SAM, is attained. Such increment is also observed in the surface plasmon, which mediated another nonlinear phenomenon such as Raman scattering and a second-harmonic generation (SHG) detected from the metal surfaces (Raether 1988; Naraoka et al. 2005).

To make a correlation between the THz field amplitude, the power of the pump laser and the angle of incidence θ, they had studied the variation of power of the reflected light and the emitted THz amplitude as a function of θ. For the taken parameters, it was noted that the point where the dip in reflected pump power is obtained (at $\theta \approx 42°$) also correlates to the angle that fulfills the resonance condition and excites the surface plasmons on the gold surface, in view of the occurrence of large THz emission at that angle. Another investigation related to the variation of THz amplitude with pump power established a linear relationship between the THz field amplitude and the incident pump power, since the THz amplitudes varied linearly with the power of pump laser.

3.9 THz SPR Near-Field Sensor

The focusing of an optical beam onto the near-field region and its use is another direct procedure to avail the benefits of (1) the relatively smaller optical beam spot size compared with THz wavelength and (2) a broad wavelength difference among the THz and optical

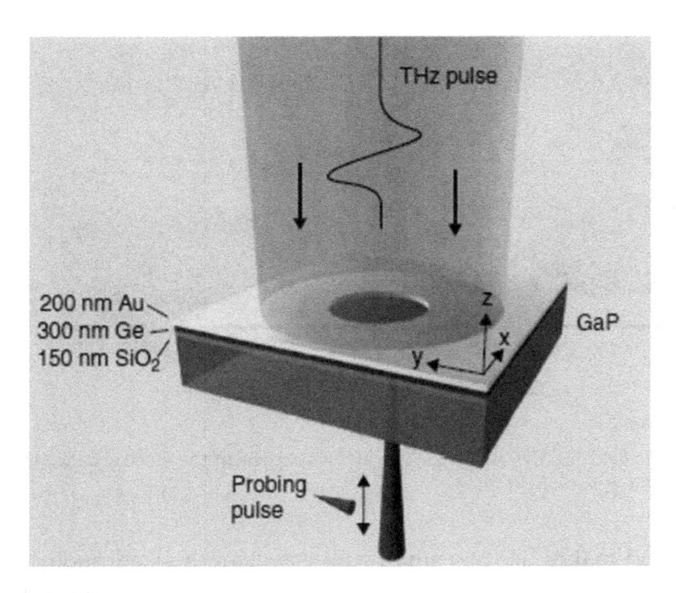

FIGURE 3.7
THz near-field microscopy setup.

regime, which controls the artifacts and interferences. In this investigation, the samples are developed on EO crystals like ZnTe and GaP for an optical beam to watch THz field.

The schematic setup of EO near-field imaging for the development of a sample on the GaP crystal is depicted in Figure 3.7. When mechanical contact between the EO crystal and the sample is made by pressing the sample on the crystal, the quasi near-field imaging is accomplished. A significant advantage of this method is an enhanced ability to probe the evolution up to the far field. The EO measurements depend on the second-order nonlinear result of polarization in EO material to an applied electric field (Planken et al. 2001; van der Valk et al. 2004). The THz electric field considerably modifies the refractive index of the EO materials. When the optical probe pulse travels through an EO detection crystal, a change in the polarization is experienced because of birefringence. After the detection crystal, a quarter-wave plate is oriented so that the probe beam passed through it will convert into circular polarization from the initial linear polarization. The signal corresponding to the electric field of THz radiation is obtained by determining the energy difference of the two orthogonal polarization direction beams with the help of two photodiodes and a Wollaston prism. As mentioned earlier, zinc blende crystals like GaP and ZnTe are preferable for EO detection. The sensitiveness of the EO detection setup on a particular component of the THz electric field vector can be chosen by modifying the orientation of the detection crystal. The crystal orientation (110) or (111) is used to measure the x- and y- components of the THz electric field, whereas crystal orientation (001) is used to measure E_z, the vertical component. These three THz field components and Maxwell's equations enable one to calculate the magnetic near field. These near-field electric and magnetic vectors (Figure 3.8) are observed behind the metallic slit with a 500 μm period and 100 μm width, when the incident ray arrives from below (Seo et al. 2007). The y-component of the magnetic field, which is determined from the calculated electric field vectors, is shown by the background grey colour scale. The positive and negative magnetic field amplitudes, respectively, are indicated by white and black. Such an operational method has been accepted in both infrared and THz regimes (Bitzer et al. 2010, 2011).

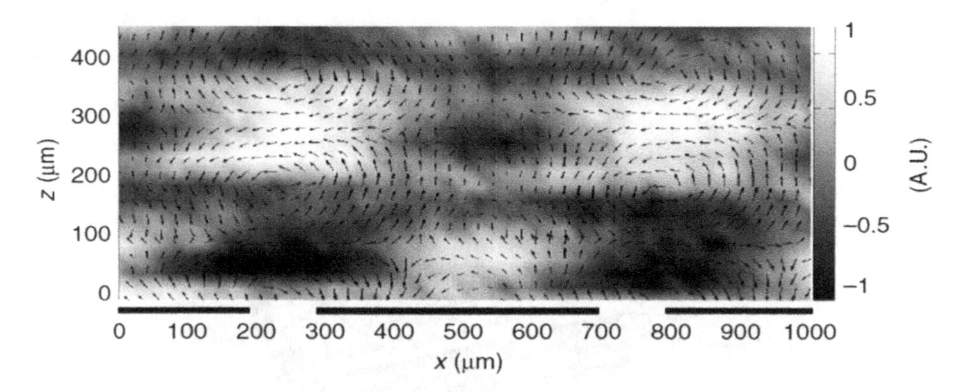

FIGURE 3.8
Electric and magnetic field profiles at 1 THz. The arrows represent the electric field, whereas the background colour represents the magnetic field.

Another method makes use of a small sub–wavelength-sized photoconductive antenna as a near-field probe, which is gated by an optical pulse of femtosecond duration (Bitzer and Walther 2008; Bitzer et al. 2010, 2011). As shown in Figure 3.9 (left side), the detector antenna is composed of an electrode structure (H-shaped) deposited on an ion-implanted silicon-on-sapphire substrate with a thickness of 500 nm. This system enables us to calculate the out-of-plane magnetic field and in-plane electric field (Figure 3.9, middle), which are near the sample with micrometer scale spatial resolution (~λ/20) at 0.5 THz. With the help of the detection method, which is based on EO sampling in a nonlinear crystal, one can determine the out-of-plane component of the electric field (Figure 3.9, right).

To investigate the quality of material and related concerns in organic light-emitting diode (OLED) productions, THz surface plasmon near-field sensors are widely adopted to achieve the demand of material characterization and sensing in the THz range. THz SPR sensors that are relevant for material sensing are being made to operate in the THz region. This is because of the benefit of the high near-field enhancement and confinement at the boundary between the dielectric layer and the SPR structure (Hailu et al. 2013).

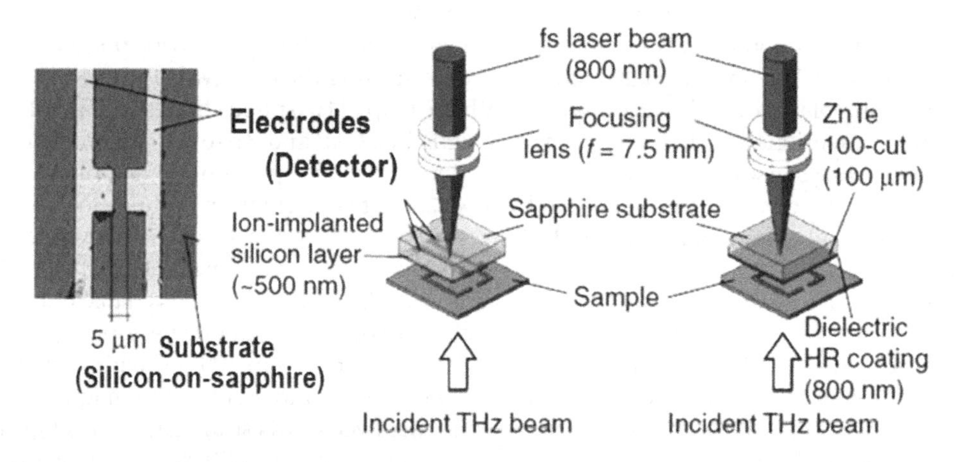

FIGURE 3.9
Photoconductive antenna of the near-field probe (left) and EO sampling in nonlinear crystal-based detection method for electric and magnetic fields.

Thick analytes in label-free sensing based on SPR using sub-wavelength sources have gained huge popularity in sensing applications and near-field microscopy (Homola et al. 1999; Swiontek et al. 2013). The sensitivity and functionality of SPR sensing could be controlled and engineered in an operating integrated waveguide by the optimization of SPR generation and by ensuring the adequate interaction between analytes and electromagnetic fields. Because the THz frequencies are quite smaller than the intrinsic plasmon frequency of metals, they cannot be enclosed by a conductor, i.e. metal surface. Due to this frequency difference, the field cannot penetrate inside the metals and surface plasmon polaritons (SPPs) are not excited.

For the generation of THz-spoof SPPs (THz-SSPPs) and consequently the enhancement of lateral confinement, metal surfaces with different structures and patterns have been explored. These include metamaterials' 2D hole arrays and periodic slits (Williams et al. 2008, 2010; Fernández-Domínguez et al. 2009; Zhu et al. 2014). The molecular sensing in the metal structured waveguide configuration can be achieved from the combination of refractive index dependence of THz-SSPPs and enhanced lateral confinement (Nylander et al. 1982; Liedberg et al. 1983).

Developments in sensor technology advance day by day so one needs to have a depth of understanding of light-matter interaction at quantum or nanosize scale, which forms the basis of nanophotonics and optoelectronics. However, these topics do not fall within the scope of this book.

3.10 Conclusions

An interaction of polarized light with the free electrons of metal film, which is in contact with the media of different refractive indices, describes the SPR phenomenon. The free electrons are excited by an evanescent field and their collective oscillations are known as surface plasmons. If one needs high-quality kinetics, real-time data acquisition, and label-free analysis, SPR is the technology to rely on. The necessary condition for the excitation of surface plasmons is that the refractive index of two media must be of opposite signs. Hence, it is essential that we choose a metal as the second medium when the first medium is a dielectric medium. The dispersion relation for surface plasmons relates the frequency of the plasmon wave and its wave number with the refractive indices of the two media. To measure the dip and the shift in the dip of reflected light, SPR instruments generally can be configured in three different optical configurations, namely Grating, Otto, and Kretschmann, by which excitation of the plasmon is achieved. There is a coupling of light beam with the surface plasmon present in the sites at the interface of metal and the dielectric medium if the component of its wave vector parallel to the interface matches with the wave vector of the surface plasmon. The phenomenon of SPR has a considerable dependency on the metal film's characteristics, wavelength and polarization state of the incident light, thickness of the metal layer, and refractive index of the prism. Towards the detection, SPR spectroscopy can detect changes of the order of $\sim 10^{-5}$ in the refractive index of the surface within a few second timescales. However, the analyte should be placed close to the surface of the SPR sensor as it responds efficiently only up to a depth of approximately 200 nm from the surface of the sensor. The phenomenon of SPR is being successfully explored for the enhancement in THz output by illuminating plain and nanostructured gold surfaces by ultrafast lasers.

3.11 Selected Problems and Solutions

PROBLEM 3.1

A metal-dielectric system is irradiated by a He-Ne laser, where the dielectric constant of the material of the coupling device (dielectric) and gold metallic layer are 2.34 and –16, respectively. Calculate the angle at which the resonance will occur. Dielectric constant of SF1 flint prism, which is used in such systems, is $\varepsilon_p = 2.89$.

SOLUTION

SPR angle (θ_{res}) can be calculated from resonance condition for SPR, given by

$$\frac{\omega}{c}\sqrt{\varepsilon_p}\sin\theta_{res} = \frac{\omega}{c}\left(\frac{\varepsilon_1\varepsilon_2}{\varepsilon_1+\varepsilon_2}\right)^{1/2}$$

$$\sin\theta_{res} = \left(\frac{\varepsilon_1\varepsilon_2}{\varepsilon_1+\varepsilon_2}\right)^{1/2} \times \frac{1}{\sqrt{\varepsilon_p}}$$

Given that $\varepsilon_1 = -16$, $\varepsilon_2 = 2.34$, $\varepsilon_p = 2.89$. Hence

$$\sin\theta_{res} = \left(\frac{-16\times 2.34}{-16+2.34}\right)^{1/2} \times \frac{1}{\sqrt{2.89}} = 0.97385$$

Or

$$\theta_{res} = 76.52°$$

PROBLEM 3.2

What happens to the angle at which surface plasmons are excited in the problem 1, if the dielectric constant of coupling device is reduced to 1.78?

SOLUTION

From SPR condition,

$$\sin\theta_{res} = \left(\frac{\varepsilon_1\varepsilon_2}{\varepsilon_1+\varepsilon_2}\right)^{1/2} \times \frac{1}{\sqrt{\varepsilon_p}}$$

and given values as $\varepsilon_1 = -16$, $\varepsilon_2 = 2.34$ but $\varepsilon_p = 1.78$, we get

$$\sin\theta_{res} = \left(\frac{-16\times 2.34}{-16+2.34}\right)^{1/2} \times \frac{1}{\sqrt{1.78}} = 1.24088$$

Or

$$\sin(\theta_{res}) > 1$$

This shows that the surface plasmons cannot be resonantly excited.

PROBLEM 3.3

A silver metallic layer is irradiated by frequency doubled Nd:YAG laser, and SPR is observed to occur at an angle of 55.76°. Calculate the dielectric constant of the coupling dielectric layer if the dielectric constant of silver is –18 and of prism is 2.89.

SOLUTION

SPR condition reads

$$\sin\theta_{res} = \left(\frac{\varepsilon_1\varepsilon_2}{\varepsilon_1 + \varepsilon_2}\right)^{1/2} \times \frac{1}{\sqrt{\varepsilon_P}}$$

Given that $\varepsilon_1 = -18$, $\varepsilon_P = 2.89$ and $\theta_{res} = 55.76°$. It means

$$\sin(55.76°) = \left(\frac{-18 \times \varepsilon_2}{-18 + \varepsilon_2}\right)^{1/2} \times \frac{1}{\sqrt{2.89}}$$

On solving we get

$$\varepsilon_2 = 1.78$$

PROBLEM 3.4

Calculate the percentage change in the angle of resonance if the silver metallic layer (dielectric constant: –18) is replaced with a gold metallic layer (dielectric constant: –16) in the arrangement where dielectric-material layers are irradiated by the He-Ne laser for surface plasmon excitation using the dielectric material of dielectric constant 2.34 and the prism of dielectric constant 2.89.

SOLUTION

For the case of gold, $\varepsilon_1 = -16$, $\varepsilon_2 = 2.34$ and $\varepsilon_P = 2.89$. Hence

$$\theta_{res} = 76.869°$$

For the case of silver $\varepsilon_1 = -18$, $\varepsilon_2 = 2.34$ and $\varepsilon_P = 2.89$. Hence

$$\theta_{res} = 74.73°$$

$$\text{Percentage change} = \frac{76.869 - 74.73}{74.73} \times 100 = 2.86\%$$

PROBLEM 3.5

Calculate the magnitude of parallel component of the wave vector of the evanescent wave when an electromagnetic wave of $\lambda = 632.8$ nm is incident at an angle of 74.73° and the dielectric constant of the coupling device and metal layer used in the arrangement are 2.34 and 2.04, respectively.

SOLUTION

Magnitude of parallel component of wave vector of evanescent wave is given by

$$k_{\text{evan,ll}} = \frac{2\pi n_p \sin(\theta)}{\lambda}$$

$$n_p = \sqrt{\varepsilon_p} = \sqrt{2.34} = 1.529$$

$$k_{\text{evan,ll}} = \frac{2\times 3.14\times 1.529\times \sin(74.73°)}{632.8\times 10^{-9}} = 14.65\ \mu\text{m}^{-1}$$

PROBLEM 3.6

An electromagnetic wave of $\lambda = 532$ nm is incident at an angle of 53.46° on a metal (silver)-dielectric interface, of refractive indices 0.155 and 1.334, respectively. Calculate the magnitude of the parallel component of the vector of the evanescent wave.

SOLUTION

Magnitude of parallel component of the wave vector of the evanescent wave is given by

$$k_{\text{evan,ll}} = \frac{2\pi n_p \sin(\theta)}{\lambda}$$

$$k_{\text{evan,ll}} = \frac{2\times 3.14\times 1.334\times \sin(74.73°)}{532\times 10^{-9}} = 12.65\ \mu\text{m}^{-1}$$

PROBLEM 3.7

Consider a single metal-dielectric interface with the relative permittivity of silver as $\varepsilon_f = -16 - 0.5j$ at $\lambda_o = 0.632\,\mu m$, and air $\varepsilon_c = 1$. Find the SPR frequency.

SOLUTION

Surface plasmon frequency reads $\omega_p = k_o\sqrt{\dfrac{\varepsilon_f\varepsilon_c}{\varepsilon_f+\varepsilon_c}}$.

Given that $\lambda_o = 0.632\,\mu m \rightarrow k_o = \dfrac{2\pi}{\lambda_o} = 9.94$ rad/μm, $\varepsilon_f = -16 - 0.5j$, and $\varepsilon_c = 1$. Hence

$$\omega_p = k_o\sqrt{\frac{\varepsilon_f\varepsilon_c}{\varepsilon_f+\varepsilon_c}} = 10.165 - 0.01075\text{i rad}/\mu\text{m}$$

PROBLEM 3.8

Calculate the propagation distance for the given surface plasmon having frequency $\omega_{\text{plasmon}} = 10.2674 - 0.0107\text{i rad}/\mu\text{m}$.

SOLUTION

The propagation distance is given by $L_z = -\dfrac{1}{\text{Im}\,(\omega_p)}$.

Putting the value $\omega_{\text{plasmon}} = 10.2674 - 0.0107\text{i rad}/\mu\text{m}$ in the above expression, we get

$$L_z = 93.5969\ \mu\text{m}$$

PROBLEM 3.9

For a gold/water interface at $\lambda = 700$ nm ($\varepsilon_{gold} = -16$; $\varepsilon_{water} = 1.770$), calculate the angular shift of the SPR dip when ε_{water} increases to 1.775. For the light-in coupling we use a semicircular glass piece with refractive index $n_{glass} = 1.5$.

SOLUTION

For the wave vector $\mathbf{k} = (k_x, k_y)$ having x- and y-components, we can write

$$k_x^2 = k_1^2 - k_{y1}^2 = k_1^2 - k_x^2 \frac{\varepsilon_1}{\varepsilon_2}$$

$$k_x = \frac{\omega}{c}\sqrt{\frac{\varepsilon_1 \varepsilon_2}{\varepsilon_1 + \varepsilon_2}} \text{ and } k_{y1} = \frac{\omega}{c}\sqrt{\frac{\varepsilon_1^2}{\varepsilon_1 + \varepsilon_2}}$$

For $\lambda = 700$ nm, $\omega = 2.69 \times 10^{15}$ rad s^{-1}, $\varepsilon_{gold} \approx -16$, $\varepsilon_{water} \approx 1.77$ and gold/water interface.

$$\frac{1}{k_{y,\text{water}}} = 238 \text{ nm and } \frac{1}{k_{y,\text{gold}}} = 26 \text{ nm}$$

Suggested Reading Material

Bitzer, A., Ortner, A., Merbold, H., Feurer, T., & Walther, M. (2011). Terahertz near-field microscopy of complementary planar metamaterials: Babinet's principle. *Optics Express, 19*(3), 2537–2545.

Bitzer, A., Ortner, A., & Walther, M. (2010). Terahertz near-field microscopy with subwavelength spatial resolution based on photoconductive antennas. *Applied Optics, 49*(19), E1–E6.

Bitzer, A., & Walther, M. (2008). Terahertz near-field imaging of metallic subwavelength holes and hole arrays. *Applied Physics Letters, 92*(23), 231101.

Fano, U. (1941). The theory of anomalous diffraction gratings and of quasi-stationary waves on metallic surfaces (Sommerfeld's waves). *JOSA, 31*(3), 213–222.

Fernández-Domínguez, A. I., Moreno, E., Martín-Moreno, L., & García-Vidal, F. J. (2009). Terahertz wedge plasmon polaritons. *Optics Letters, 34*(13), 2063–2065.

Gao, Y., Chen, M. K., Yang, C. E., Chang, Y. C., Yin, S., Hui, R., ... & Luo, C. (2009). Analysis of terahertz generation via nanostructure enhanced plasmonic excitations. *Journal of Applied Physics, 106*(7), 074302.

Garwe, F., Schmidt, A., Zieger, G., May, T., Wynne, K., Hübner, U., ... & Meyer, H. G. (2011). Bi-directional terahertz emission from gold-coated nanogratings by excitation via femtosecond laser pulses. *Applied Physics B, 102*(3), 551–554.

Gladun, A. D., Leiman, V. G., & Arsenin, A. V. (2007). On the mechanism of generation of terahertz electromagnetic radiation upon irradiation of a nanostructured metal surface by femtosecond laser pulses. *Quantum Electronics, 37*(12), 1166.

Hailu, D. M., Alqarni, S., Cui, B., & Saeedkia, D. (2013). Terahertz surface plasmon resonance sensor for material sensing. In photonics North 2013. *International Society for Optics and Photonics, 8915*, 89151G.

Homola, J., Yee, S. S., & Gauglitz, G. (1999). Surface plasmon resonance sensors. *Sensors and Actuators B: Chemical, 54*(1–2), 3–15.

Kadlec, F., Kužel, P., & Coutaz, J. L. (2004). Optical rectification at metal surfaces. *Optics Letters, 29*(22), 2674–2676.

Kadlec, F., Kužel, P., & Coutaz, J. L. (2005). Study of terahertz radiation generated by optical rectification on thin gold films. *Optics Letters, 30*(11), 1402–1404.

Kretschmann, E., & Raether, H. (1968). Radiative decay of non radiative surface plasmons excited by light. *Zeitschrift Für Naturforschung A, 23*(12), 2135–2136.

Liedberg, B., Nylander, C., & Lunström, I. (1983). Surface plasmon resonance for gas detection and biosensing. *Sensors and Actuators, 4*, 299–304.

Naraoka, R., Kaise, G., Kajikawa, K., Okawa, H., Ikezawa, H., & Hashimoto, K. (2002). Nonlinear optical property of hemicyanine self-assembled monolayers on gold and its adsorption kinetics probed by optical second-harmonic generation and surface plasmon resonance spectroscopy. *Chemical Physics Letters, 362*(1–2), 26–30.

Naraoka, R., Okawa, H., Hashimoto, K., & Kajikawa, K. (2005). Surface plasmon resonance enhanced second-harmonic generation in Kretschmann configuration. *Optics Communications, 248*(1–3), 249–256.

Nylander, C., Liedberg, B., & Lind, T. (1982). Gas detection by means of surface plasmon resonance. *Sensors and Actuators, 3*, 79–88.

Otto, A. (1968). Excitation of nonradiative surface plasma waves in silver by the method of frustrated total reflection. *Zeitschrift für Physik A Hadrons and Nuclei, 216*(4), 398–410.

Planken, P. C. M., Nienhuys, H.-K., Bakker, H. J., & Wenckebach, T. (2001). Measurement and calculation of the orientation dependence of terahertz pulse detection in ZnTe. *JOSA B, 18*(3), 313–317.

Polyushkin, D. K., Hendry, E., Stone, E. K., & Barnes, W. L. (2011). THz generation from plasmonic nanoparticle arrays. *Nano Letters, 11*(11), 4718–4724.

Raether, H. (1977). Hass G., Francombe M., and Hoffman R, eds. *Physics of thin films*. Academic Press: New York. (Raether, H. (1977). Physics of Thin Films, Ed. Hass G., Francombe M. and Hoffman R. Academic Press: New York)

Raether, H. (1988). Surface plasmons on smooth surfaces. *Surface plasmons on smooth and rough surfaces and on gratings*. Cham, Switzerland: Springer, pp. 4–39.

Ramakrishnan, G., Kumar, N., Planken, P. C., Tanaka, D., & Kajikawa, K. (2012). Surface plasmon-enhanced terahertz emission from a hemicyanine self-assembled monolayer. *Optics Express, 20*(4), 4067–4073.

Ramakrishnan, G., & Planken, P. C. (2011). Percolation-enhanced generation of terahertz pulses by optical rectification on ultrathin gold films. *Optics Letters, 36*(13), 2572–2574.

Rayleigh, L. (1907). On the dynamical theory of gratings. *Proceedings of the Royal Society of London. Series A, Containing Papers of a Mathematical and Physical Character, 79*(532), 399–416.

Seo, M. A., Adam, A. J. L., Kang, J. H., … Kim, D. S. (2007). Fourier-transform terahertz near-field imaging of one-dimensional slit arrays: Mapping of electric-field-, magnetic-field-, and Poynting vectors. *Optics Express, 15*(19), 11781–11789.

Swiontek, S. E., Pulsifer, D. P., & Lakhtakia, A. (2013). Optical sensing of analytes in aqueous solutions with a multiple surface-plasmon-polariton-wave platform. *Scientific Reports, 3*, 1409.

van der Valk, N. C. J., Wenckebach, T., & Planken, P. C. M. (2004). Full mathematical description of electro-optic detection in optically isotropic crystals. *JOSA B, 21*(3), 622–631.

Welsh, G. H., Hunt, N. T., & Wynne, K. (2007). Terahertz-pulse emission through laser excitation of surface plasmons in a metal grating. *Physical Review Letters, 98*(2), 026803.

Welsh, G. H., & Wynne, K. (2009). Generation of ultrafast terahertz radiation pulses on metallic nanostructured surfaces. *Optics Express, 17*(4), 2470–2480.

Williams, C. R., Andrews, S. R., Maier, S. A., Fernández-Domínguez, A. I., Martín-Moreno, L., & García-Vidal, F. J. (2008). Highly confined guiding of terahertz surface plasmon polaritons on structured metal surfaces. *Nature Photonics, 2*(3), 175–179.

Williams, C. R., Misra, M., Andrews, S. R., … Martin-Moreno, L. (2010). Dual band terahertz waveguiding on a planar metal surface patterned with annular holes. *Applied Physics Letters, 96*(1), 11101.

Wood, R. W. (1902). On a remarkable case of uneven distribution of light in a diffraction grating spectrum. *Proceedings of the Physical Society of London, 4*(21), 396–402.

Wood, R. W. (1912). XXVII. Diffraction gratings with controlled groove form and abnormal distribution of intensity. *The London, Edinburgh, and Dublin Philosophical Magazine and Journal of Science, 23*(134), 310–317.

Zawadzka, J., Jaroszynski, D. A., Carey, J. J., & Wynne, K. (2001). Evanescent-wave acceleration of ultrashort electron pulses. *Applied Physics Letters, 79*(14), 2130–2132.

Zhu, J., Ma, Z., Sun, W., Ding, F., He, Q., Zhou, L., & Ma, Y. (2014). Ultra-broadband terahertz metamaterial absorber. *Applied Physics Letters, 105*(2), 021102.

4

THz Radiation Using Gases/Plasmas

4.1 Introduction

The neighboring terahertz (THz) frequency regions, i.e. infrared (IR) and microwave regions, have been extensively investigated and researched for possible applications in science and technology. THz technology has aroused immense interest among the concerned fraternity because of its exceptional properties such as (1) the penetration of THz radiation through numerous non-polar and non-metallic materials, e.g. woods, paper, textiles and plastics; (2) absorption lines for different chemical substances and molecules, such as oxygen and carbon monoxide, corresponding to their rotational and vibrational levels also fall in the THz range; (3) absorption of THz radiation by water molecules ; (4) reflection by metals; and (5) primarily, its non-ionizing behavior quenches the harmfulness of living cells. These distinctive traits of THz radiation can be tailored according to the need of technology for their applications in numerous diversified fields.

THz radiation can be generated by a number of methods and techniques such as optical THz generation, quantum cascade laser, optical rectification, solid-state electronic devices, accelerator-based sources, and sources based on laser plasma interaction. Some of these methods have already been discussed in the previous chapters. Most of these methods have certain drawbacks, viz. lower damage limit of certain media restricts the efficiency of emitted THz energetic pulses to be not very high. To overcome these drawbacks, a non-linear medium, i.e. plasma, is utilized in several techniques involving strong laser-plasma interaction. This occurs because plasma can withstand very high-power lasers and has no damage limit.

4.2 Tunnel Ionization

Ionization involves an electron escape from the atom (or molecule) via crossing the potential barrier. Classically, an electron must have a higher amount of energy compared with that of the potential barrier and only then it is allowed to leave the atom (or molecule). But with the application of an external field, the height of the potential barrier can be suppressed, and the electron can easily tunnel through the potential instead of going all over the way due to its wave nature. This distortion of potential barrier, due to intense electric field, permits the electron to escape from the atom. This quantum mechanical phenomenon where ionization happens due to quantum tunneling, which is forbidden by the classical laws, is known as tunnel ionization. Tunnel ionization allows the particle to escape from the distorted Coulomb potential barrier with non-zero probability. The probability of an electron's tunneling through the barrier with the width of the potential barrier drops off exponentially. Therefore, an electron with higher energy encounters a thinner potential barrier, which further increases the tunneling probability.

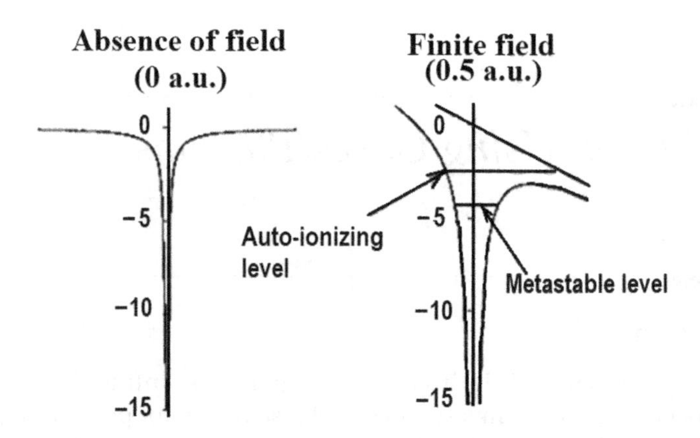

FIGURE 4.1
Schematic of tunnel ionization of atoms in external electric fields.

Let us consider that the direct current (DC) electric field interacts with an atom. As shown in Figure 4.1, the external field distorts the Coulomb potential barrier drastically and the bound electrons can tunnel through the barrier to ionize the atom. If the field is strong enough, the zero-field stable energy levels can become lower in energy than the asymptotic value of the potential. On the other hand, for an alternating current (AC) field, the tunneled electron may come back to its parent ion after the half-period of the field as the direction of the electric field reverses. This may let the electron recombine with the nucleus (nuclei) and the resultant kinetic energy is released as light (high-harmonic generation).

In the weak field regime, only one photon is absorbed by the electron. Under intense field conditions, the absorption of multiple numbers of photons from the laser field ionizes the electrons bound to an atom; this is termed multiphoton ionization (MPI). In the multiphoton regime, the probability of n-photon absorption is proportional to I^n, where I is the intensity of the incident laser field. The lifetime of the system is governed by the Heisenberg uncertainty principle. Keldysh (1965) did not take into account the details of atomic structure to determine the ionization probability and perturbation of the ground state while modeling the MPI process. He modeled only the excitation of the ground state electrons to the Volkov states in an atom; hence, the details related to the Coulomb interaction for the excited state of the electron have not been taken into account. However, Perelomov (2012) has derived a short-range potential Perelomov-Popov-Terent'ev (PPT) model with the inclusion of Coulomb interaction at larger internuclear distance in quasi-classical action. Larochelle et al. (1998) have set up an experiment by interacting a Ti:sapphire laser with rare gas atoms and compared theoretically ion-intensity curves. They successfully verified the total ionization rate for all the rare gases as predicted by the PPT model in the intermediate regime of Keldysh parameter. Malik et al. (2010) have anticipated the generation of the tunable THz radiation where high plasma density and larger residual current were achieved due to the tunnel ionization of nitrogen and hydrogen (Figure 4.2). They proposed two lasers for achieving the required value of the electric field for the quick tunnel ionization. It is clear from Figure 4.2 that there is a specific value of the field for the maximum rate of tunnel ionization for different gases. For example, the optimum field is 4.16×10^{11} V/m for the nitrogen gas, whereas it is 3.12×10^{11} V/m for the hydrogen gas. In their proposal, they could reduce the phase difference between launching of the lasers and the creation of plasma by tunnel ionization.

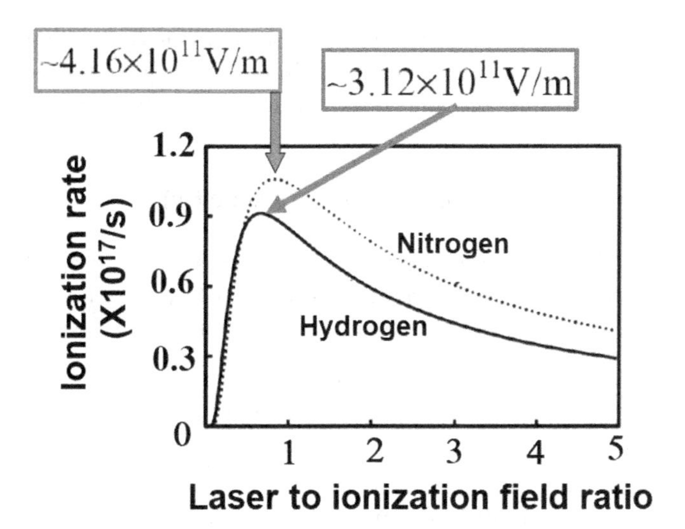

FIGURE 4.2
Dependence of ionization rate of nitrogen and hydrogen, showing the highest rate at particular values of the electric field.

4.2.1 Keldysh Parameter

Processes like MPI and tunnel ionization can occur under a high-frequency electromagnetic field, i.e. an alternating sinusoidal field. Hence, the barrier against ionization in any given direction is lowered only for the duration of a single cycle, which implies that only levels that can tunnel ionize on a timescale shorter than the inverse frequency of the radiation can undergo tunnel ionization. Thus, low-frequency radiation at high intensities is suited for such processes. MPI, on the other hand, requires that individual photons have a large energy, so that a given process can take place with a fewer number of photons. The larger the energy is, the higher the frequency of the incident laser radiation. The two processes can take place in different regimes of the frequency spectrum. A quantitative measure of this division had been proposed by Keldysh, who suggested that the adiabaticity parameter 'G' (now known as the Keldysh parameter), a quantitative indicator, compares the ionization potential of the system in free-field (E_i) with ponderomotive energy (U_P), which evolves due to the laser field. That is

$$G = \sqrt{\frac{E_i}{2U_P}}$$

The Keldysh parameter tells about the regime in which the processes might take place. MPI mechanism, where absorption of more than one photon is responsible for ionization, occurs when G > 1, i.e. IP > $E_i \gg h\nu_0$, where IP stands for the ionization potential. When the fields are very intense and involve higher ponderomotive energy such that $E_i < U_p$, then G < 1 is the result. This is tunnel ionization arising due to optical field ionization, which is particularly preferred for near-IR strong laser pulses. Recent studies have revealed that for tunnel ionization one requires G < 0.5. The Keldysh parameter for H-like atoms is calculated as

$$G^2 = (0.73Z^2)/[(1+a^2)\ I_{14}\ \lambda_{\mu m}^2]$$

Here the intensity I_{14} is in 10^{14} W/cm^2 and wavelength, λ, is in microns units.

Interaction of atoms or molecules with high laser fields is dependent on the ionization energy and the quiver velocity of electrons in the high-frequency field. We consider the electric field of an electromagnetic wave, which imparts an oscillatory energy to the electrons, as

$$E = E_0\ (\cos\ \omega t\ \hat{x} + a\ \sin\ \omega t\ \hat{y})$$

The polarization of the laser light is determined by the constant a, which is zero ($a = 0$) for linearly polarized light and is 1 ($a = \pm 1$) for circularly polarized light. The oscillation energy of electrons, also referred to as ponderomotive energy, is given by

$$U_P = \frac{e^2 E_0^2}{4m\omega^2}(1 + a^2)$$

To estimate its significance more appropriately, we take different laser fields: one with micrometer wavelength and 10^{14} W/cm^2 intensity that results in $U_P = 9.33 I_{14}\ \lambda_{\mu m}^2$ and another field for an Nd:glass laser of 1-mm wavelength, which focussed to an intensity of $I \sim 10^{16}$ W/cm^2 and gives KeV order ponderomotive energy.

4.2.2 Frequency of Oscillations

When laser light is focussed along the optic axis, by using axicon lens, the plasma is generated in a cylindrical shape (Figure 4.3) in the region of another side of the lens and where gas is taken to ionize. Assuming the stationary background of ions due to their low mobility compared with the electrons, natural transverse oscillations of electrons against this state can be found that are due to residual momentum imparted by the laser light. Hence, the cylindrical geometry of the plasma channel reflects the involvement of oscillating dipoles at frequency ω. Considering the internal damping as γ_i and taking the dipole moment per unit length as Γ, the oscillations can be related as per the relation

$$\frac{d^2\vec{\Gamma}}{dt^2} + 2\gamma_i \frac{d\vec{\Gamma}}{dt} + \omega^2\vec{\Gamma} = 0$$

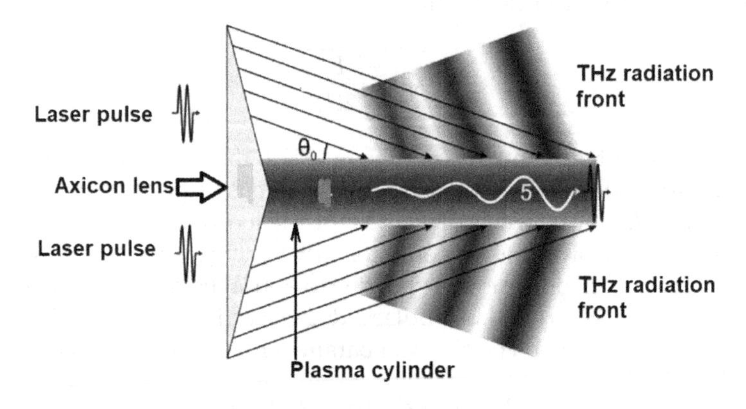

FIGURE 4.3
Creation of plasma cylinder whose electrons oscillate due to residual momentum imparted by lasers on the plasma electrons and emit THz radiation due to dipole oscillations.

In a realistic situation, the plasma created in the cylindrical form will not carry the uniform density, though the density will reduce in the radial direction and be maximum on the axis of the cylinder. To maintain the non-uniformity of plasma density in the cylinder, we divide the cylinder into two portions with an inner radius a_1 and outer radius a $(a_1 < a)$. These two concentric cylinders have uniform density individually, but the density varies relatively such that the inner cylinder is denser than the outer one. For the calculation of plasma dipole oscillation frequency, consider the electric field as $\vec{E} = -\vec{\nabla}\phi$, where potential $\phi = A(r)e^{-i(\omega t+\theta)}$ together with θ as the azimuthal coordinate. The density perturbations arise due to nonlinear forces originating in the plasma by the laser field, which can be evaluated by using equations of continuity and momentum as follows

$$n_1 = \frac{\vec{\nabla}\cdot\left(ne\vec{\nabla}\phi\right)}{m\omega^2}$$

With the help of Poisson's equation $\nabla^2\phi = 4\pi e n_1$ we get (CGS system)

$$\left(1-\frac{\omega_p^2}{\omega^2}\right)\nabla^2\phi - \vec{\nabla}\left(\frac{\omega_p^2}{\omega^2}\right)\cdot\vec{\nabla}\phi = 0$$

In cylindrical polar coordinate system, this can be written as

$$\frac{1}{r}\frac{\partial}{\partial r}\left\{\left(1-\frac{\omega_p^2}{\omega^2}\right)r\frac{\partial\phi}{\partial r}\right\}\left(1-\frac{\omega_p^2}{\omega^2}\right)\frac{\phi}{r^2} = 0$$

Plasma density is considered for three different regions, partitioned by inner and outer boundaries, with plasma frequency ω_{p_1}, ω_{p_2}, and ω_{p_3}, respectively.

$$\omega_{p_1} = \omega_{pm} \text{ when } 0 \le r \le a_1$$

$$0 < \omega_{p_2} < \omega_{pm} \text{ when } a_1 \le r \le a \text{ and}$$

$$\omega_{p_3} = 0 \text{ when } r \ge a \text{ (outside the cylinder)}$$

We integrate these equations from a_1^- (just below boundary) and from a_1^+ (just above boundary) for the first boundary, i.e.

$$\left(1-\frac{\omega_{p1}^2}{\omega^2}\right)r\frac{\partial\phi}{\partial r}\bigg|_{r=a_1^-} = \left(1-\frac{\omega_{p2}^2}{\omega^2}\right)r\frac{\partial\phi}{\partial r}\bigg|_{r=a_1^+}$$

Similarly, the second boundary condition yields

$$\left(1-\frac{\omega_{p2}^2}{\omega^2}\right)r\frac{\partial\phi}{\partial r}\bigg|_{r=a^-} = r\frac{\partial\phi}{\partial r}\bigg|_{r=a^+}$$

Since $\nabla^2\phi = 0$ for all three regions, we have

$$\omega_{p1} = \omega_{pm} = \text{constant} => \frac{\partial}{\partial r}\left(\frac{\omega_p^2}{\omega^2}\right) = 0 \text{ when } 0 \le r \le a_1$$

$$0 < \omega_{p_2} (= \text{constant}) < \omega_{pm} => \frac{\partial}{\partial r}\left(\frac{\omega_p^2}{\omega^2}\right) = 0 \text{ when } a_1 \le r \le a, \text{ and}$$

$$\omega_{p_3} = 0 => \frac{\partial}{\partial r}\left(\frac{\omega_p^2}{\omega^2}\right) = 0 \text{ when } r \ge a$$

To design a model for all the regions, we keep in mind that the Laplace equation should be satisfied in all the regions and then choose the corresponding potential distribution as

$$\phi = C_1 r \text{ when } 0 \le r \le a_1$$

$$\phi = C_2 r + \frac{C_3}{r} \text{ when } a_1 \le r \le a, \text{ and}$$

$$\phi = \frac{C_4}{r} \text{ when } r \ge a$$

Using the boundary conditions, we get the following equation in ω,

$$4\omega^4 - 2(\omega_{p_1}^2 + 2\omega_{p_2}^2)\omega^2 + \left\{\left(1 + \frac{a_1^2}{a^2}\right)\omega_{p_1}^2 + \left(1 - \frac{a_1^2}{a^2}\right)\omega_{p_2}^2\right\}\omega_{p_2}^2 = 0 \tag{4.1}$$

It contains the allowed frequency of the oscillating dipoles considered in the plasma cylinder. These oscillating dipoles will radiate at the allowed frequency, which is the solution of Eq. (4.1). Solving this equation, the frequencies of radiation are obtained as

$$\omega_1 = \frac{1}{2}\left[\omega_{p_1}{}^2 + 2\omega_{p_2}{}^2 \pm \sqrt{\left\{\left(\omega_{p_1}{}^2 + 2\omega_{p_2}{}^2\right)^2 - 4\omega_{p_2}{}^2\left\{\left(1 + \frac{a_1{}^2}{a^2}\right)\omega_{p_1}{}^2 + \left(1 - \frac{a_1{}^2}{a^2}\right)\omega_{p_2}{}^2\right\}\right\}}\right]^{\frac{1}{2}} \tag{4.2}$$

4.2.3 Emission of THz Radiation by Dipole Oscillations

The average power of the emitted THz radiation, calculated based on the concept of oscillating dipoles, gives rise to (Malik et el. 2010)

$$\langle \Pi \rangle = \frac{\pi L \omega^3 \Gamma_0^2}{8c^2}$$

where Γ_0 is the maximum amplitude of dipole moment per unit length, and L is the length of the plasma cylinder.

The spatial distribution of plasma density in the plasma cylinder significantly affects the frequency of emitted THz radiation. The relative size of both the concentric cylinders and the total volume of evolved plasma affect the frequency in view of the contribution of resonant dipole oscillations. Eq. (4.2) shows that the THz frequency increases with the increment in plasma frequency and it is enhanced as the value of ratio a_1/a (taken as a_r in Figure 4.4) increases. For a specific value of a_1/a, pressure of the gas increases which further enhances the plasma density and hence the frequency and

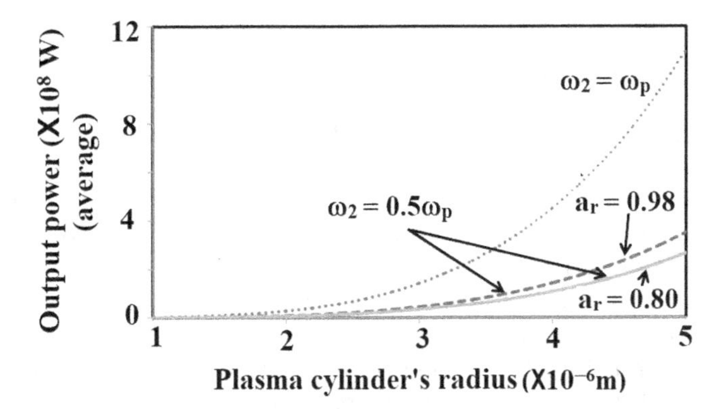

FIGURE 4.4
Dependence of output THz power on the plasma cylinder's radius. Here a_r is the ratio a_1/a.

power of the emitted radiation are increased. The frequency saturates at smaller values of ω_{p_2} for larger values of a_1/a, i.e. when the size of the inner cylinder is larger. On the other hand, the line $\omega_2 = \omega_p$ in Figure 4.4 reveals that the highest power THz output is achieved when the whole plasma cylinder carries the same density as the inner one. Under this situation, all the dipoles oscillate resonantly and the highest radiation is emitted by them.

The interesting part of the tunnel ionized plasma and the emitted THz radiation (Malik et al. 2010) is that the direction of the radiation can be controlled based on the initial phase difference of the two lasers used for quick tunnel ionization (Figure 4.5). Here, it is seen that the THz radiation can be made to emit almost along the axis of the plasma cylinder when there is a phase difference of π radian. However, the intensity of such radiation is weaker than the radiation emitted at some finite obliqueness, though the obliqueness is less than 0.1 rad.

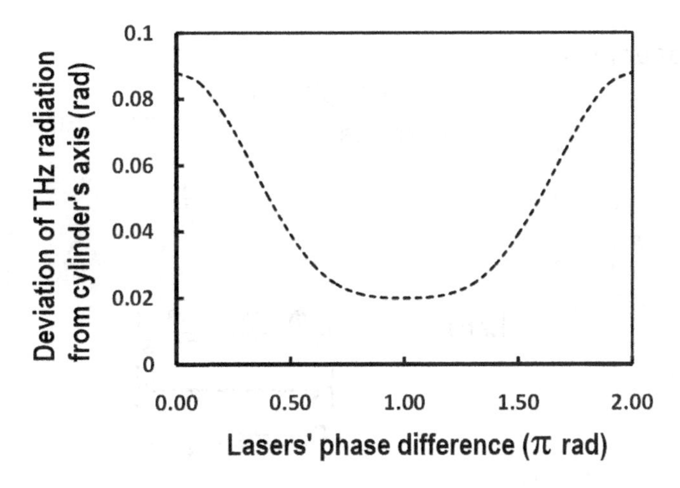

FIGURE 4.5
THz radiation deviation versus lasers' phase difference.

4.3 Laser Beating Process

The superposition of two waves that are identical in every aspect, except one wave parameter, shows some peculiar phenomenon. For example, waves travelling in opposite directions form a standing wave, waves differing in phase result in an interference, and the waves having slightly different frequencies cause beating. The light surrounding us, mostly, is incoherent in nature, which is not applicable to form beats on macroscopic time or distance scales. But in the case of coherent sources like a laser, we can observe beat notes which are different from the audio beat notes. The most accurate method to obtain beats is by splitting the monochromatic source light into two to obtain coherent beams and to shift the frequency of one of the beams by using an acousto-optic modulator and interfere them again. Efficient THz radiation can also be generated by the laser beating process in corrugated plasma or gases at different frequencies. THz sources based on beating can also be scaled to high peak powers with good tunability, efficiency, and directionality. The basic mechanism behind the process (Figure 4.6) can be understood based on the following mathematical treatment of the problem where two Gaussian laser beams are employed.

Two spatial-Gaussian laser beams are considered, which are polarized along the y-direction (propagating along the x-direction) and have different frequencies and wave numbers. The electric fields for these kinds of lasers are given by

$$\vec{E}_1 = \hat{y}E_0 e^{-y^2/a_0^2} e^{i(k_1 x - \omega_1 t)}, \; \vec{E}_2 = \hat{y}E_0 e^{-y^2/a_0^2} e^{i(k_2 x - \omega_2 t)}$$

Using the equation of motion, we can calculate the oscillatory velocity, $v_j = eE_j/mi\omega$ of plasma electrons resulted by laser beating, where $j = 1, 2$ represents the cases of first and second laser beams. The beating of two laser beams imparts ponderomotive force $\vec{f}_p^{NL}$ at

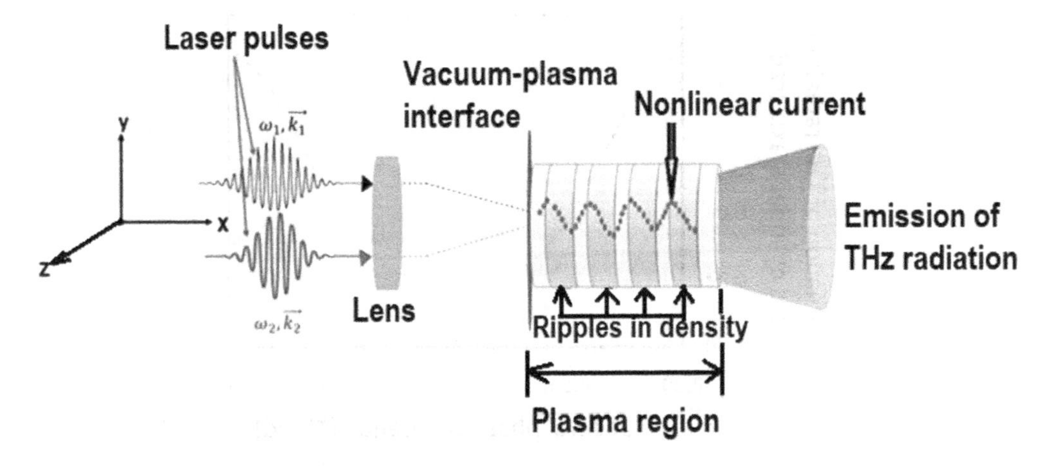

FIGURE 4.6
Schematic diagram of laser beating for THz radiation generation.

beating frequency $\omega = \omega_1 - \omega_2$ and wave number $k' = k_1 - k_2$. The components of the ponderomotive force can be expressed as

$$f_{px}^{NL} = -\frac{ie^2 E_0^2 k'}{2m\omega_1\omega_2} e^{-2y^2/a_0^2} e^{i(k'x-\omega' t)}$$

$$f_{py}^{NL} = \frac{e^2 E_0^2}{2m\omega_1\omega_2 a_0} \left(\frac{4y}{a_0}\right) e^{-2y^2/a_0^2} e^{i(k'x-\omega' t)}$$

The ponderomotive force induces nonlinear perturbations (say n^{NL}) in the electron density which produce a space charge field of potential φ, and self-consistently linear density perturbations (say n^L). With the help of the equation of motion and continuity equation, the linear density perturbations can be obtained as

$$n^L = \frac{\chi_e \varphi}{4\pi e k^2}, \ n^{NL} = \left[n_0 k / m\omega\left(\omega^2\right)\right] f_{py}$$

Now combining the effect of both the linear and nonlinear density perturbations, the electrostatic space-charge potential φ can be estimated in terms of the ponderomotive force $\vec{f}_p^{NL}$ by the following expression:

$$\vec{\nabla}\varphi = \left(\frac{4\pi e n_0}{m\omega'^2(1+\chi_e)}\right)\vec{f}_p^{NL}$$

where the electric susceptibility of the plasma is defined as $\chi_e = -\omega_p^2/\omega^2$. We can calculate the resultant nonlinear electron velocity by using the equation of motion as

$$\frac{\partial \vec{v}^{NL}}{\partial t} = \frac{e\vec{\nabla}\varphi}{m} + \frac{\vec{f}_p^{NL}}{m}$$

$$\vec{v}^{NL} = \frac{i\omega' \vec{f}_p^{NL}}{m\left(\omega'^2 - \omega_p{}^2\right)}$$

These oscillations excite the nonlinear current $\vec{J}^{NL} = -\left(\frac{1}{2}\right) n_\alpha e^{i\alpha x} \vec{v}^{NL}$ in the corrugated plasma, which has density ripples given by $n_\alpha e^{i\alpha x}$. Hence, the current density's components are obtained as

$$J_x^{NL} = -\frac{1}{4}\frac{e^3 n_\alpha \omega E_0^2 k'}{m^2\omega_1\omega_2(\omega^2-\omega_p^2)} e^{-\frac{2y^2}{a_0^2}} e^{i(kx-\omega t)}$$

$$J_y^{NL} = -\frac{1}{4}\frac{ie^3 n_\alpha E_0^2}{m^2\omega_1\omega_2}\frac{\omega}{a_0(\omega^2-\omega_p^2)}\left(\frac{4y}{a_0}\right) e^{-\frac{2y^2}{a_0^2}} e^{i(kx-\omega t)}$$

The above expressions show that the oscillatory current is generated at the beating frequency (same as of the ponderomotive force) and beating wave number $k = k_1 - k_2 + \alpha$ (different from the ponderomotive force). This oscillating current can be thought of as an antenna for the emission of efficient THz radiation.

4.3.1 Resonance Condition for THz Emission

THz radiations can be generated through current density $\vec{J}^{NL}$ oscillating at the frequency ω' and the wave number k'. Hence, the resonance occurs when tuning of the two wave numbers k and k' takes place. This can be done based on the wave number of ripples α, i.e. $k = k' + \alpha$ with the introduction of space-periodically modulated density $n = n_0 + n'$, where, $n' = n_\alpha e^{i\alpha x}$ together with n_α as the amplitude. We can generate these periodic density ripples with the help of a machining beam (Hazra et al. 2004; Kuo et al. 2007). Actually, slow wave structure is required for the resonant excitation of radiation. When the condition of resonance is met, the strongest THz field is radiated due to the maximum energy and momentum transfer from the lasers (via ponderomotive force) to the oscillating nonlinear current. To calculate the field of emitted THz radiation, the following wave equation is used, which is obtained from Maxwell's equations

$$-\nabla^2 \vec{E}_{\text{THz}} + \vec{\nabla}\left(\vec{\nabla} \cdot \vec{E}_{\text{THz}}\right) = -\frac{4\pi i\omega}{c^2}\vec{J}^{NL} + \frac{\omega^2}{c^2}\varepsilon\vec{E}_{\text{THz}} \tag{4.3}$$

The exact phase-matching condition is related to the dispersion relation of the wave. The coefficient of one of the terms of Eq. (4.3) yields this, for example, $k^2 - \omega^2\varepsilon/c^2 = 0$ in the present case. From this, the resonance condition is obtained (Malik et al. 2011a) as

$$\frac{\alpha c}{\omega_p} = \frac{\omega}{\omega_p}\left\{\left(1 - \frac{\omega_p^2}{\omega^2}\right)^{\frac{1}{2}} - 1\right\} \tag{4.4}$$

The periodicity of the density ripples is given by $\alpha c/\omega_p$, which is normalized with the plasma frequency. Here $\varepsilon = 1 - \omega_p^2/\omega^2$ is the electric permittivity of the plasma at the frequency $\omega \equiv \omega'$. At resonance condition, $\omega = \omega_p$ and the maximum momentum and energy transfer takes place, as there is a phase-matching between the wavenumbers of the ripples and the beat wave. It can be seen based on the relation (4.4) that the parameter α is required to be smaller for the larger beating frequency. However, this parameter attains higher values when $\omega = \omega_p$ (Figure 4.7). Because the wavelength of the density ripples is, $\lambda_{\text{ripples}} = \frac{2\pi}{\alpha}$, the resonance condition will be satisfied when these ripples are constructed closer to each other. Also, the higher ripples in the density are expected to create the stronger THz radiation.

4.3.2 Frequency and Power of THz Radiation

The solution of Eq. (4.3) yields the following expression for the THz field E_{THz}

$$\left|\frac{E_{THz}}{E_0}\right| = \frac{n_\alpha}{n_0} \frac{e\omega_p{}^2\omega^2 |E_0 y|}{m\omega_1\omega_2(\omega^2 - \omega_p{}^2)^2 a_0{}^2} e^{-\frac{2y^2}{a_0{}^2}} \tag{4.5}$$

This expression represents the normalized field of the THz radiation, whose behavior with the laser frequency can be seen in Figure 4.7. The maximum amount of energy is seen to transfer at the resonance condition, i.e. when $\omega/\omega_p \approx 1$ (or $\omega \approx \omega_p$). The THz field falls down when the beating frequency (normalized with ω_p, i.e. ω/ω_p) is far from

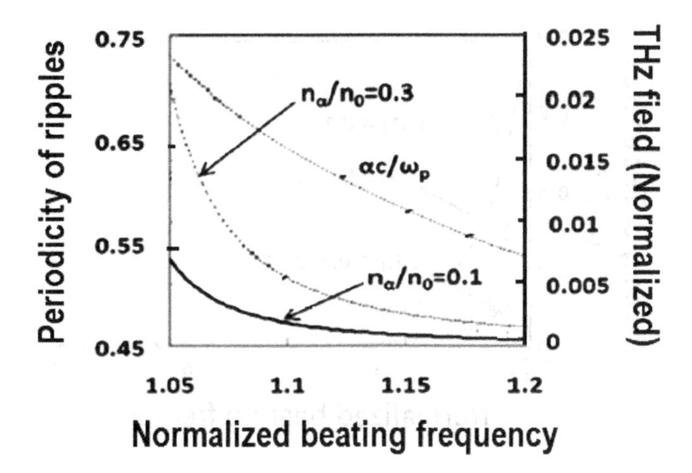

FIGURE 4.7
Variation of periodicity of density ripples ($\alpha c/\omega_p$) and THz field with normalized beating frequency.

the resonance condition. Another important result is the decrement in the periodicity of ripples for higher values of beating frequency ω. Hence, the construction of spatial-modulated ripples at closer distances or with large amplitude would lead to the emission of strong and efficient THz radiation. It can be understood as follows. When the normalized amplitude of the density ripples, i.e. n_α/n_0, increases, electrons in greater number take part in the nonlinear oscillating current and hence increase the amount of THz radiation emission. This effect is also found to be more profound near the resonance condition, i.e. when ω is close to ω_p.

In addition to the above results, one sees the inverse proportionality (square) of the normalized amplitude of the THz field on the beam width a_0 from Eq. (4.5). The THz field will increase in large proportion for the smaller beam width of the lasers because the ponderomotive force becomes stronger for the reduced beam width owing to the great increment in the gradient of the lasers intensity. The increased ponderomotive force leads to the enhanced nonlinear current, which leads to the emission of stronger radiation. For example, the THz field is found to be doubled when the beam width is reduced from 0.10 to 0.05 mm [Eq. (4.5)].

4.3.3 Efficiency of Mechanism

The ratio of energy densities of the THz radiation and the incident lasers determine the efficiency of the THz radiation generation. The average electromagnetic energy density (stored in unit volume) of the electric and magnetic fields (Rothwell and Cloud 2009), in general, is given by the relations $\langle W_E\rangle = \left(\frac{1}{8\pi}\varepsilon\frac{\partial}{\partial\omega}\left[\omega\left(1-\frac{\omega_p^2}{\omega^2}\right)\right]\right)\langle|E_E|^2\rangle$ and $< W_B > = \frac{1}{8\pi\mu_0}\left(\frac{k\langle E_E\rangle}{\omega}\right)^2$.

Based on this, the total average energy density of the emitted THz radiation and that of the incident lasers (pump) are calculated, the ratio of which yields efficiency η of the THz radiation

$$\eta = \left(\frac{n_\alpha}{n_0}\right)^2 \frac{e^2\omega_p{}^4\omega^4E_0{}^2}{16\sqrt{2}m^2\omega_1{}^2\omega_2{}^2(\omega^2-\omega_p{}^2)^4a_0{}^2} \tag{4.6}$$

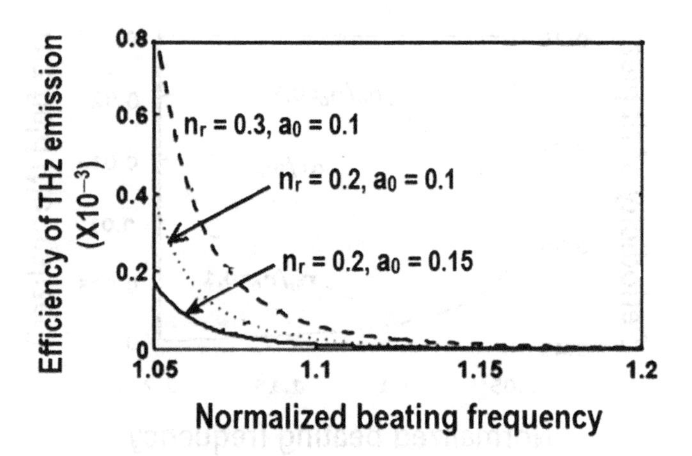

FIGURE 4.8
Efficiency of THz emission versus beating frequency, showing the effect of $n_r = \frac{n_\alpha}{n_0}$ and beam width a_0 (in millimeters).

The dependence of efficiency of THz emission on the beating frequency is shown in Figure 4.8. The effect of the density amplitude of the density ripples and beam width of the lasers is also investigated and the efficiency is found to decrease as we go away from the resonance condition. For the fixed ripple amplitude, the magnitude of efficiency decreases with the increment of beam width. However, the efficiency is enhanced with the higher amplitudes of the density ripples for other fixed parameters.

From the above results, it is clear that the field of THz radiation can be tuned with the help of laser and plasma parameters. For example, the decrement of laser beam width and an increment in the ripples' amplitude enhance the THz field significantly and the mechanism becomes very efficient.

4.3.4 Multiple Resonance-Led THz Radiation

In this section we are talking about the laser beating for the THz generation, but in addition to the laser's electric field there is also a periodic electric field with polarization in the same direction as that of the laser's field. This static electric field is periodic in the direction of propagation of the beams, in which direction the density ripples ($n_\alpha e^{i\alpha z}$) are also created in plasma in order to meet the resonance condition via matching the wave numbers (Figure 4.9). The periodic electric field is taken as

$$\vec{E}_S = E_{0S} e^{i\delta z} \hat{y} \tag{4.7}$$

where δ is its periodicity, given by $\delta = 2\pi/2s_b$ together with S_b as the half wavelength of the field. The total electric field exerted on the electrons is the sum of the electrostatic field and the laser's field, $\vec{E}_{jL}$; hence, $\vec{E} = \vec{E}_S + \vec{E}_{jL}$. In the case of the super-Gaussian profile (detailed description given later (Section 4.5.3)) of the laser beams, the electric field polarized in the y-direction is written as

$$\vec{E}_{jL} = E_{0L} e^{-\left(\frac{y}{a_0}\right)^q} e^{i(k_j z - \omega_j t)} \hat{y} \tag{4.8}$$

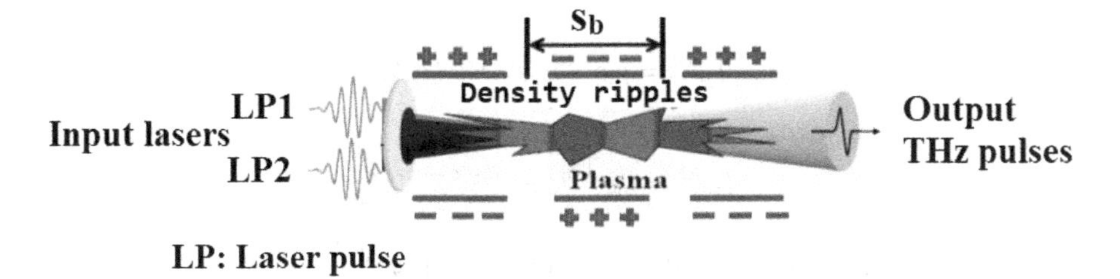

FIGURE 4.9
Schematic for multiple resonance-led THz emission, showing the application of periodic electrostatic field (half wavelength as S_b) and the density ripples.

where a_0 is the beam width of the lasers and q is the index of the lasers, called the sG index for the super-Gaussian (sG) beams. The expression for $q = 2$ will correspond to the beams of Gaussian profile and the expression for $q > 2$ represents the super-Gaussian profile.

The coupling of the laser's field and the electrostatic field takes place; hence, now the ponderomotive force is also realized at modified frequency (ω_1, ω_2) and wave number (k'_2, k'_3), obtained in terms of k_1, k_2, and periodicities α and δ of the density ripples and the electrostatic field, in addition to the usual beat frequency $\omega = \omega_1 - \omega_2$ and wave number $k = k_1 - k_2$. This is in view of the following form of the ponderomotive force (Singh and Malik 2020)

$$\vec{f}_p^{\ NL} = \frac{e^2}{2m(i\omega_1 - \nu)(i\omega_2 + \nu)} \vec{\nabla}[\vec{E}_1 \cdot \vec{E}_2^* + \vec{E}_1 \cdot \vec{E}_S^* + \vec{E}_2^* \cdot \vec{E}_S + \vec{E}_S \cdot \vec{E}_S^*] \tag{4.9}$$

In the above expression, the first term of the square bracket in the right-hand side (RHS) represents the usual laser beating in plasma, the second term is due to the coupling of laser 1 with the electrostatic field, the third term is due to the coupling of laser 2 with the electrostatic field, and the last term is superfluous being non-oscillatory in nature (field E_S is the static field). The ponderomotive force caused by the coupling of the first and second lasers' fields with the electrostatic field is called as external field–induced (EFI) ponderomotive force. Only the following three components / terms will be responsible for the emission of THz radiation

$$f_{py}^{\ NL} = (f_{py}^{\ NL})_{\text{beating}} + (f_{py_1}^{\ NL})_{\text{EFI}} + (f_{py2}^{\ NL})_{\text{EFI}} \tag{4.10}$$

In what manner do the last two terms of the RHS of Eq. (4.10) differ? Actually, the magnitudes of these terms will be equal but these will oscillate at different frequencies and wave numbers. The magnitude of the EFI ponderomotive force would be much smaller than that of the beating-enabled ponderomotive force due to the smaller magnitude of the electrostatic field compared with the laser's field. The oscillations of the electrons under the effect of these forces become nonlinear, creating density perturbations due to the redistribution of the electrons in plasma. Based on this different kinds of nonlinear currents will be generated, which will radiate the EFI THz (or EFIT radiation) in addition to the beating-enabled THz radiation. The following three kinds of the currents are observed in this situation (Singh and Malik 2020)

$$J_{1y}^{\ NL} = 2J_0 \frac{E_{0L}}{E_{0S}} e^{-2\left(\frac{y}{a_0}\right)^q} e^{i\{(k+\alpha)z - \omega t\}} \tag{4.11}$$

$$J_{2y}{}^{NL} = J_0 e^{-\left(\frac{y}{a_0}\right)^q} e^{i\{(k_1+\alpha-\delta)z-\omega_1 t\}} \tag{4.12}$$

$$J_{3y}{}^{NL} = J_0 e^{-\left(\frac{y}{a_0}\right)^q} e^{-i\{(k_2-\alpha-\delta)z-\omega_2 t\}} \tag{4.13}$$

Here $J_0 = -\frac{iN_\alpha e^3 E_{0L} E_{0S}}{4m^2(i\omega_1 - \nu)(i\omega_2 + \nu)} \frac{q}{a_0} \frac{\omega}{[\omega(\omega + i\nu) - \omega_p{}^2]} \left(\frac{y}{a_0}\right)^{q-1}$.

Consistent with the nature of EFI ponderomotive forces, the magnitude of the current densities J_{2y}^{NL} and J_{3y}^{NL} is equal and is much less than that of the current J_{1y}^{NL} generated by the usual laser beating. Also, all these currents oscillate at different frequencies and wave numbers, as has been the case of ponderomotive forces. For example, the current density J_{1y}^{NL} oscillates at the frequency $\omega = \omega_1 - \omega_2$ and the wave number $k + \alpha = k_1 - k_2 + \alpha\ (= k_{T1})$. In view of α as the wave number of the density ripples, this current can be generated resonantly at the beating frequency by tuning the wave numbers of the lasers and the density ripples. However, the current densities J_{2y}^{NL} and J_{3y}^{NL} are observed to be generated at frequencies ω_1 and ω_2 and wave numbers $k_1 + \alpha - \delta\ (= k_{T2})$ and $k_2 - \alpha - \delta\ (= -k_{T3})$, respectively. Owing to the terms α and δ in their wave numbers, these currents can be generated resonantly by tuning the periodicities of the electrostatic field and the density ripples. However, these currents oscillate at the frequency of the respective laser (ω_1 or ω_2). In short, the lasers beat together when they are incident on plasma; additionally both the lasers also produce ponderomotive force via their interaction with the periodic electrostatic field. Hence, the plasma electrons oscillate under the influence of the laser's field and the external electric field.

In view of the magnitude, the current generated by the laser beating is called the primary current, whereas the EFI currents are understood to be secondary in nature. There are phase-matching conditions for all these currents and based on which we can excite or suppress any current in the plasma. Since these currents are the sources for the radiation, three kinds of THz radiations are observed. These are termed as beat wave–enabled THz radiation and EFI THz radiation (two types). In the case of the beat wave–enabled THz radiation, following is the phase matching condition in collisional plasma with ν as the electron-neutral collision frequency

$$\left(\frac{\alpha c}{\omega_p}\right) = \frac{\omega}{\omega_p}\left[\left(1 - \frac{\omega_p^2}{\omega(\omega + i\nu)}\right)^{1/2} - 1\right] \tag{4.14}$$

In the case of the EFI-enabled THz radiation, the following are the phase matching conditions

$$\left(\frac{\alpha c}{\omega_p}\right) = \frac{\omega}{\omega_p}\left[\left(1 - \frac{\omega_p^2}{\omega(\omega + i\nu)}\right)^{1/2} - (k_1 - \delta)\frac{c}{\omega}\right] \tag{4.15}$$

$$\left(\frac{\alpha c}{\omega_p}\right) = \frac{\omega}{\omega_p}\left[\left(1 - \frac{\omega_p^2}{\omega(\omega + i\nu)}\right)^{1/2} - (-k_2 + \delta)\frac{c}{\omega}\right] \tag{4.16}$$

The beat wave–enabled primary THz field has the following amplitude

$$E_{0THzLB} = \frac{4\pi\omega z J_0}{k_{T1}c^2}\frac{E_{0L}}{E_{0S}}e^{-2\left(\frac{y}{b_w}\right)^p} \tag{4.17}$$

The EFI-enabled secondary THz fields have the following amplitudes

$$E_{0THzEFI1} = \frac{2\pi\omega z J_0}{k_{T2}c^2}e^{-\left(\frac{y}{b_w}\right)^p} \tag{4.18}$$

$$E_{0THzEFI2} = \frac{2\pi\omega z J_0}{k_{T3}c^2}e^{-\left(\frac{y}{b_w}\right)^p} \tag{4.19}$$

Figures 4.10 and 4.11 make a comparative study of the beat wave–enabled THz radiation field and the EFI THz radiation field. Both the fields are found to be reduced if the lasers of larger beam width are used in the said mechanism (Singh and Malik 2020). Actually, weaker ponderomotive force is exerted on the plasma electrons when the lasers have a larger beam width due to the reduced intensity gradient of the lasers. This leads to the generation of weaker current because of which the emission of a weak THz field is obtained. If we look at the magnitudes of these two THz fields, it is found that the beat wave–enabled THz field is about 18 times higher than the EFI THz field. The point of interest is that one can suppress any THz field (out of the three) by varying the parameters concerning lasers, plasma, and electrostatic field. For example, we can further enhance the EFI THz field by either increasing the strength of electrostatic field or by reducing its periodicity. The beat wave–enabled THz field can be reduced by using the lasers with frequencies having a little more difference or by reducing the sG index.

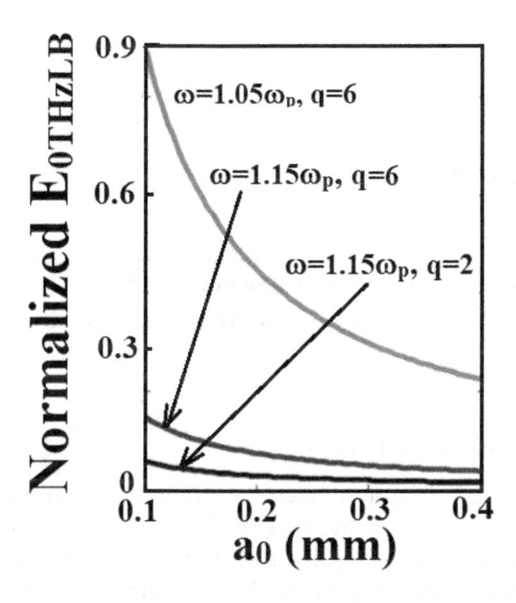

FIGURE 4.10
Beat wave–enabled THz field (normalized) versus laser beam width, showing the effect of the sG index and beating frequency.

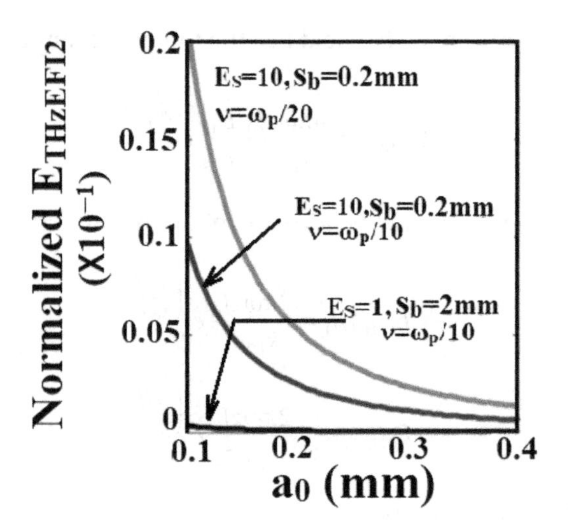

FIGURE 4.11
EFI THz field (normalized) versus laser beam width, showing the effect of strength of the electrostatic field (E_S), periodicity of electrostatic field (S_b), and electron-neutral collision frequency ν. E_S carries the unit kV/cm.

With the concept of this multiple resonance-led THz radiation, we can create three kinds of THz fields using the same mechanism of laser beating. This way multifocal THz radiation can be achieved, which will be quite useful in cancer diagnosis and its treatment. For further application, efficiency of mechanisms, and other related aspects, the readers are advised to go through the recently published article from Singh and Malik (2020).

4.4 Role of Magnetic Field

To investigate the role of the magnetic field in the emission of THz radiation by gases / plasma, many experiments have been performed. Wu et al. (2007) have found that the magnetic field enhances the efficiency of the emitted field. McLaughlin et al. (2000) also realized continuous increment in the THz emitted field when they increased the magnetic field up to 8 T.

4.4.1 Tunnel Ionization–Based Emission

The mechanism of tunnel ionization discussed in an earlier section (Section 4.2) is taken further as an example to uncover the role of the external magnetic field B_0 (Malik and Malik 2011). The magnetic field is applied along the axis of the plasma cylinder. This will add another motion to the electrons due to their cyclotron motion; hence, the dipole oscillations will also be modified accordingly. This will lead to a control on the frequency of oscillations and hence the THz frequency by varying the magnetic field. Under this situation, Poisson's equation reads

$$\frac{1}{r}\frac{\partial}{\partial r}\left\{\left(1-\frac{\omega_p^2}{\omega^2-\omega_c^2}\right)r\frac{\partial\phi}{\partial r}\right\}-\frac{\omega_c}{\omega(\omega^2-\omega_c^2)}\frac{\phi}{r}\frac{\partial\omega_p^2}{\partial r}-\left(1-\frac{\omega_p^2}{\omega^2}\right)\frac{\phi}{r^2}=0$$

where ω_c stands for the electron cyclotron frequency. The plasma cylinder will not carry uniform density, as has been in the case of unmagnetized plasma. A similar kind of approach is used for calculating the frequency of oscillations and for finding the following equation in the oscillation frequency ω

$$\begin{aligned}&4a^2\omega^6-2(4\omega_c^2+2\omega_{p_2}^2+\omega_{p_1}^2)a^2\omega^4-2(\omega_c\omega_{p_1}^2a^2)\omega^3\\&+[2\omega_c^2a^2(2\omega_c^2+2\omega_{p_2}^2+\omega_{p_1}^2)+\omega_{p_1}^2\omega_{p_2}^2(a^2+a_1^2)+\omega_{p_2}^2(a^2-a_1^2)]\omega^2\\&+2\omega_{p_1}^2\omega_c a^2(\omega_{p_1}^2+\omega_c^2)\omega+\omega_{p_2}^2\omega_c^2[\omega_{p_1}^2-\omega_{p_2}^2(a^2-a_1^2)]=0\end{aligned} \tag{4.20}$$

When we solve this equation numerically, the frequency comes out to be in the THz range. For an unmagnetized case, the frequency is obtained as $\frac{\omega_p}{\sqrt{2}}$. The variation of frequency of oscillations (THz frequency) with the magnetic field is depicted in Figure 4.12 for different plasma densities and sizes of the plasma cylinder. The tuning of THz frequency is possible with the help of the magnetic field and the density produced in the plasma cylinder. Higher density in the outer region of cylinder, when the size of the cylinder remains fixed, results in higher frequency, and the maximum frequency is achieved when $\omega_{p2}=\omega_p$ ($\equiv \omega_{pm}$) and $a_r = a_1/a = 1$. Hence, we conclude that the extension of plasma volume raises the THz frequency and confirms its dependence on the spatial distribution of plasma. Also, the THz frequency varies linearly with the magnetic field B_0, when the whole cylinder carries uniform plasma density or the oscillating dipoles.

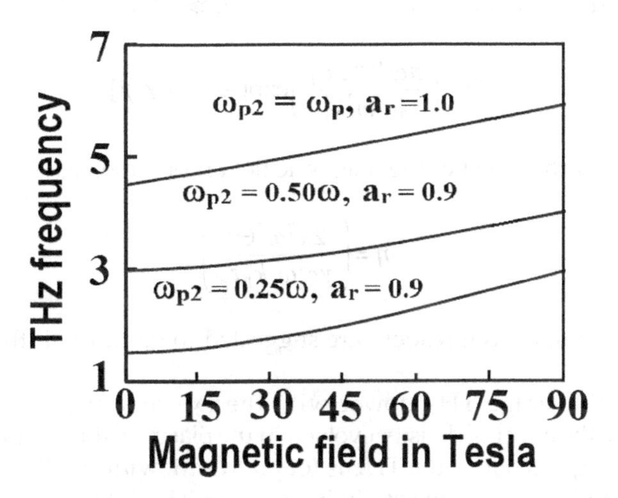

FIGURE 4.12
Tuning of emitted THz frequency with the help of magnetic field B_0.

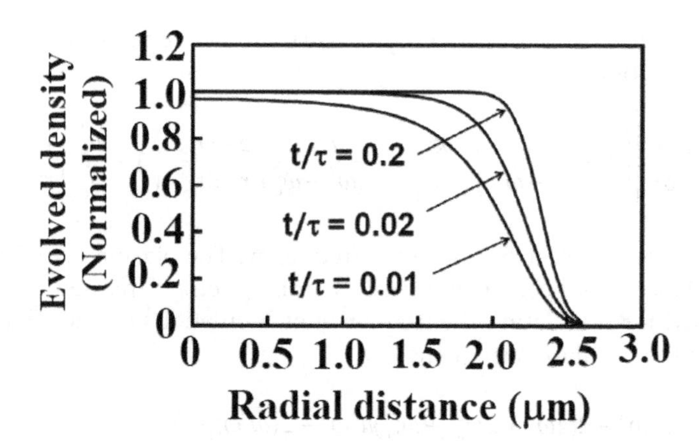

FIGURE 4.13
Evolution of plasma density with normalized time in the plasma cylinder in the radial direction.

It is interesting to see the evolution of plasma density with time in the plasma cylinder in the radial direction (Figure 4.13). In smaller times ($t/\tau = 0.01$), the plasma density does not stay uniform in the cylinder and a gradient in the density starts after a 1-μm radial distance that keeps increasing. However, with sufficient time of evolution ($t/\tau = 0.2$) the density becomes uniform until a 2.25-μm radial distance, i.e. up to 90% of the size of the cylinder. The main purpose of the mechanism discussed in Malik and Malik (2011) was to reduce the time of plasma density evolution and to enhance the resonance between the ponderomotive force and the nonlinear current for efficient THz emission.

To determine the power of emitted THz radiation, the length of plasma cylinder is taken as $L \equiv w/\tan\alpha_0$, where w is the initial spot size of the laser beam. The radiative damping γ_r is also included in the equation, which governs the dipole moment per unit volume Γ. For the damping to be weak and the ionization to be mild, the total emitted power is calculated (Malik and Malik 2011) in the far zone, i.e. $L \gg c/\omega$. The total power reads

$$\Pi = \left(\frac{\pi\omega^5\Theta^2 L E_0^2}{4c^2\omega_L^2 k_\perp^4}\right)\exp[-2(\gamma_i + \gamma_r)t]$$

The efficiency under the effect of the magnetic field is obtained as

$$\eta = \left(\frac{2\pi^2\omega^5\Theta^2}{\gamma c^4\omega_L k_\perp^4 \tau_p}\right)$$

For details of the symbols the readers are suggested to go through the article Malik and Malik (2011).

The variation of average THz power with the external magnetic field is shown in Figure 4.14, where the effect of density evolved in the plasma cylinder and the size of the cylinder can be investigated. Here, the THz is found to emit with smaller power if the plasma cylinder is small and when the density in it is smaller. The THz power is increased as soon as the density increases in the cylinder and, finally, the highest power is obtained when the whole plasma cylinder carries uniform density. Under this situation ($\omega_{p2} = \omega_p$, $a_r = 1$), THz power increases with the magnetic field linearly until 45 T and then the enhancement rate is

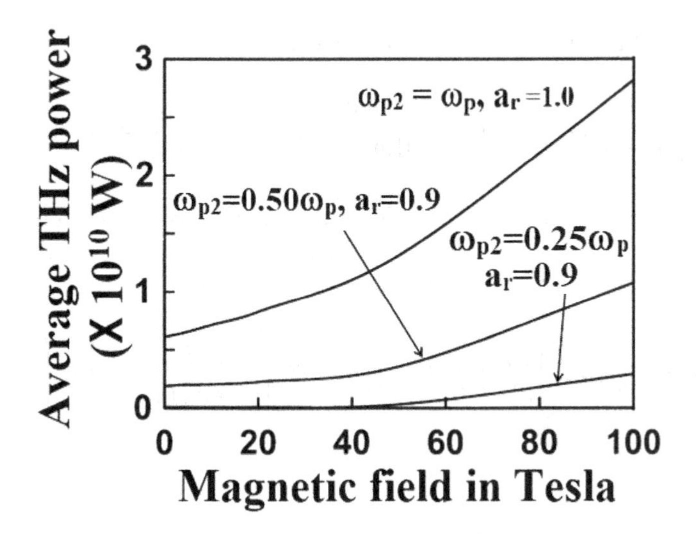

FIGURE 4.14
Variation of average power of THz radiation with the magnetic field B_0.

suddenly doubled when the magnetic field is larger than ≈60 T. In accordance to the variation of THz power, the efficiency of the mechanism is also increased with the magnetic field. Clearly, the role of the magnetic field in tunnel-ionized plasma is to tune the frequency and power of the emitted THz radiation and the efficiency of the emitted radiation.

4.4.2 Laser Beating–Based Emission

Consider two spatial-Gaussian lasers (Malik et al. 2012) co-propagating in plasma with different frequencies and wave numbers in the presence of the magnetic field. The electric field of these lasers polarized in the y-direction is written as

$$\vec{E}_j = \hat{y}E_0 e^{-y^2/a_o^2} e^{i(k_j x - \omega_j t)} \text{ with } j = 1,2$$

Since the lasers propagate in the x-direction and their electric field is in the y-direction, it is advisable to apply the external magnetic field in the z-direction. The Lorentz force will act on the electrons in the x-y plane and, hence, will help increasing the impact of the electric field of the lasers in the y-direction, i.e. the transverse direction. Using the equation of motion $\frac{\partial \vec{v}_1^{NL}}{\partial t} = \frac{e\vec{\nabla}\phi_p}{m} - (\vec{v}_1^{NL} \times \vec{\omega}_c)$, where $\vec{\omega}_c = \frac{eB_0}{mc}\hat{z}$, and continuity equation $\frac{\partial n_1^{NL}}{\partial t} = -n_0 \vec{\nabla}\cdot \vec{v}_1^{NL}$, nonlinear density and nonlinear velocity are calculated for obtaining the nonlinear current density $\vec{J}^{NL}$.

The current density is put in the following wave equation, which is obtained based on the Maxwell's equations. The point of observation is the existence of dielectric tensor $\bar{\varepsilon}$ in the wave equation, which is due to the anisotropy created in the plasma by the external magnetic field (discussed in Chapter 1).

$$-\nabla^2 \vec{E}_{\text{THz}} + \vec{\nabla}\left(\vec{\nabla}\cdot\vec{E}_{\text{THz}}\right) = -\frac{4\pi i\omega}{c^2}\vec{J}^{NL} + \frac{\omega^2}{c^2}\bar{\bar{\varepsilon}}\vec{E}_{\text{THz}} \tag{4.21}$$

The components of the dielectric tensor $\bar{\bar{\varepsilon}}$ are as under

$$\varepsilon_{xx}=\varepsilon_{yy}=1-\frac{\omega_p^2}{\left(\omega^2-\omega_c^2\right)},\ \varepsilon_{yx}=-\varepsilon_{xy}=\frac{i\omega_c\omega_p^2}{\omega\left(\omega^2-\omega_c^2\right)},\ \varepsilon_{zz}=1-\frac{\omega_p^2}{\omega^2},\text{ and }\varepsilon_{xz}=\varepsilon_{zx}=\varepsilon_{zy}=\varepsilon_{yz}=0$$

The modified electric permittivity of the medium (or the dielectric tensor) is expected to affect the emitted THz field amplitude and the efficiency of the mechanism. The above wave equation in terms of tensor components is written as

$$-\nabla^2\vec{E}_{THz}+\vec{\nabla}(\vec{\nabla}\cdot\vec{E}_{THz})=-\frac{4\pi i\omega}{c^2}(\hat{x}J_x{}^{NL}+\hat{y}J_y{}^{NL})+\frac{\omega^2}{c^2}\left[\begin{pmatrix}\varepsilon_{xx}E_{THzx}\\+\varepsilon_{xy}E_{THzy}\end{pmatrix}\hat{x}+\begin{pmatrix}\varepsilon_{yx}E_{THzx}\\+\varepsilon_{yy}E_{THzy}\end{pmatrix}\hat{y}\right] \tag{4.22}$$

The solution of this equation gives rise to the following THz field

$$\left|\frac{E_{0y}}{E_0}\right|=\left|\frac{\lambda_1 x}{2k}\right|\left[\lambda_2\left(\frac{4y}{a_0^2}-\lambda_3 k'\right)+\lambda_4\left(k'+\frac{4y\lambda_3}{a_0^2}\right)\right]e^{-\frac{2y^2}{a_0^2}} \tag{4.23}$$

where the coefficients λ_1, λ_2, λ_3, and λ_4 are given as

$$\lambda_1=\frac{in_\alpha e\omega\omega_p{}^2E_0}{4n_0mc^2\omega_1\omega_2(\omega^2-\omega_h{}^2)(\omega^2-\omega_c{}^2)},\ \lambda_2=\frac{\omega^2(\omega^2-\omega_c^2)+\omega_c^2\omega_p^2}{\omega},\ \lambda_3=\frac{\omega_c\omega_p^2}{\omega(\omega^2-\omega_h^2)}$$

and $\lambda_4=\omega_c(\omega^2+\omega_p^2-\omega_c^2)$.

To understand the role of magnetic field on the emitted THz radiation using spatial-Gaussian laser beams (Malik et al. 2012), the normalized THz field is plotted for its transverse profile and its variation with normalized cyclotron frequency (Figure 4.15). Here, a peak is observed at particular values of y/a_0 and ω_c/ω_p. The reason behind this peculiar behavior can

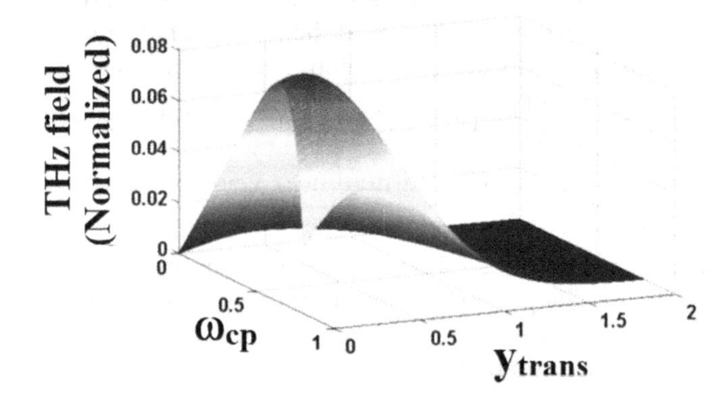

FIGURE 4.15
Transverse profile ($y_{trans}=y/a_0$) of the normalized THz field and its variation with normalized cyclotron frequency $\omega_{cp}=\omega_c/\omega_p$.

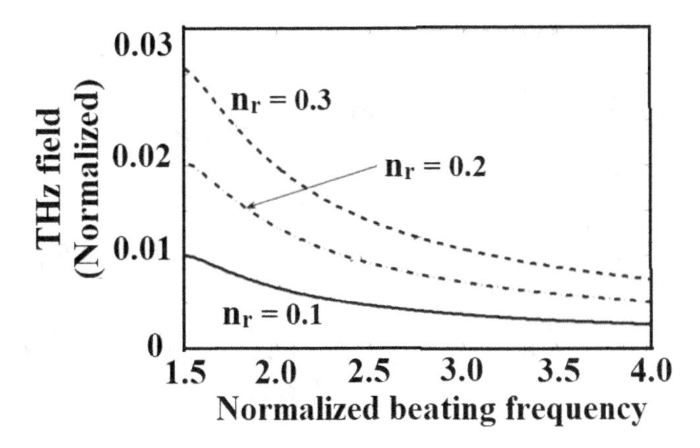

FIGURE 4.16
Normalized emitted THz field versus normalized beating frequency $\omega = \omega/\omega_p$, showing the impact of the ripple's amplitude $n_r = n_\alpha/n_0$.

be understood based on the resonance condition. The resonance condition $\omega \approx \omega_h = \sqrt{\omega_p^2 + \omega_c^2}$ can be satisfied for a particular value of ω_c/ω_p keeping the plasma frequency constant. At that value, the maximum energy transfer can take place, leading the THz amplitude to peak. On the other hand, the ponderomotive force at a particular value of y/a_0 becomes strongest, creating the largest nonlinear current; hence, the THz field peaks at the same value of y/a_0.

Figure 4.16 shows the variation of THz field amplitude with the beating frequency. Here the effect of the density ripple's amplitude on the THz field is also presented. As the beat wave frequency gets closer to the plasma frequency (i.e. towards resonance condition $\omega \approx \omega_h$), the THz field amplitude is enhanced, provided $k^2 > 0$. Furthermore, when the beating frequency is far from ω_h, the THz amplitude decreases. In addition, the THz field is enhanced for the larger values of n_r. Since n_r represents the amplitude of density ripples, the THz radiation attains larger field amplitude for the larger density ripples. The ripples are related to the density of electrons. Increasing ripple amplitude means the electrons are taking part in excitation of the nonlinear current in larger numbers. It leads to a larger current because of which the higher amplitude THz radiation is emitted. The difference between the values of the THz field for different values of n_r becomes more pronounced towards the beating frequency close to 1.5 ω_p, i.e. towards the resonance condition.

In addition to the THz field amplitude, the efficiency of the mechanism is calculated based on the energy densities of the pump lasers and the emitted THz radiation. This is given as

$$\eta = \left(\frac{\lambda_1 x}{2k}\right)^2 \left[\frac{(\lambda_2 + \lambda_3\lambda_4)^2}{a_0^2} + \frac{(\lambda_2\lambda_3 - \lambda_4)^2\omega^2}{4c^2}\right] \tag{4.24}$$

The dependence of efficiency of THz emission on the applied magnetic field (via the frequency ratio $\omega_{cp} = \omega_c/\omega_p$) is shown in Figure 4.17 along with the effect of beating frequency via $\omega_{nf} = \omega/\omega_p$. Here, the concept of resonant excitation of the THz field is also exposed, though the peaks are attained by the THz field at different values of ω_{cp} for different beating frequencies. The figure shows that the THz field increases linearly with ω_{cp} (or the amplitude of the magnetic field) until a peak is achieved; then the field falls down. This is due to the additional cyclotron motion of the electrons because of which the nonlinear current oscillates at hybrid frequency ω_h, i.e. larger than the plasma frequency ω_p. Since the

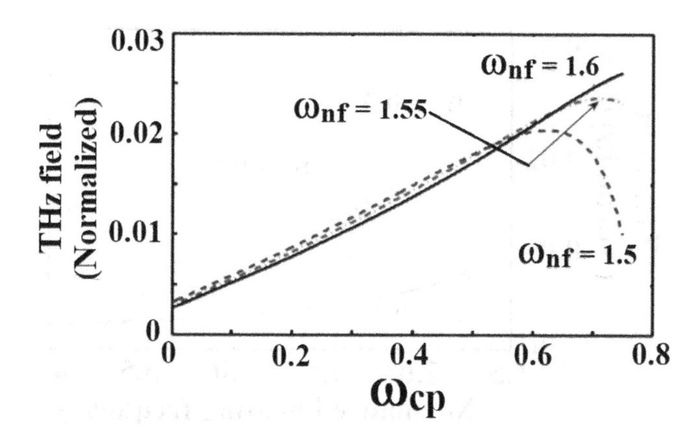

FIGURE 4.17
Variation of normalized emitted THz field with normalized cyclotron frequency $\omega_{cp} = \omega_c/\omega_p$. Here $\omega_{nf} = \omega/\omega_p$.

upper hybrid frequency $\omega_h = \sqrt{\omega_p^2 + \omega_c^2}$ is the frequency at which the resonance is met, the resonance is expected to take place at a particular value of ω_{cp} for the fixed value of plasma frequency ω_p (or ω_{nf}, as $\omega_{nf} = \omega/\omega_p$). At this value, the THz field is emitted with the largest magnitude due to the maximum energy transfer to the electrons and the excitation of the strongest nonlinear current in the plasma.

4.5 Role of Laser Pulse Profile

In laser-plasma interaction, while propagating through the medium, the laser pulse modifies various properties of the medium. Along with the intensity and energy effect, spatial and temporal profiles of the incident laser beams also affect the dynamics of laser-plasma interaction. If we talk about the THz emission, this radiation also encounters a significant change in its profile and efficiency of the scheme due to different laser profile parameters like beam order, index, skewness, and so forth. These parameters change the shape of the laser pulses, which means the laser beams with particular shape can be achieved by tailoring these parameters. The particular shape of laser pulses can be created based on several techniques, involving optical spectral shapers (Chou et al. 2003), adaptive optics (Cherezova et al. 1998), frequency-to-time mapping (Ye et al. 2011), and fibre-optic shaping method (Wang et al. 2010). The laser pulses create a modified nonlinear current due to the modified form of nonlinear ponderomotive force acting on the electrons. Since this current is responsible for the emission of THz radiation, the role of the laser pulses having a particular shape is vital in creating the desired nonlinear current via ponderomotive force suitable for efficient generation of THz radiation.

4.5.1 Temporal and Spatial-Gaussian Profiles of Laser Beams

Pulsed Gaussian beam can be represented in frequency domain as (Wang et al. 1997)

$$U(r,z,\omega) = \tilde{A}(\omega)\frac{\sigma_0}{\sigma(z)}\exp\left\{i\varphi(z) - ik\left[z + \frac{r^2}{2q(z)}\right]\right\} \tag{4.25}$$

At the center frequency of the spectrum, the beam parameters are as follows:

$$q(z) = \frac{1}{R(z)} - i\frac{2}{k_0\sigma^2(z)}$$

$$\sigma^2(z) = \sigma_0^2\left[1 + \left(\frac{z}{z_0}\right)^2\right]$$

$$R(z) = z\left[1 + \left(\frac{z}{z_0}\right)^{-2}\right]$$

$$\varphi(z) = \tan^{-1}\left(\frac{z}{z_0}\right)$$

$$z_0 = \frac{\pi\sigma_0^2}{\lambda_0}$$

where $\tilde{A}(\omega - \omega_0)$ is the spectral amplitude of the pulse, $\sigma_0 = W_0\sqrt{k/k_0}$, W_0 is the waist size, given by $W_0 = \sqrt{L/k}$ together with L as the length of the cavity, and k_0, λ_0 are the wave number and wavelength, respectively.

Using inverse Fourier transform, the time-domain pulsed field can be obtained from Eq. (4.25) as

$$U(r,z,t) = \frac{1}{\sqrt{2\pi}}\int_{-\infty}^{+\infty} U(r,z,\omega)e^{i\omega t}d\omega \tag{4.26}$$

By using a complex temporal variable

$$\tau(r,z,t) = t - \frac{\left(z + \frac{r^2}{2q(z)}\right)}{c}$$

$$= t - \frac{\left(z + \frac{r^2}{2R(z)}\right)}{c} + i\frac{r^2}{\omega_0\sigma^2(z)}$$

we obtain the spatial as well as the temporal profile of Gaussian laser pulse in free space as

$$U(r,z,t) = \frac{\sigma_0}{\sigma(z)}A(\tau)e^{i[\omega_0\tau + \varphi(z)]} \tag{4.27}$$

Specifically the electric field of linearly polarized spatial-Gaussian laser beams can be expressed as (Malik et al. 2011a)

$$\vec{E}_j = \hat{y}E_0e^{-y^2/a_0^2}e^{i(k_jz - \omega_jt)} \tag{4.28}$$

4.5.2 Spatial-Triangular Laser Beams

Many techniques including optical spectral shapers, frequency-to-time mapping, fibre-optic shaping method, and adaptive optics can generate pulses with a triangular shape. The electric field of linearly polarized spatial-triangular laser beams can be expressed as (Malik et al. 2011b)

$$\vec{E}_j = \hat{y}E_0\left(1-\left|\frac{y}{a_0}\right|\right)e^{i(k_jx-\omega_jt)} \text{ for } \left|\frac{y}{a_0}\right| < 1$$
$$= 0 \text{ otherwise.}$$

The ponderomotive force, evolved due to the intensity gradient of these laser beams in the beating process, can be expressed as

$$\vec{f}_p^{NL} = -\frac{e^2E_0^2}{2m\omega_1\omega_2}\left(1-\left|\frac{y}{a_0}\right|\right)\left\{ik\left(1-\left|\frac{y}{a_0}\right|\right)\hat{x}-\frac{2}{a_0}\hat{y}\right\}e^{i(kx-\omega t)} \tag{4.29}$$

The emitted THz field in the above process (Malik et al. 2011b) can be found as

$$\left|\frac{E_{0_y}}{E_0}\right| = \frac{n_\alpha}{2n_0}\frac{\omega_p^2\omega^2\,|\vec{v}_2^*|}{\omega_1(\omega^2-\omega_p^2)^2a_0}\left(1-\left|\frac{y}{a_0}\right|\right), \text{ where } \vec{v}_2^* = \frac{e\vec{E}_2^*}{mi\,\omega_2} \tag{4.30}$$

THz radiation generation, due to beating process, by using spatial-triangular beams has been found by Malik et al. (2011b). They have proved that a spatial-triangular laser has an advantage over the spatial-Gaussian profile for the emission of efficient THz radiation. To study in detail the effect of the laser profile on the emitted field, they considered two co-propagating triangular lasers in the z-direction with the polarization along the perpendicular y-direction. A steep variation in the intensity profile of these spatial-triangular pulses in transverse direction put forth a stronger nonlinear ponderomotive force. Hence, a highly nonlinear current oscillating at the beating frequency of the incident laser beams is responsible for stronger THz radiation (Figure 4.18). Higher amplitude of the emitted

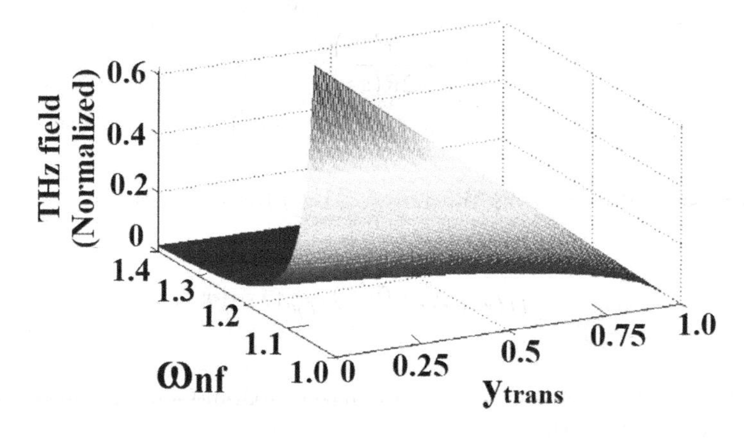

FIGURE 4.18
Transverse profile ($y_{trans} = y/a_0$) of normalized THz field and its variation with normalized beating frequency $\omega_{nf} = \omega/\omega_p$.

field results in higher efficiency, which can also be enhanced by lowering beam width a_0 and increasing n_α. A better collimation quality along the axis due to higher intensity amplitude is also realized in the scheme based on spatial-triangular laser pulses.

4.5.3 Super-Gaussian Laser Beams

The electric field for linearly polarized super-Gaussian laser beams (Malik and Malik, 2012) (for $q > 2$) can be expressed as

$$\vec{E}_j = \hat{y}E_0 \exp\left\{-\left(\frac{y}{a_0}\right)^q\right\} e^{i(k_j x - \omega_j t)} \tag{4.31}$$

The profile of this super-Gaussian laser beam is shown in Figure 4.19, where a comparison of the same is also made with the Gaussian profile of the beams. The laser index 2, i.e. $q = 2$, in Eq. (4.34) gives rise to the Gaussian profile, where the whole field or the intensity is localized on the axis (Figure 4.19). However, the maximum field or the intensity stays for larger space around the axis in a super-Gaussian profile of the beams. This region of maximum intensity / field can be further enhanced by increasing the laser index q. A comparison of the graphs with $q = 4$ and 6 reveals that the gradient in the field or the intensity for the beams having larger values of q stays stronger. This property of the super-Gaussian laser beams is exploited to achieve stronger ponderomotive force for the excitation of strong THz radiation by the larger nonlinear current produced in the plasma.

The THz electric field amplitude obtained in the process of laser beating can be obtained as (Malik and Malik 2012)

$$\left|\frac{E_{0y}}{E_0}\right| = \frac{n_\beta}{n_0}\frac{q}{2}\frac{\omega_p^2\omega^2\left|\vec{v}_2^{\,*}\right|}{\omega_1 a_0\left(\omega^2-\omega_p^2\right)^2}\left(\frac{y}{a_0}\right)^{q-1}\exp\left\{-2\left(\frac{y}{a_0}\right)^q\right\} \tag{4.32}$$

Here $\left|\vec{v}_2^{\,*}\right| = \left|\frac{e\vec{E}_2^{\,*}}{mi\omega_2}\right|$.

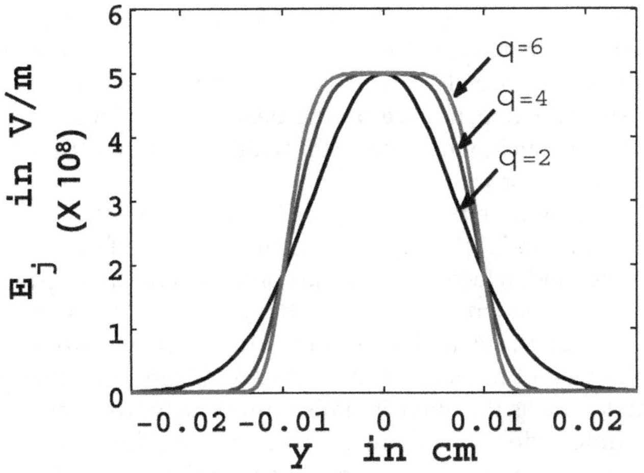

FIGURE 4.19
Comparison of super-Gaussian laser beams ($q = 4, 6$) with Gaussian beam ($q = 2$) for $E_0 = 5 \times 10^8$ V/m.

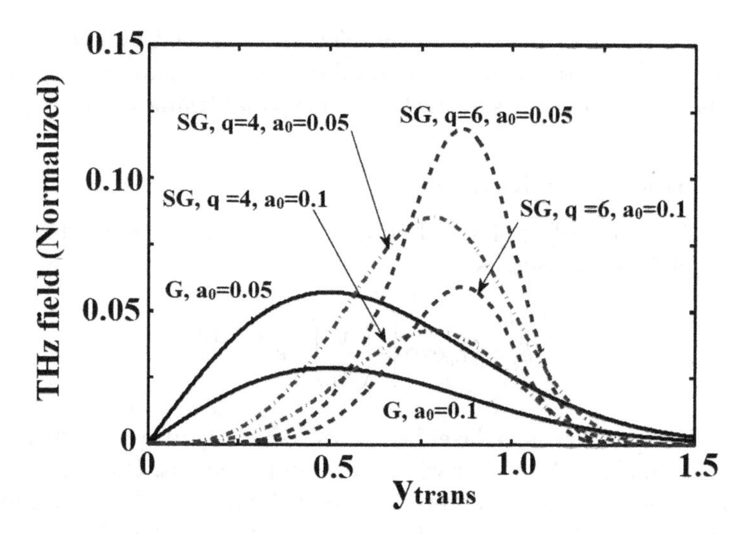

FIGURE 4.20
Transverse profile ($y_{\text{trans}} = y/a_0$) of the THz field, showing the effect of beam width a_0. Here SG is the super-Gaussian laser beam with q as its index and G is the Gaussian beam.

The role of laser index q of the super-Gaussian beams for the generation of strong and collimated THz radiation is uncovered from Figure 4.20, when a comparison is made between the Gaussian beam (marked as G) and super-Gaussian beams (marked as SG), and for the graphs prepared for different values of q, i.e. $q = 4$ and 6 in the case of super-Gaussian beams. The THz field and hence the radiation becomes more collimated due to its larger confinement in the smaller transverse region for the case of $q = 6$. The focus of the radiation changes its place when the value of q is changed. For example, the radiation for the case of super-Gaussian beams changes its position from $y/a_0 = 0.78$ to $y/a_0 = 0.86$ when the value of q is changed from 4 to 6; on the other hand, the radiation is focussed at $y/a_0 = 0.5$ for the Gaussian beam. Another benefit of using super-Gaussian lasers is that the field of emitted THz radiation stays symmetrical about the point of focus, otherwise it is asymmetrical when the Gaussian beams are used. Also, the stronger THz field is achieved for super-Gaussian lasers with a larger index. It means the emitted THz field and the position of its focus can be tuned by varying the laser index for super-Gaussian beams. In addition, if we examine the significance of beam width, we find that the lasers with a higher index parameter show stronger dependence on the beam width. This is in view of the larger enhancement in the THz field when the beam width is reduced from 0.1 to 0.05 mm in the case of $q = 6$ than the case of $q = 4$.

Why is the THz field symmetrical for super-Gaussian lasers? Why is the THz field asymmetrical for Gaussian lasers? Actually, the intensity gradient of the incident lasers modifies the electron dynamics and, hence, the way the nonlinear current is generated. The intensity gradient and hence the ponderomotive force changes its direction when the electrons face the rise of the laser pulse (front) and fall of the pulse (rear). Accordingly, the electrons have to change their direction of motion. For the Gaussian beams, this change has to take place immediately or suddenly as the intensity falls immediately from the axis. The electrons do not find sufficient time to do that and are unable to traverse their original path. This leads to an asymmetry in the current oscillations and, hence, in the THz field. On the other hand, the maximum intensity of the super-Gaussian lasers stays for a larger space where the gradient is zero and, hence, the ponderomotive force vanishes for this point of time. Before and after

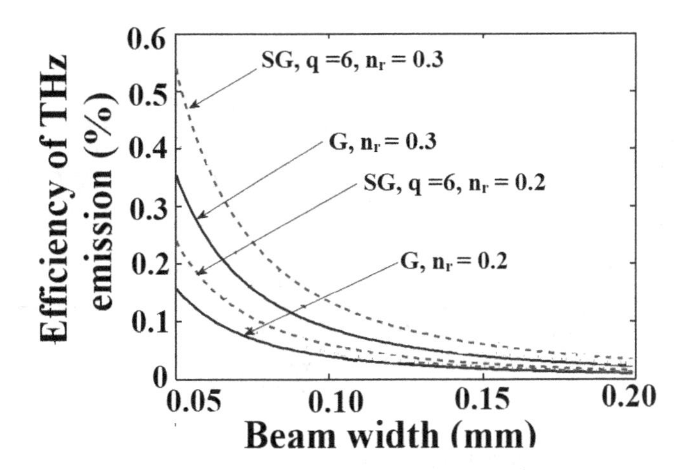

FIGURE 4.21
Efficiency of THz emission with a_0 of Gaussian lasers (G) and super-Gaussian lasers (SG, with index q). Here n_r represents the ratio of n_α and n_0.

this, the direction of the force is opposite. In view of this, the electrons get sufficient time to revert back to their original path and hence cause symmetrical oscillations of the nonlinear current. Finally, a symmetrical distribution of the THz field is realized.

Malik and Malik (2012) also have calculated the efficiency of the THz emission, the variation of which is shown with the beam width in Figure 4.21. To prove the supremacy of the super-Gaussian beams over the Gaussian beams, the graphs are made for both the lasers; it is shown that the efficiency is always higher for super-Gaussian lasers. However, for the super-Gaussian and Gaussian beams, the efficiency of the THz emission decreases for the larger beam width; this effect seems to be prominent in the case of super-Gaussian beams and that too for the beams which have larger index parameters. In addition, the effect of the density ripple's amplitude n_α is discussed for both types of the lasers. Clearly, the efficiency is enhanced with the higher amplitude of the ripples for both the laser beams, which is consistent with the variation of THz field. However, the percentage increase in the efficiency for super-Gaussian beams, when n_r is increased from 0.2 to 0.3, is quite larger than the one for the Gaussian beams. This way, the super-Gaussian profile of the laser proves to be better than the Gaussian profile for the efficient THz generation.

4.5.4 cosh-Gaussian and Skew cosh-Gaussian Laser Beams

The coshyperbolic-Gaussian beams (chG beams) are also called decentered Gaussian beams. In general, their electric field is represented by $E = E_0 \cosh(\Omega_0 x)\exp\left(-\frac{x^2}{a_0^2}\right)$. The parameter Ω_0 associated with the cosh part controls the shape of these beams (Lü et al. 1999; Konar et al. 2007; Patil et al. 2012). With a suitable choice of parameter $\Omega_{0,}$ their profile can resemble closely to the flat-top field distributions (Singh and Malik 2018). On the other hand, the laser beams with an electric field in the form $\vec{E} = E_{0L} \cosh^n\left(\frac{ys}{a_0}\right) e^{-\left(\frac{y}{a_0}\right)^2} e^{i(kz-\omega t)}\hat{y}$ are called skew-chG beams, polarized in the y-direction (Malik 2020). The parameters s and n are called the skewness parameter and the order of the beam. Here $s > 0$, otherwise the expression for $s = 0$ will correspond to the Gaussian beam. These parameters decide the

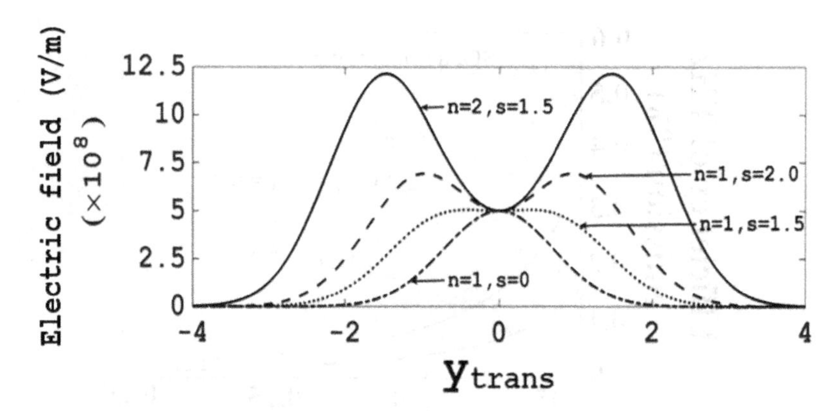

FIGURE 4.22
Electric field profile of skew-coshyperbolic Gaussian beams, controlled by the parameters s and n.

shape of these laser beams (Figure 4.22). For example, a specific combination of s and n ($n = 1$, $s = 2.0$ and $n = 2$, $s = 1.5$) creates an appreciable dip at the peak laser field. The parameter $s = 1.5$ can create a huge skewness for $n = 2$ and the skewness can be further enhanced by raising the values of n. Actually, Hermite-sinusoidal–Gaussian (HSG) beams are one of the solutions of the paraxial wave equation in the rectangular coordinate system. The HSG beams represent the more general beams such as cosh (sinh) beams, cosine (sine) beams, Gaussian beams, and Hermite-Gaussian beams, which can be thought of as their special cases (Casperson et al. 1997; Tovar and Casperson 1998).

4.5.4.1 Nonrelativistic Case

Malik (2020) has obtained bifocal THz radiation and unifocal THz radiation based on these skew-cosh-Gaussian beams. The most interesting part of the study is that there is a mechanism to convert the unifocal radiation into bifocal radiation and the vice-versa (Figure 4.23). Also, the peaks of the emitted THz field can be either enhanced or

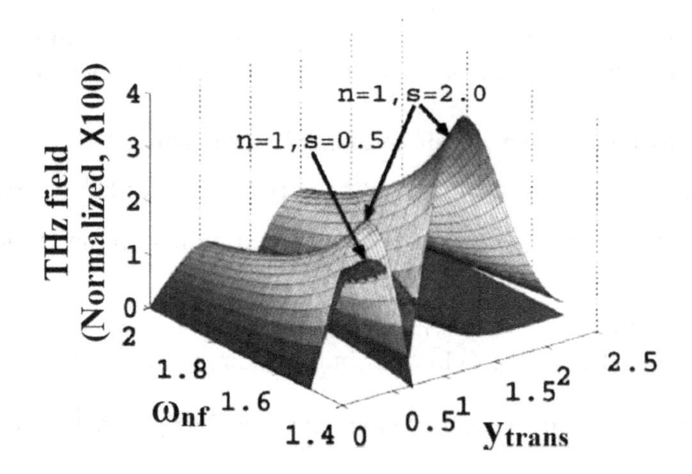

FIGURE 4.23
Transverse profile ($y_{trans} = y/a_0$) of the THz field and its variation with normalized beating frequency $\omega_{nf} = \omega/\omega_p$. The conversion of bifocal to unifocal THz radiation is done based on the parameter s.

suppressed. Another part of this important result is that one can find the position of the THz peak, and its position can be optimized as per the requirement based on the skewness parameter s and order of the beams n. This position is governed by the following expression

$$\frac{y}{a_0} = -\frac{1}{3as}\left(b + \sqrt[3]{\frac{\Delta_1 + \sqrt{\Delta_1^2 - 4\Delta_0^3}}{2}} + \Delta_0\left\{\sqrt[3]{\frac{\Delta_1 + \sqrt{\Delta_1^2 - 4\Delta_0^3}}{2}}\right\}^{-1}\right) \tag{4.33}$$

where $\Delta_0 = b^2 - 3ac$ and $\Delta_1 = 2b^3 - 9abc + 27a^2d$ together with $a = 16n - \frac{8}{3s^2}$, $b = -4n$, $c = 2 - ns^2$, and $d = 4n$. This is clear from the above relation that the foci of the THz field and radiation can be controlled by the parameters s and n.

4.5.4.2 Relativistic Case

Gill et al. (2017) have considered the relativistic case for the THz emission, which means the intensity of the lasers is taken such that the motion of the electrons becomes relativistic and there is a variation in their mass. They have done the calculations based on the beating of two linearly polarized cosh-Gaussian beams whose electric fields are polarized along the y-axis, i.e. when $\vec{E}_j = E_0 \cosh\left(\frac{ys}{b}\right)\exp\left(-\left(\frac{ys}{b}\right)^p\right)\exp\left(i\left(k_j z - \omega_j t\right)\right)\hat{y}$ together with skewness parameter s, laser index p, beam width b, and $j = 1, 2$ for two different lasers. The beams profile is cosh-Gaussian when $p > 2$.

Here, the relativistic effects are considered due to high intensity of the incident laser beams. Relativistic ponderomotive force for an intense electromagnetic beam can be expressed as $\vec{F}_p = -mc^2\vec{\nabla}(\gamma - 1)$ with $\gamma = \left(1 - \frac{|v_1 + v_2|^2}{c^2}\right)^{-1/2}$, which yields

$$\vec{F}_p^{NL} = \frac{mA}{\omega_1\omega_2}\left(1 - \frac{2A}{\omega_1\omega_2}\cos(kz - \omega t)\right)^{-3/2}$$
$$\times\left[\left(2p\left(\frac{ys}{b}\right)^{p-1}\frac{s}{b} - \frac{s}{b}\tanh\left(\frac{ys}{b}\right)\right)\cos(kz - \omega t)\hat{y} + k\sin(kz - \omega t)\hat{z}\right] \tag{4.34}$$

where m is the rest mass of the electron, ω_p is the plasma frequency, and v_1 and v_2 are oscillatory velocities of the electrons.

The wave equation, discussed earlier, can be used to obtain the field of the emitted THz radiation as

$$E_{THzy} = \frac{-\pi i\omega mzen_\alpha Ae^{i\alpha z}}{c^2k'\omega_1\omega_2}\left[1 - \frac{2A}{\omega_1\omega_2}\cos(kz - \omega t)\right]^{-\frac{3}{2}} CD\cos(kz - \omega t) \tag{4.35}$$

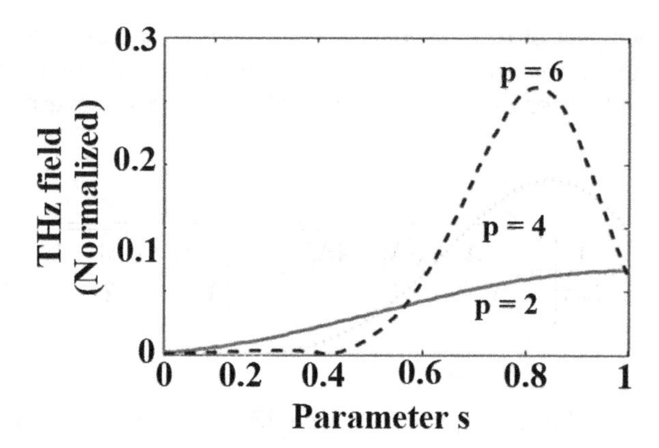

FIGURE 4.24
Role of skewness parameter s in controlling the emitted THz field in plasma under relativistic effects. The field maximizes at a particular value of s and its position can be controlled based on the index p.

where $A = \frac{e^2 E_0^2}{m^2}\cosh^2\left(\frac{ys}{b}\right)\exp\left(-2\left(\frac{ys}{b}\right)^p\right)$

$$C = \left(2p\left(\frac{ys}{b}\right)^{p-1}\frac{s}{b} - \frac{s}{b}\tanh\left(\frac{ys}{b}\right)\right)$$

$$D = \frac{i\omega}{m(\omega^2 - \omega_p^2)}$$

Figure 4.24 is plotted to uncover the role of skewness parameter s and laser index p. The skewness is a characteristic of the cosh-Gaussian laser profile and the field of emitted THz radiation varies with it. For a particular combination of skew parameter and laser index, i.e. when $p = 6$ and $s = 0.8$, the THz field peaks, which means the stronger THz radiation is possible for a particular combination of p and s. Interestingly, the position of the peak can be controlled based on the index p.

The transverse profile of the THz field is shown in Figure 4.25, where the effect of laser index p is investigated and the role of cosh-Gaussian beams is evident in producing collimated and more focussed radiation. When the index p is enhanced, the THz field is found to be enhanced very significantly and is more focussed, which means the field is confined to a small region. Hence, the role of p is to produce stronger and more collimated THz radiation, when the intensity of lasers is high enough to produce relativistic effects.

4.6 Control of Polarization of THz Radiation

So far we have discussed THz radiation generation and the roles of laser pulse shape and external magnetic field in controlling the frequency, power, and focus of the radiation. Since the control of polarization of the emitted THz radiation is also a burning issue

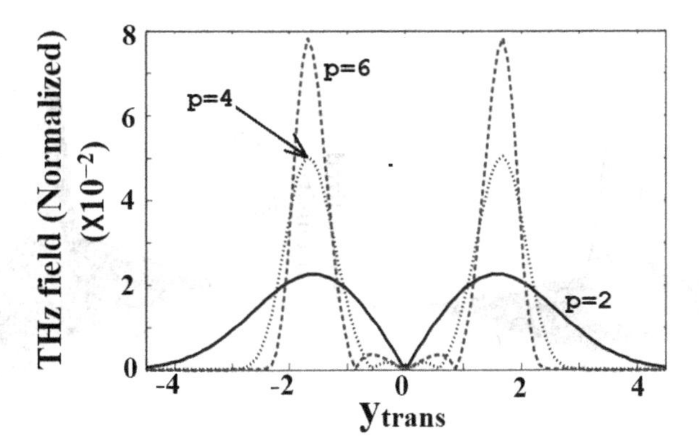

FIGURE 4.25
Transverse profile of the THz field (normalized), showing the effect of the laser index in producing collimated and more focussed radiation.

these days, we will discuss an interesting method to control the polarization (Punia and Malik 2019). In this section, we review the laser beating process used by the laser beams with a doughnut profile in plasma and modulated / periodic density. The control of polarization is based on the simultaneous application of two kinds of magnetic fields, helical solenoidal and wiggler. These types of fields can be produced by the superposition of the axial field on the wiggler field pattern. This is done with the help of the solenoid, which produces a magnetic field in the axial direction. For this reason, this field is also called a solenoidal field. The electrons will start revolving around the direction of propagation of the laser beams due to the influence of the transverse magnetic field under the helical wiggler field. This kind of magnetic field not only extends the time duration for the particles (electrons) and laser interaction, but also causes them to retain a significant amount of energy due to which intense radiation is emitted in view of the larger current produced in plasma.

The dark hollow Gaussian beams (Figure 4.26) are taken with their electric field as (Punia and Malik 2019)

$$\vec{E}_j = E_0\left(\frac{y^2}{\beta^2}\right)^{\mathcal{O}_j} \exp(-\left(\frac{y^2}{\beta^2}\right))e^{i(\xi_j z-\omega_j t)}\ \hat{y} \tag{4.36}$$

where β is the beam width, $\mathcal{O}$ is the beam order, ξ is the wave number, ω is the angular frequency, and E_0 is the amplitude of the field of the laser beams. In laser beams with the above profile (possessing a dark spot at the centre, i.e. minimum intensity there, and whose spot size increases with the order of the beam), the reduced interaction area of electrons with the laser pulse (than that of plane-wave approximation) is realized owing to the Gaussian nature of the beams. With the application of the solenoidal field, better collimation of the electrons is achieved, which dominates the process of acceleration. The periodic nature of the wiggler magnetic field and its controllable strength fine-tunes the polarization of emitted THz radiation. The configuration of such a magnetic field has also been employed for controlling gyratory motion of the electrons in free electron lasers (El-Bahi et al. 2002).

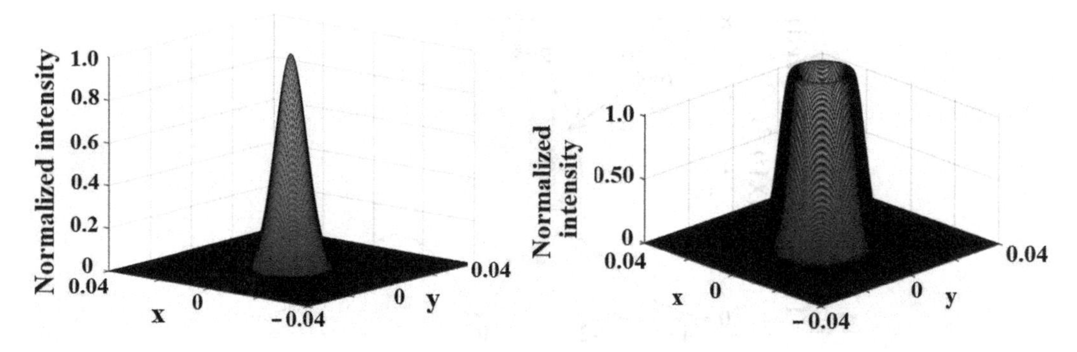

FIGURE 4.26
Intensity profile of Gaussian beam (left) and dark hollow Gaussian beam (right).

Considering the wiggler period as δ_w, the wiggler wavelength λ_w can be given by $\lambda_w = \frac{2\pi}{\delta_w}$. Further, taking the maximum strength of the solenoidal and helical fields as B_s and B_w, the combined magnetic field can be expressed as

$$\vec{B} = B_s\hat{z} + B_w\left[\cos(\delta_w z)\,\hat{x} + \sin(\delta_w z)\hat{y}\right] \tag{4.37}$$

This expression shows the circularly polarized nature of the wiggler magnetic field, which will enable violent acceleration of the electrons by providing them helical motion contrary to the static magnetic field. The role of the solenoidal field, applied in the z-direction, is in controlling the transverse velocity / motion of the electrons. The magnetic field, defined by Eq. (4.37), will lead the electrons to make a spiral trajectory, where their helical motion is provided by the wiggler field and their transverse motion is controlled by the solenoidal field that also guides beam transportation.

Following Punia and Malik (2019), the impact of combined wiggler and solenoidal magnetic fields on the emitted radiation (normalized THz field) is shown in Figure 4.27. Clearly, the increasing solenoidal field or the enhanced solenoidal cyclotron frequency of the electrons leads to the THz emission with a stronger field. The same result, i.e. the stronger THz emission with the increasing strength of the wiggler magnetic field, through the increased wiggler cyclotron frequency of electrons Ω_w, is seen in Figure 4.27. The figure also shows that increasing both the magnetic fields leads to very strong THz emission. The enhancement in the THz field with the magnetic fields is due to the greater enhancement of the nonlinear current generated due to the laser beating in plasma.

In the said mechanism, the THz radiation field stays in the x-y plane. The polarization angle is defined as the ratio of amplitudes of the THz field in the y-direction and that in the x-direction, i.e. the ratio E_{0THzy}/E_{0THzx}. The role of wiggler period or the wiggler wavelength in exciting stronger THz radiation is uncovered in Figure 4.28. Since the pattern of helical wiggler field turns out to be spiral geometry around the axis of the beam propagation, the pitch of this geometry can be manipulated by changing the wiggler wavelength or the period. This will affect very significantly the dynamics of the electrons in plasma. Increasing the wavelength of the wiggler field causes electrons to be efficient in

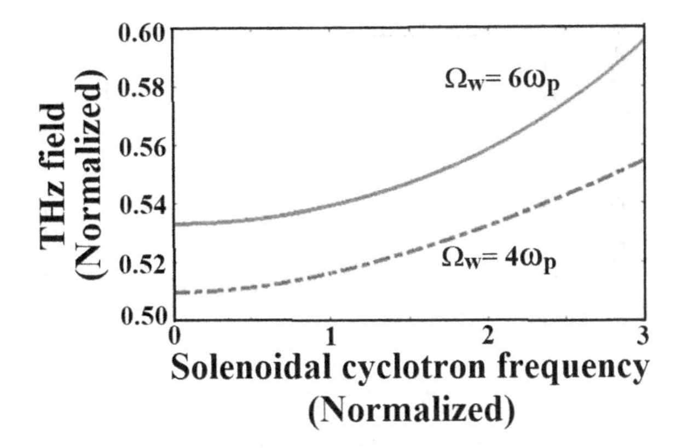

FIGURE 4.27
THz field (normalized) versus solenoidal cyclotron frequency, showing the effect of wiggler cyclotron frequency.

the x-direction, whereas the electrons tend to be more in the y-direction on the reduction of this wavelength. This way, one can control the polarization of the THz field by manipulating the wiggler field through a control on the nonlinear current excited by the nonlinear motion of the electrons. When the electrons ought to be in the x-direction, the angle of polarization decreases from 90° to 0° due to the increasing value of the x-component of the current density, which leads to the enhanced E_{0THzx}. Hence, the THz radiation attains polarization in the y-direction (x-direction) when the wiggler is formed at closer (larger) distances.

It is not only the wiggler wavelength that changes the polarization of the emitted radiation. The wiggler cyclotron frequency or simply the wiggler field strength also plays a vital role in controlling the THz polarization. Figure 4.29 unveils the importance of the same (wiggler field) for different values of wiggler wavelength or the wiggler period.

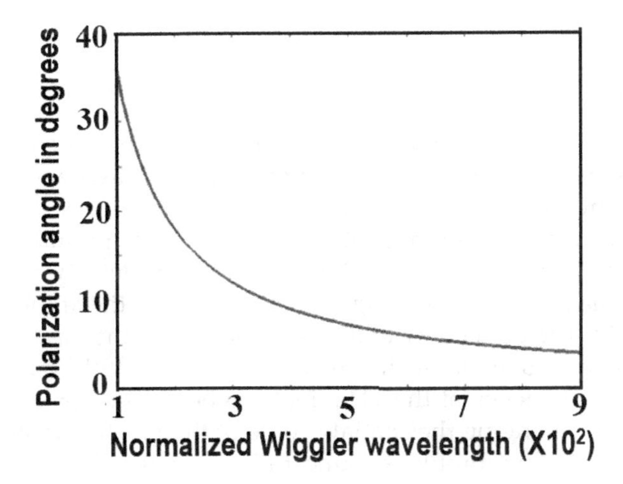

FIGURE 4.28
Control of the polarization angle of the THz field by wavelength of the wiggler field.

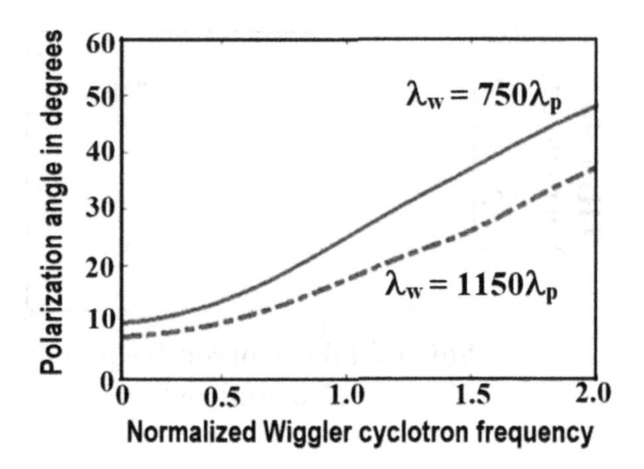

FIGURE 4.29
Control of polarization angle of the THz field by the cyclotron frequency and wavelength of wiggler field.

The increasing wiggler cyclotron frequency leads to the enhanced polarization angle of the emitted THz radiation for both the wiggler wavelengths, which means the direction of polarization of the THz field moves towards the *y*-axis with the increased wiggler field. Actually, with the larger cyclotron motion, the electrons experience an effect similar to that of the shorter wiggler wavelength and should be towards the *y*-axis, leading to the larger value of the *y*-component of the current density. This increases the *y*-component of the THz electric field E_{0THzy}, leading to the polarization towards the *y*-axis. Theoretically, the *x*- and *y*-components of the THz field depend on the wiggler cyclotron frequency and the *x*-component decreases with its increasing values. Opposite dependence of these THz field components is seen on the wiggler wavelength; hence, the polarization angle decreases with the increased wavelength of the wiggler field.

4.7 Conclusions

Laser beating mechanism allows the generation of reliable and efficient THz radiation whose frequency and power can be tuned with the help of laser and plasma parameters. At resonance condition, the beating frequency matches well with the plasma frequency (upper hybrid frequency) in unmagnetized (magnetized) plasma and the maximum transfer of energy takes place. The higher amplitude of density ripples and choosing the laser with a smaller beam width significantly enhance the amplitude of the emitted THz radiation field. The periodicity of density ripples also increases the THz field when the ripples are generated at closer distances, in view of fulfilling the resonance condition. Moreover, the efficiency of the THz field can be tailored by an externally applied magnetic field or by changing the spatial profile of the incident laser pulse. The spatial symmetry of the THz field profile is found to be secured with super-Gaussian beams compared with the other beams, and the collimated and strong radiation is achieved with the super-Gaussian lasers having a higher index. The application of an electrostatic field periodic in the direction of laser propagation and the density ripples produces and controls three kinds of THz radiation fields. The polarization of the emitted THz field can be controlled with the application of a magnetic field obtained by the combination of solenoidal and wiggler fields.

4.8 Selected Problems and Solutions

PROBLEM 4.1

For the generation of THz radiation by the laser beating process, two Gaussian laser beams (beam width $b = 100\ \mu$m) are co-propagating into plasma with different frequencies and wave numbers but the same amplitude $E_0 = 2.0 \times 10^{10}\ \text{Vm}^{-1}$. Calculate the ponderomotive force raised during laser-matter interaction for beating frequency $\omega_b = 4.5 \times 10^{13}$ rad/s at $y = b$.

SOLUTION

The ponderomotive force is given by $F_p = -\frac{e^2}{4m\omega^2}\nabla\left(E^2\right)$

The electric field profile of Gaussian laser pulses in one-dimension can be taken as

$$E = E_0 e^{-\left(\frac{y^2}{b^2}\right)} e^{i(k\,z-\omega\,t)}$$

Then $F_p = \frac{e^2 y}{b^2 m\omega^2} E_0^2 e^{-2\left(\frac{y^2}{b^2}\right)} = \frac{e^2}{bm\omega_b{}^2} E_0^2 e^{-2}$

$$F_p = \frac{(1.6\times10^{-19})^2 \times (2\times10^{10})^2 (0.13533)}{\left(100\times10^{-6}\right)(9.1\times10^{-31})(4.5\times10^{13})^2}$$

$$F_p = 7.52\times10^{-12} N$$

PROBLEM 4.2

The dark hollow Gaussian laser beam with an amplitude $E_0 = 5.0 \times 10^{10}\ \text{Vm}^{-1}$ and pulse duration of a few *ps* with beam width 10 μm is incident on plasma for the generation of THz radiation. Calculate the corresponding power emitted in the form of THz radiation.

SOLUTION

The power can be estimated using $P = I \cdot A$. The laser's intensity is determined by $I = \frac{1}{2}c\epsilon_0 E_0^2 = 3.31\times10^{14}\ \text{W}/\text{cm}^2$.

The electric field profile of dark hollow Gaussian laser pulse in one-dimension can be considered as

$$E = E_0\left(\frac{y^2}{b^2}\right)^n e^{-\left(\frac{y^2}{b^2}\right)} e^{i(k\,z-\omega\,t)}$$

Hence, the area under the curve can be calculated as $A = \int_{-\infty}^{+\infty} \frac{y^2}{b^2} e^{-\frac{y^2}{b^2}} .dy = \frac{b}{2}\sqrt{\pi}$ where b is the beam width of the laser. Hence, the emitted power can be calculated using the formula

$$P = I \cdot \int_{-\infty}^{+\infty} \frac{y^2}{b^2} e^{-\frac{y^2}{b^2}} .dy$$

$$\cong 29 \text{ TW (for beam order } n = 1)$$

Suggested Reading Material

Casperson, L. W., Hall, D. G., & Tovar, A. A. (1997). Sinusoidal-gaussian beams in complex optical systems. *JOSA A, 14*(12), 3341–3348.

Cherezova, T. Y., Chesnokov, S. S., Kaptsov, L. N., & Kudryashov, A. V. (1998). Super-Gaussian output laser beam formation by bimorph adaptive mirror. *Optics Communications, 155*(1–3), 99.

Chou, J., Han, Y., & Jalali, B. (2003). Adaptive RF-photonic arbitrary waveform generator. *IEEE Photonics Technology Letters, 15*(4), 581–583.

El-Bahi, R., Rhimi, M. N., & Cheikhrouhou, A. W. (2002). Electron trajectories in a helical free-electron laser. *Brazilian Journal of Physics, 32*(3), 790–797.

Gill, R., Singh, D., & Malik, H. K. (2017). Multifocal terahertz radiation by intense lasers in rippled plasma. *Journal of Theoretical and Applied Physics, 11*(2), 103–108.

Hazra, S., Chini, T. K., Sanyal, M. K., Grenzer, J., & Pietsch, U. (2004). Ripple structure of crystalline layers in ion-beam-induced Si wafers. *Physical Review B, 70*(12), 121307.

Keldysh, L. V. (1965). Diagram technique for nonequilibrium processes. *Sov. Phys. JETP,* 20(4), 1018–1026.

Konar, S., Mishra, M., & Jana, S. (2007). Nonlinear evolution of Cosh-Gaussian laser beams and generation of flat top spatial solitons in cubic quintic nonlinear media. *Physics Letters A, 362*(5–6), 505–510.

Kuo, C. C., Pai, C. H., Lin, M. W., Lee, K. H., Lin, J. Y., Wang, J., & Chen, S. Y. (2007). Enhancement of relativistic harmonic generation by an optically preformed periodic plasma waveguide. *Physical Review Letters, 98*(3), 033901.

Larochelle, S., Talebpour, A., & Chin, S. L. (1998). Non-sequential multiple ionization of rare gas atoms in a Ti: Sapphire laser field. *Journal of Physics B: Atomic, Molecular and Optical Physics, 31*(6), 1201.

Lü, B., Ma, H., & Zhang, B. (1999). Propagation properties of Cosh-Gaussian beams. *Optics communications, 164*(4–6), 165–170.

Malik, H. K. (2020). Generalized treatment of skew-Cosh-Gaussian lasers for bifocal terahertz radiation. *Physics Letters A,* 384(15), 126304.

Malik, H. K., & Malik, A. K. (2011). Tunable and collimated terahertz radiation generation by femtosecond laser pulses. *Applied Physics Letters, 99*(25), 251101.

Malik, H. K., & Malik, A. K. (2012). Strong and collimated terahertz radiation by super-Gaussian lasers. *Europhysics Letters, 100*(4), 45001.

Malik, A. K., Malik, H. K., & Kawata, S. (2010). Investigations on terahertz radiation generated by two superposed femtosecond laser pulses. *Journal of Applied Physics, 107*(11), 113105.

Malik, A. K., Malik, H. K., & Nishida, Y. (2011a). Terahertz radiation generation by beating of two spatial-Gaussian lasers. *Physics Letters A, 375*(8), 1191–1194.

Malik, A. K., Malik, H. K., & Stroth, U. (2011b). Strong terahertz radiation by beating of spatial-triangular lasers in a plasma. *Applied Physics Letters, 99*(7), 071107.

Malik, A. K., Malik, H. K., & Stroth, U. (2012). Terahertz radiation generation by beating of two spatial-gaussian lasers in the presence of a static magnetic field. *Physical Review E, 85*(1), 016401.

McLaughlin, R., Corchia, A., Johnston, M. B., Chen, Q., Ciesla, C. M., Arnone, D. D., ... & Pepper, M. (2000). Enhanced coherent terahertz emission from indium arsenide in the presence of a magnetic field. *Applied Physics Letters, 76*(15), 2038–2040.

Patil, S. D., Takale, M. V., Navare, S. T., Fulari, V. J., & Dongare, M. B. (2012). Relativistic self-focusing of cosh-Gaussian laser beams in a plasma. *Optics & Laser Technology, 44*(2), 314–317.

Perelomov, A. (2012). *Generalized coherent states and their applications*. Springer Science & Business Media.

Punia, S., & Malik, H. K. (2019). Controlling polarization of THz radiation in pair plasma. *Plasma Sources Science and Technology, 28*(11), 115018.

Rothwell, E. J. and Cloud, M. J. (2009). *Electromagnetics*. CRC Press.

Singh, D., & Malik, H. K. (2018). Generation of terahertz radiations by flat top laser pulses in modulated density plasmas. *Plasma and Fusion Science*, Apple Academic Press, p. 385–395.

Singh, D., & Malik, H. K. (2020). Terahertz emission by multiple resonances under external periodic electrostatic field. *Physical Review E, 101*(4), 043207.

Tovar, A. A., & Casperson, L. W. (1998). Production and propagation of Hermite–sinusoidal-gaussian laser beams. *JOSA A, 15*(9), 2425–2432.

Wang, H., Latkin, A. I., Boscolo, S., Harper, P., & Turitsyn, S. K. (2010). Generation of triangular-shaped optical pulses in normally dispersive fibre. *Journal of Optics, 12*(3), 035205.

Wang, Z., Zhang, Z., Xu, Z., & Lin, Q. (1997). Space-time profiles of an ultrashort pulsed gaussian beam. *IEEE Journal of Quantum Electronics, 33*(4), 566–573.

Wu, H. C., Sheng, Z. M., Dong, Q. L., Xu, H., & Zhang, J. (2007). Powerful terahertz emission from laser wakefields in inhomogeneous magnetized plasmas. *Physical Review, 75*(1), 016407.

Ye, J., Yan, L., Pan, W., Luo, B., Zou, X., Yi, A., & Yao, S. (2011). Photonic generation of triangular-shaped pulses based on frequency-to-time conversion. *Optics, 36*(8), 1458–1460.

5

Terahertz Radiation and Its Detection

5.1 Introduction

The topic of terahertz (THz) radiation continues to attract interest from the scientific community, not just because it remains the only part of the electromagnetic spectrum that has not been covered in its entirety, but because of its numerous application prospects in both pure and applied sciences. Ultrafast spectroscopy of semiconductors and atoms (Beard et al. 2002; Segschneider et al. 2002; Hangyo et al. 2005), nonlinear spectroscopy (Gaal et al. 2006), material characterization and counterfeit detection (Ahi and Anwar 2016), medical diagnostics (Sun et al. 2011), wireless transmission (Ishigaki et al. 2012), and THz imaging (Jiang and Zhang 1998l; Dobroiu et al. 2006) are only a few examples of their use. These applications arise out of its unique placement between infrared (IR) and microwave bands, allowing for high-resolution characterization without the risk of ionization.

Though there are many sources of THz radiation, they occur naturally in the blackbody spectrum of any object at a temperature above 2 K. As dicussed in earlier chapter, there are reports of semiconductors generating THz waves through the Josephson effect and through the photo-Dember effect in InAs (Liu et al. 2006); high band gap semiconductors like ZnSe, ZnTe, GaSe, and $LiNbO_3$ produce these waves on interacting with lasers (Faure et al. 2004); and optical rectification of laser pulses in second-order nonlinear crystals (Löffler et al. 2005) is considered a standard method in low-power applications like THz pump-probe spectroscopy. None can be considered suitable for handling high-intensity waves. Thus, an efficient method of THz generation, which is affordable yet compact and able to support high-power applications, is difficult to find. Plasma is thus the obvious substitute, since it has a high nonlinearity and is already broken down, so there is little or no concern about material damage in high-power applications such as detailed above.

The use of plasma as a THz source can be first credited to the pioneering work of Hamster et al. (1993, 1994) who took advantage of wakefields produced through the nonlinear ponderomotive force to generate THz pulses. Other research groups built on their efforts and soon a variety of plasma-based methods emerged, such as producing THz radiation through Cherenkov wakes in magnetized plasma (Yoshii et al. 1997), through tunnel ionization of a gas jet (Malik et al. 2010), through beating of Gaussian beams in spatially periodic plasma (Malik et al. 2011), through synchrotron radiation (Byrd et al. 2006) and through nanoparticle-based media (Sharma et al. 2020). Some other groups focussed on the optical nonlinearity of plasma to produce broadband THz pulses through four-wave rectification (FWR; Cook and Hochstrasser 2000), which was soon extended to produce broadband radiation using an external direct current (DC) bias to the plasma region (Löffler et al. 2000; Löffler and Roskos 2002), and using two-colour lasers [alternating current (AC) bias] in a laser-induced plasma (Cook and Hochstrasser 2000; Li et al. 2016).

Today, the research on THz radiation detectors (Sizov 2018) is conducted for the use of graphene plasmons to detect resonant THz (Bandurin et al. 2018), detection of THz in two-dimensional materials (Murphy et al. 2018), use of THz radiation to detect the methylation in blood cancer DNA (Cheon et al. 2019), the sensitive and fast detection of THz with the help of an antenna-integrated graphene pn-junction (Castilla et al. 2019), emission of THz by multiple resonances under the periodic external electrostatic field (Singh and Malik 2020), and generation of bifocal THz radiations (Malik 2020).

5.2 Production of THz Radiation

There are basically three different kinds of methods to produce THz radiation through laser-plasma interaction. Though all three utilize high-intensity lasers, their effect on the plasma and hence on the type of radiation produced are quite different. These methods are described in detail in the following subsections.

5.2.1 THz Generation Through Laser-Wakefield Oscillations in Plasma

This is the first method of plasma-based THz generation, and it relies on wakefields produced by high-intensity laser pulses. It was first reported by Hamster et al. (1993, 1994), and is, incidentally, also the first account of THz generation by laser-plasma interaction. They shot femtosecond pulses with mJ energy at both solid and gaseous targets to first generate a plasma, which then interacted with the intense laser field to generate a wakefield that oscillated at the plasma frequency. This oscillation led to transient THz pulses lasting a few picoseconds in the time domain (resonant with the plasma frequency), which corresponded to broadband THz radiation in the frequency domain.

In the first stage of experimentation, the team shot 0.5 J laser pulses with 120 fs pulse width at 0.8 μm into He gas taken at various pressures. The objective was to confirm the presence of a strongly resonant time-domain signal when the condition $\omega_P \tau_0 \gg 2$ was satisfied. Here, ω_P represents the oscillation frequency of the electron plasma density wakefield, which is created due to the perturbation caused by the ponderomotive force of the laser, and τ_0 is the full-width at half maximum (FWHM) of the laser pulse. As expected, a few-cycle time-domain signal (~1.7 ps) at the plasma frequency was observed for smaller electron plasma densities, but non-resonant signals centred around 1.5 THz were noticed for higher densities (Figure 5.1), which were attributed to nonlinear mixing of the ω_0 and $\omega_0 \pm \Delta\omega$ side bands, where $\Delta\omega$ is the laser bandwidth.

The second stage of experimentation was basically the same as above, but He was replaced by a solid target of Al-coated glass. The incident pulse energy was 200 mJ and the resultant THz pulse power was in the excess of 1 MW. Incidentally, the generation of THz radiation was strongly correlated with the generation of X-rays and MeV energy electrons as well, which proved the presence of a laser wakefield. The experiments conducted by Hamster et al. (1994) proved that laser-produced wakefields also could be used as THz sources, and added to the already well-known application in particle acceleration; the subject attracted a great deal of attention from the scientific community.

Löffler et al. (2000) improved on this design by introducing a high-voltage DC bias at the site of plasma formation. They found that by applying a bias field of around 10.6 kV/cm, THz pulses can be emitted with almost the same intensity as those radiated by large-area intrinsic GaAs emitters. Also, the frequency-domain spectra peaked at a slightly higher frequency compared with the biased large-area GaAs emitter.

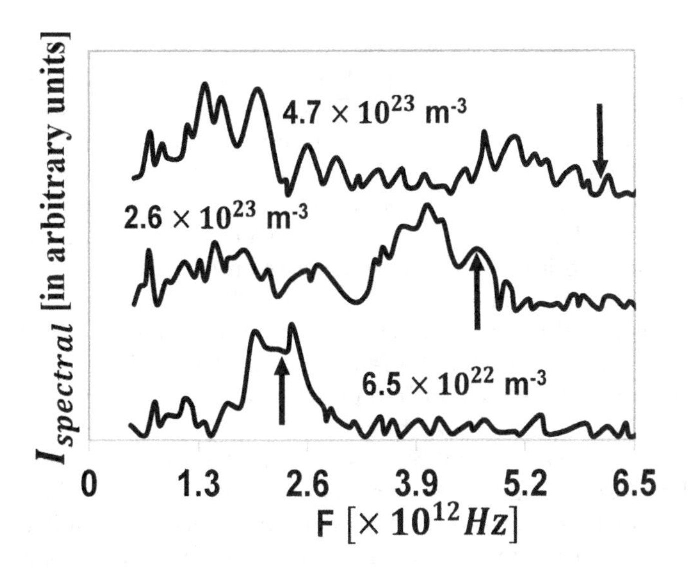

FIGURE 5.1
The arrows point at the plasma frequency for every density taken. Clearly, for lower density (6.5 × 10^{22} m^{-3}) the THz transients are resonant with the electron plasma frequency, while for higher density (4.7 × 10^{23} m^{-3}), the non-resonant side bands centred at 1.5 THz are prominent.

Löffler et al. (2000) conducted a comparative study of air-plasma emitters with large-area GaAs emitters. They observed that the emitted THz signals have 60% more peak amplitude compared with the case of air plasma. The biasing of GaAs emitters with 1 kV/cm revealed that the amplitude of emitted signals is higher in one order of magnitude than the radiations produced from unbiased GaAs and biased air (Figure 5.2). The essential

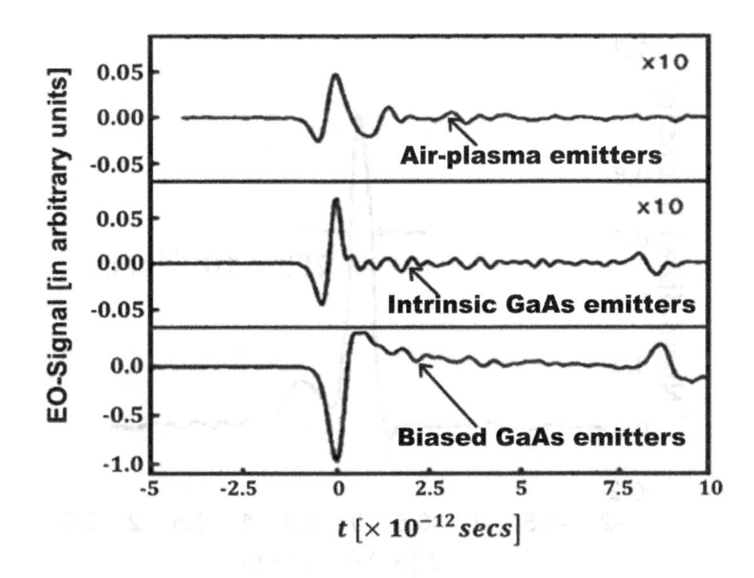

FIGURE 5.2
The time-domain signal for high-voltage DC-biased (10.6 kV/cm) air-plasma emitters, intrinsic GaAs emitters, and biased GaAs emitters.

measurement for the adoption of emitters in imaging or spectroscopy systems is the signal-to-noise ratio (SNR) in Fourier spectrum. Therefore, the same group also investigated the amplitude of SNR as a function of THz frequency for these three cases, i.e. biased air, intrinsic GaAs emitters, and biased GaAs emitters. In the frequency range lower than 1.5 THz, there was an increment in the SNR for biased GaAs emitters. The SNR characteristics of plasma emitters were found to be similar to biased GaAs emitters in the frequency spectrum of 1–3 THz; however, the SNR values were lower by a factor of 3 for plasma emitters. The plasma emitters and intrinsic GaAs emitters are comparable in the lower frequency range (for more details, see Löffler et al. 2000).

The method of generating few-cycle THz radiation resonant with the plasma frequency was novel and useful, but it suffered a few basic problems. The first problem is obvious: the method failed to produce the required radiation at higher electron plasma density. The second, and more important issue with the method was that since it relied on the plasma density for operation, it was susceptible to significant losses on part of the radiated waves, especially if the plasma length was more than the skin depth. Although this is exactly what made the process appealing with respect to tunability, it prohibited one from collecting energy from a larger plasma volume (Hamster et al. 1993). This problem was taken care of by Yoshii et al. (1997), who used a static and homogeneous magnetic field in conjunction with a high-intensity laser in plasma to create a Cherenkov wake that not only was both electromagnetic and electrostatic ($\vec{k} \cdot \vec{E} \neq 0$ and $\vec{k} \times \vec{E} \neq 0$), but also moved at non-zero group velocity so the THz radiation could gain energy by covering a greater volume. In addition, since the same basic principle was being used here, the tunability was not lost. And finally, unlike the previous case where the energy was being transmitted into a general 2π angle, here the radiation focussed in a small area centred on the axis. The experimental verification was made by Yugami et al. (2002), and the typical time-domain waveform obtained is shown in Figure 5.3.

It is interesting to note that Hamster et al. (1994) could not properly explain the occurrence of non-resonant THz pulses in their experiment with a solid target, or the observed

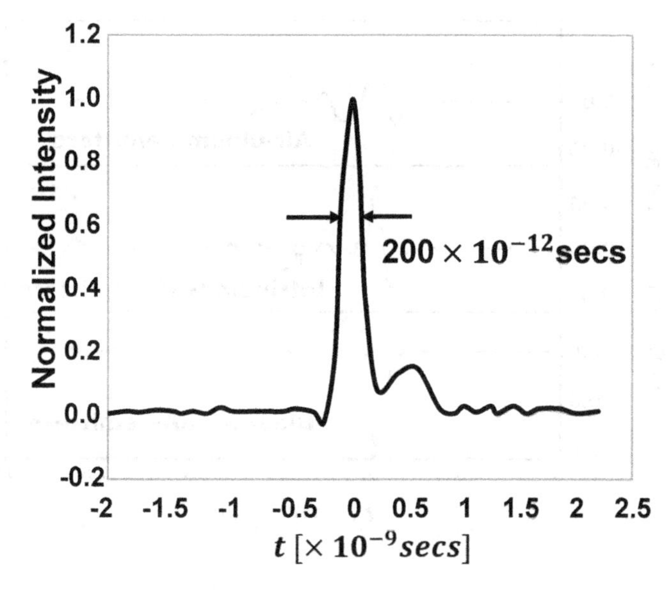

FIGURE 5.3
Typical radiation waveform of radiation obtained from Cherenkov wakes detected by a crystal detector.

polarization of the radiation. Both of these were partially explained by Sheng et al. (2005), when they analyzed the generation of THz pulses through similar means, but slightly different than what Hamster used. They assumed the presence of inhomogeneous plasma and a laser pulse incident obliquely on the plasma, and they proved through linear mode conversion that the wakefield produced by the ponderomotive force of the intense laser could generate THz pulses at the plasma frequency. They confirmed the validity of their results by comparing them with 2D PIC simulations. They also proved that the radiation was tunable and could be scaled up to high-power applications simply by raising the laser field strengths to a few GV/m. Figure 5.4 shows the time-domain waveform for different angles of incidence [Figure 5.4(I)], and emission spectra of the radiation for different angles of incidence [Figure 5.4(II)]. Both the analytical model and PIC simulation plots are shown in this figure and a good agreement between them is observed. The scale of emission spectra is proportional to the fourth power of laser intensities. At higher intensities, the peaks in the emission spectrum are smaller than the scaling of the fourth power of laser intensities because there is no longer proportionality of wakefield amplitude to the second power of laser intensities. In other words, breaking of plasma waves occurred at very high intensities. Consequently, a complicated spectrum resulted (for more details, see Sheng et al. 2005).

Through their results, Sheng et al. (2005) further validated the importance of laser wakefields in both particle acceleration and THz emission, and described a method to measure the wakefield using the emitted radiation. This provided a definitive link between plasma-based particle acceleration and the generation of THz radiation.

5.2.2 Terahertz Generation in Plasma With Density Perturbations

The explanation given by Sheng et al. (2005) about the Hamster et al. (1994) results using an inhomogeneous plasma density paved the way for a long line of methods that improved on the latter's original paper by assuming the action of the ponderomotive force on inhomogeneous / corrugated plasma. They now fall under an entirely different class of plasma-based THz emission processes that are mediated by a laser ponderomotive force. The presence of a perturbed density profile is the characteristic of all these techniques.

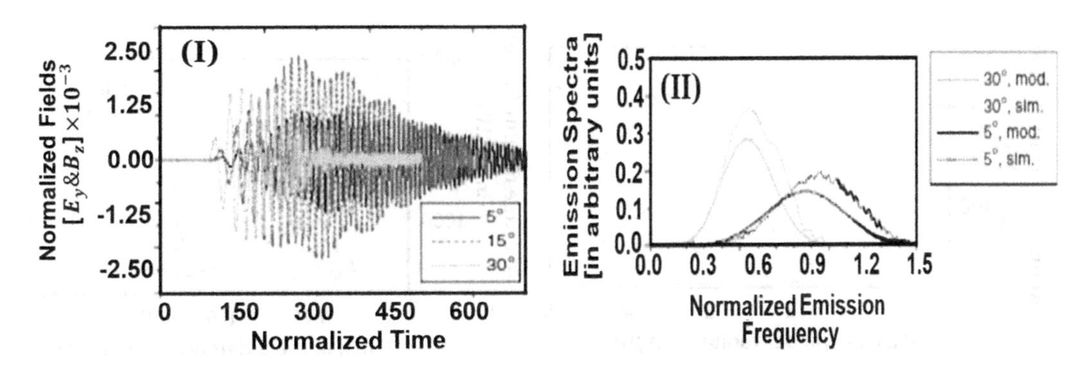

FIGURE 5.4
(I) Time-domain waveform for different angles of incidence and (II) comparison of frequency-domain emission spectra between particle-in-cell (PIC) simulations and analytical model for different angles of incidence. Here, mod. and sim. correspond to analytical model and PIC simulations, respectively.

In many later reports (Malik et al. 2010, 2011), further developments were made by replacing intense pulses of a single kind with superposed pulses. Bystrov et al. (2005) proposed an analytical model where THz pulses were generated due to tunnel ionization of air by a single intense laser after passing through an axicon, and then Malik et al. (2010) built on this approach by considering two superposed femtosecond pulses with a phase delay as the source of the ponderomotive force. They found that this phase delay plays an important role in optimizing the rate of tunnel ionization, evolution of plasma density, and, consequently, in residual current due to dipole oscillations. The directionality of the emitted radiation, too, depends on the initial phase difference, and the THz power depends on the incident laser field strength. More studies were made by assuming different polarization combinations of the input pulses, and the results of the THz power as a function of plasma radius and initial phase delay for three polarization combinations were shown (Figure 5.5).

Malik et al. (2011) also developed models in which THz radiation was generated by beating of two spatial-Gaussian lasers in spatially periodic plasma, and concluded that a transverse nonlinear current density produced by the ponderomotive force and which propagated through the plasma was the source of the radiation. The periodic plasma functioned similarly to periodically poled $LiNbO_3$ crystals used for phase matching in nonlinear optics. Other improvements also were made to this approach, like adding a DC magnetic field (Malik et al. 2012) and assuming magnetized collisional plasma (Singh and Malik 2015), all adding to the efficiency of conversion. Though little experimental verification has been made for these methods, they hold great promise for becoming the technological basis of yet another class of standard tabletop THz emitter.

THz emission from plasma with perturbed density profiles accounts for the second method of plasma-based THz generation. However, owing to its great similarity to the first method, it is often considered a subclass of THz generation methods that are mediated by ponderomotive forces of intense lasers.

5.2.3 THz Generation Through Nonlinear Effects in Plasma

This is the third, and final, method of plasma-based THz generation devised by Cook and Hochstrasser (2000) in their pioneering work. Instead of wakefields and ponderomotive forces, they used higher-order nonlinearities of plasma to generate THz radiation (Roskos et al. 2007).

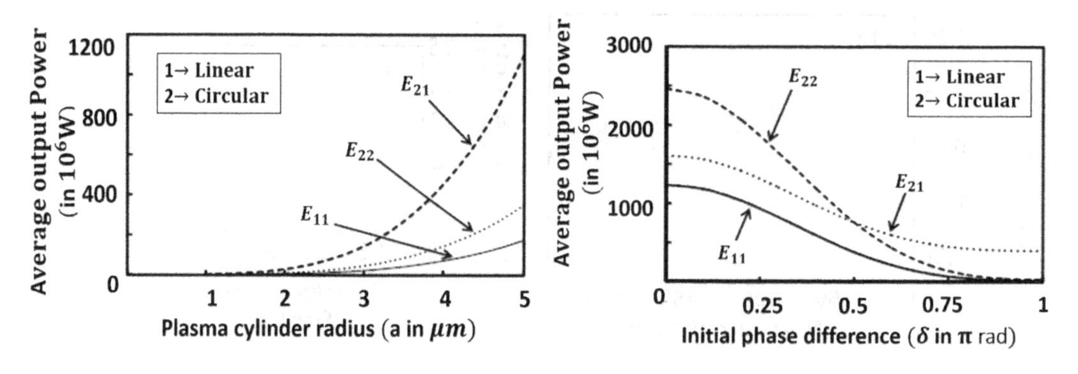

FIGURE 5.5
THz power as a function of the radius of the plasma cylinder (left) and initial phase difference (right) for three polarization combinations: both pulses are linearly polarized, one linearly and other circularly polarized, and both are circularly polarized.

The method was straightforward: intense 65 fs laser pulses at a fundamental wavelength of 800 nm and 1 kHz repetition rate were shot at a beam splitter that divided them into two separate beams. One of the beams was forced to pass through a second-order nonlinear crystal (like beta barium borate crystal), which promptly converted it into 400 nm through second-harmonic generation (SHG). This 400 nm pulses were focussed onto a gas like N or Ar or even pure air using off-axis paraboloid mirrors, and the other fundamental beam from the beam splitter was focussed onto the same point (Figure 5.6). The high intensity of the beams caused plasma to form, and in the presence of third-order nonlinearity $\chi^{(3)}$ in plasma, a sub-picosecond time-domain signal was obtained. Their Fourier transform into the frequency domain, broadband THz radiation covering the entire THz gap, and more were observed by Cook and Hochstrasser (2000).

Mathematically, the nonlinear polarization that led to the THz–time-domain signal can be expressed as

$$P_i^{(3)}(t) \propto \varepsilon_0 \chi_{ijkl}^{(3)} E_j^{2\omega_0}(t) E_k^{\omega_0 *}(t) E_l^{\omega_0 *}(t) \tag{5.1}$$

where the *indices i, j, k,* and l represent the polarization directions of the involved photons, and the electric fields are functions of time *t*, to show that the mixing happens in the time domain. Clearly, since the interacting photons lie in the UV-visible range, the third-order polarization barely has any periodicity, because THz frequencies are close to static when compared with the other frequencies involved.

The group (Cook and Hochstrasser 2000) further studied the power spectrum of the generated THz radiation. Since two photons of the fundamental frequency were used for every THz photon created, the THz power scale quadratically with the 800 nm input

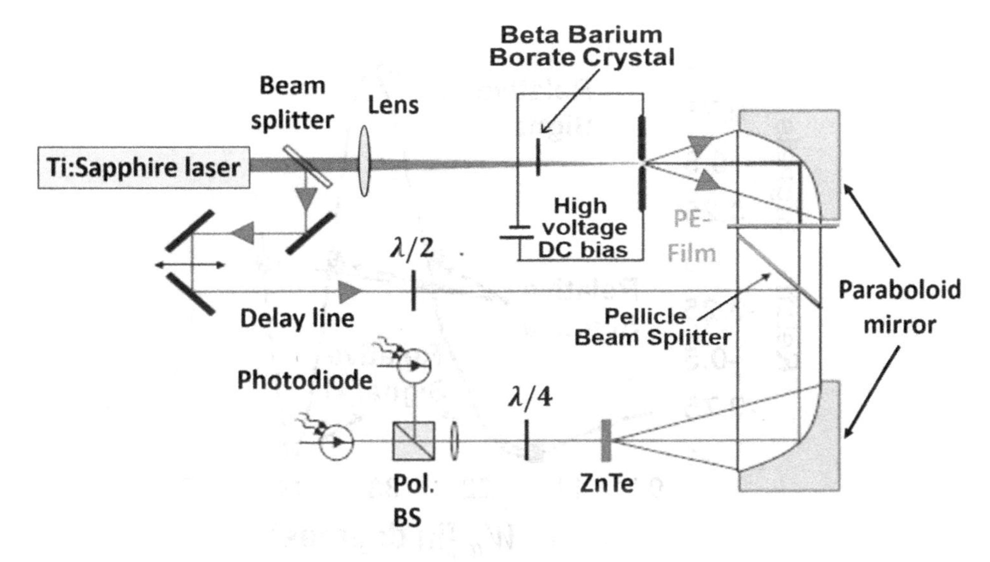

FIGURE 5.6
Typical experimental setup for THz generation using FWR in plasma. The experiment described in this section does not have a high-voltage DC bias for the SHG in the BBO crystal. The delay line is also replaced by a simple microscope coverslip with varying angles of incidence to simulate a phase delay between the fundamental and second-harmonic pulses.

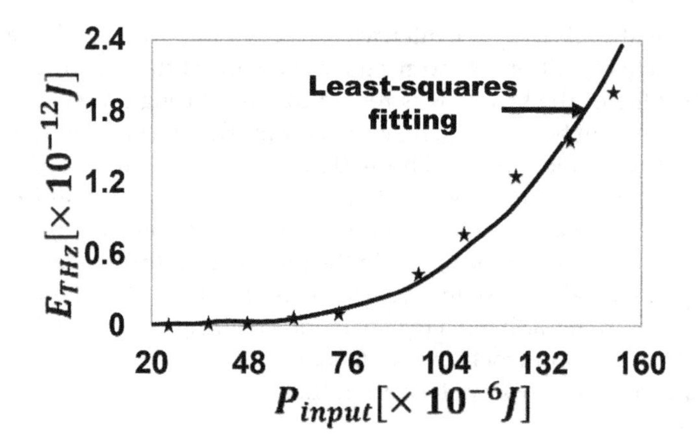

FIGURE 5.7
The solid line is the least-squares fit to the data obtained from power studies (starred points). The experimental data are in perfect agreement with the theoretical predictions.

power was indeed observed, seemingly confirming the THz generation process as FWR in air (Figure 5.7).

As another confirmation of the FWR model, a microscope coverslip was introduced between the BBO crystal and the off-axis paraboloid mirror focussing the beams; and phase delays between the fundamental and second-harmonic beams were simulated by varying the angle of incidence between the $2\omega_0$ pulses and the coverslip (Figure 5.8). As expected, a strong THz signal was received only when the incidence angle was such that the phase delay was 0. The time dependence of the electric field of the THz spectrum is shown in Figure 5.9 (Cook and Hochstrasser 2000).

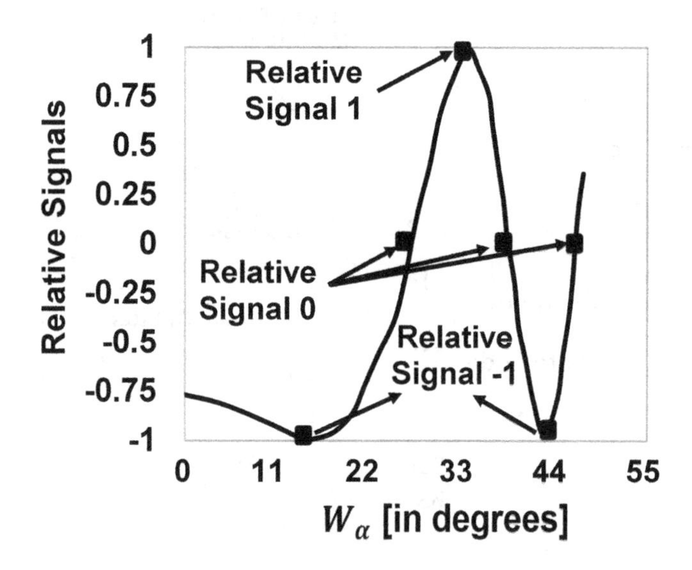

FIGURE 5.8
The solid curve is a least-squares fit to the phase-dependent experimental data (rectangular boxes). The angles correlated with zero crossings (relative signal 0), minima (relative signal –1) and maxima (relative signal 1) were detected.

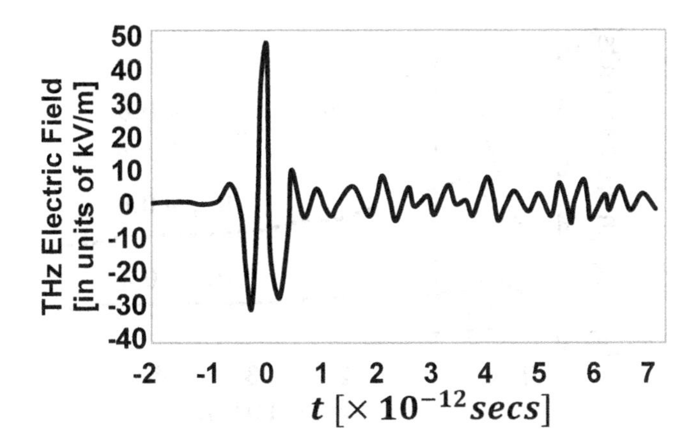

FIGURE 5.9
Behaviour of THz electric field as a function of time of the picosecond order.

Strictly speaking, Cook and Hochstrasser (2000) did not guess the formation of an air/gas plasma on focussing the lasers (though they found indications of the same); they merely thought the FWR was taking place due to the third-order nonlinearity of air. Only later did Kress et al. (2004) conclusively prove that the FWR process could not proceed without the formation of plasma by showing that no signal is received before a threshold input pump energy is reached (Figure 5.10). In analogy to modification (as was seen in the previous section), this process is named the AC-bias method or $\omega - 2\omega$ method. Many more improvements have been made to this process (Blanchard et al. 2009), and it remains the standard tabletop THz emitter technology with wide uses, especially in THz–time-domain spectroscopy (TDS) and ultrafast spectroscopy.

5.2.4 THz Emission Using Field Effect Transistors

In the previous sections, we reported that the laser-induced nonlinear oscillatory current in gaseous plasma or solid targets acts as a source for the THz emission. However, the field-effect transistor (FET) channels can also work as a cavities for the propagation of

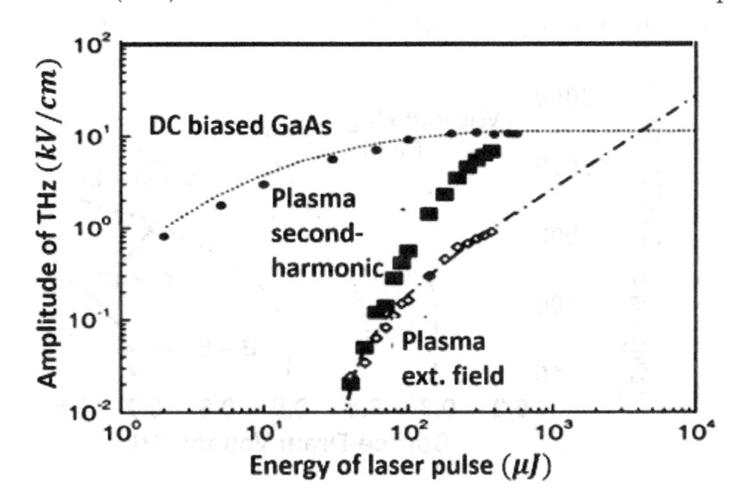

FIGURE 5.10
Comparative study of the amplitude of THz as a function of energy of laser pulse for different THz sources.

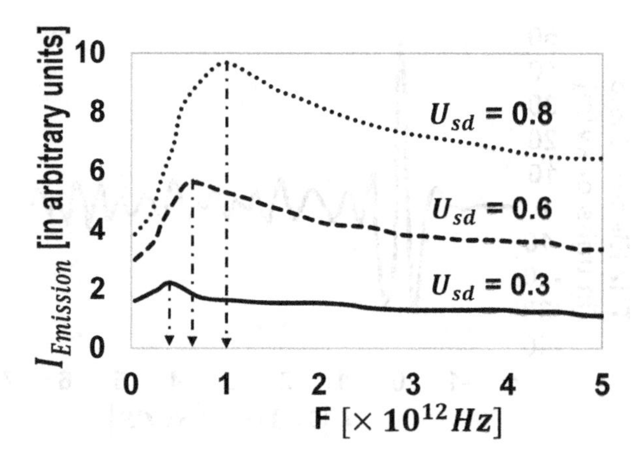

FIGURE 5.11
THz emission spectra for InGaAs HEMT for different source-drain voltages (U_{sd}). The maxima in the emission intensity (I_{Emission}) is marked by arrows.

plasma waves and hence for the emission of THz radiation. At the beginning of the 1990s, the spontaneous generation of electromagnetic radiation was proposed theoretically due to steady current flow in an asymmetric FET channel (Dyakonov and Shur 1993). This FET resonator can deliver the typical velocity of plasma waves in $\sim 10^6\,\text{m} / \text{s}$ range corresponding to a micron gate length, which may cause a fall in the resonance frequency, f, in the THz range. Moreover, it predicts that the transistor channel utilizes the nonlinear properties of two-dimensional plasma for the mixing of THz radiation (Dyakonov and Shur 1996).

To determine the efficiency and tunability of the emitted THz radiation by FET, many experiments have been performed. Owing to high carrier mobility that may raise the quality factor for THz emission, Knap et al. (2004) chose InGaAs high electron mobility transistors (HEMTs). With the parameters such as a gate length of 60 nm and 1.3 μm drain-source separation, it observes THz emission due to plasma wave instability or when the drain current crosses a specific critical value, as shown in Figure 5.11. At the threshold voltage, emission intensity appeared and this threshold value for drain-source voltage rises with an increment in an external applied magnetic field (Dyakonova et al. 2005) as magnetic fields instigate the channel resistance (Figure 5.12).

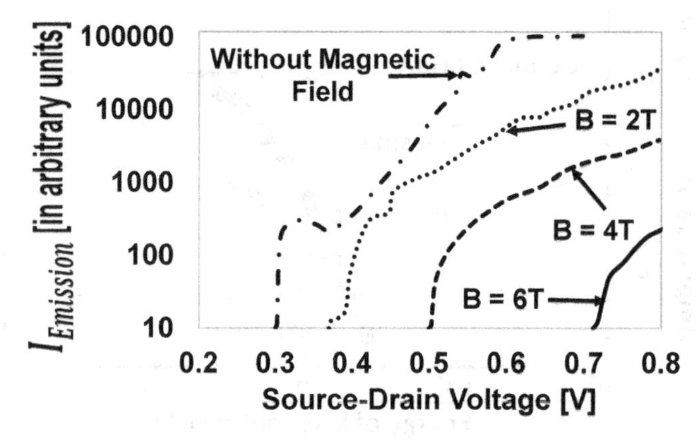

FIGURE 5.12
Variation of THz intensity with source-drain voltage under several magnitude of magnetic fields for the InGaAs HEMT emitter.

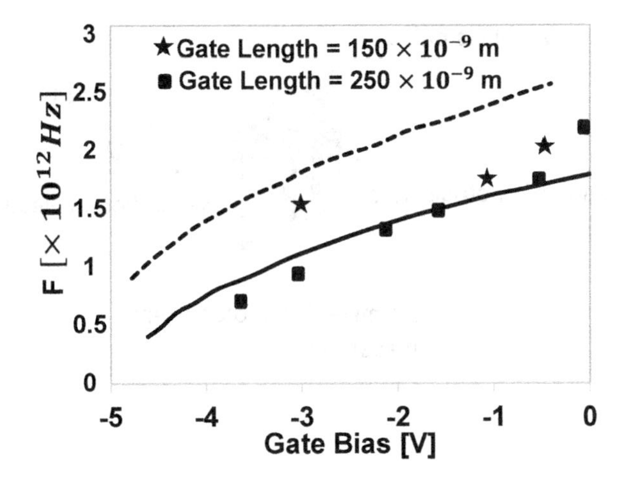

FIGURE 5.13
Comparative study of experimental results (star and rectangular points) and theoretical calculations (solid and dashed lines) of THz frequency with gate bias for AlGaN transistor with gate lengths of 250 and 150 nm.

Furthermore, the frequency of emitted radiation always falls in the THz range irrespective of the gate length.

Apart from this, El Fatimy et al. (2010) created an additional metal layer between the drain and gate terminals (field plate) of transistors to enhance the amplitude of emitted radiation. It has been estimated that the transistor gate voltage adjusts the maxima of emitted amplitude because of the modified velocity of plasma waves, as shown in Figure 5.13.

Recently, research has been carried out in this field for the enhancement and betterment of the THz power (achievable up to ~ 0.1 μW) and to engineer the efficient THz emitters to be tunable by the gate voltage. Most importantly, the elements associated with FET-based THz emitters could be quickly and cheaply integrated into arrays by using modern technology that leads to the generation of high-power sources.

5.3 Detection of Terahertz Radiation

Developing a reliable means of detecting THz radiation is just as important as generating them, and sometimes even more so because inaccuracies in detection mechanisms may lead one to misjudge the validity of the generation process. Just as with the sources of generation, coming up with suitable detection methods that can perform satisfactorily under all conditions has been quite challenging. There are various techniques available like photoconductive sampling, electro-optic sampling, liquid He-cooled bolometers, sum-frequency generation (SFG), and THz–air-biased coherent detection (ABCD). The following subsections will describe them in detail.

5.3.1 Photoconductive Detection/Sampling

Photoconductive detection uses the same phenomenon as photoconductive generation, except here the external bias voltage applied between the antenna leads is due to the THz pulse alone, and a femtosecond visible pulse is focussed synchronously onto the antenna

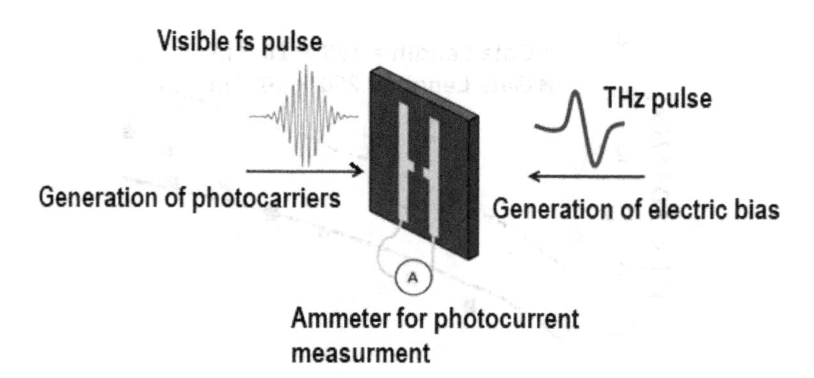

FIGURE 5.14
Photoconductive sampling.

gap (Figure 5.14). Due to this current is generated across the antenna leads, which is amplified using a low-bandwidth amplifier. This amplified current measures the strength of the generated THz field, as the current induced across the electrodes of the antennae is proportional to the electric field at the time when the conductivity is high. Further, the carriers present in the semiconducting substrate have an instantaneous lifetime. Hence, the THz electric field strength is only measured for an extremely small portion of the complete electric field waveform (approximately for femtoseconds).

5.3.2 Electro-Optic Detection/Sampling

Electro-optic detection / sampling is a method in which THz pulse induces a birefringence in an electro-optic medium (such as ZnTe) that is proportional to the electric field of the THz pulse. These changes in the birefringence can be measured by observing the change in the polarization state of a probe pulse. The change in the polarization state also changes the optical power at a polarizer, which can be measured easily.

The nonlinear materials used for generation of THz radiation by optical rectification can also be used for its detection by using the Pockels effect, wherein certain crystalline materials become birefringent in the presence of an electric field, due to which the optical polarization of the detection pulse changes and is proportional to the strength of THz electric field. With the help of suitable polarizers and photodiodes, this polarization change can be measured (Figure 5.15). The bandwidth of the emitted THz radiation depends on the crystal thickness, material characteristics, and laser pulse duration.

The THz electric field at the output of the detector is given by

$$\frac{\Delta P}{P_{\text{probe}}} = \frac{\omega n^3 E_{\text{THz}} r_{14} L}{c} \tag{5.2}$$

where r_{14} is the electro-optic coefficient of the ZnTe detector, L is the length of detector, n is the near-IR refractive index for the ZnTe detector, ω is the frequency, and E_{THz} is the electric field of the THz radiation.

5.3.3 Terahertz Detection Using Sum-Frequency Generation in Nonlinear Crystals

SFG detection is a frequency up-conversion technique that uses the second order nonlinearity of non-centrosymmetric crystals to produce sum-frequency pulses from incident

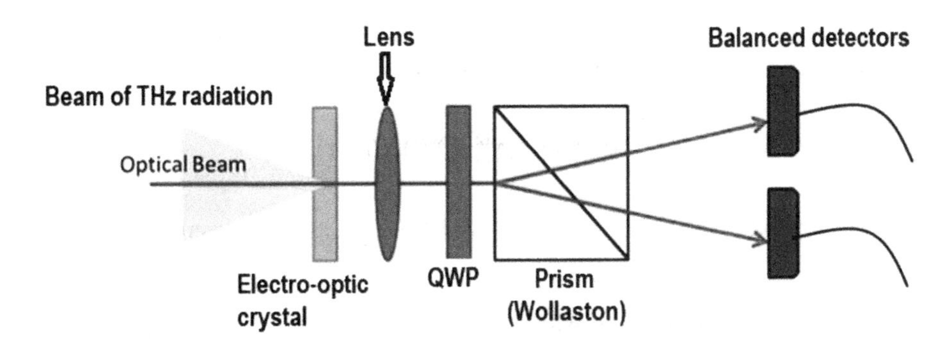

FIGURE 5.15
Electro-optic sampling.

THz and optical pulses at the fundamental frequencies. By analysing the SFG spectrum, one can get a direct estimate of the THz spectrum and bandwidth, and thus the THz field.

The technique was first devised by Baiz and Kubarych (2011) to detect mid-IR pulses, and it was later adapted to the THz regime by Blank et al. (2013). The team used a 150 fs, 800 nm pulse as the fundamental optical pulse, and tjeu allowed it to mix with the generated THz signal (after a controlled delay) in a 0.5-mm-thick non-centrosymmetric <100> ZnTe crystal pre-optimized for phase matching. An ultrabroadband THz spectrum, extending well into the 125 THz range, was detected (Figure 5.16).

The mathematical paradigm basically involved writing the second-order nonlinear polarization in time domain (since the signals are all pulses) and then using it to obtain the final time-domain field. This could be Fourier transformed into the frequency domain

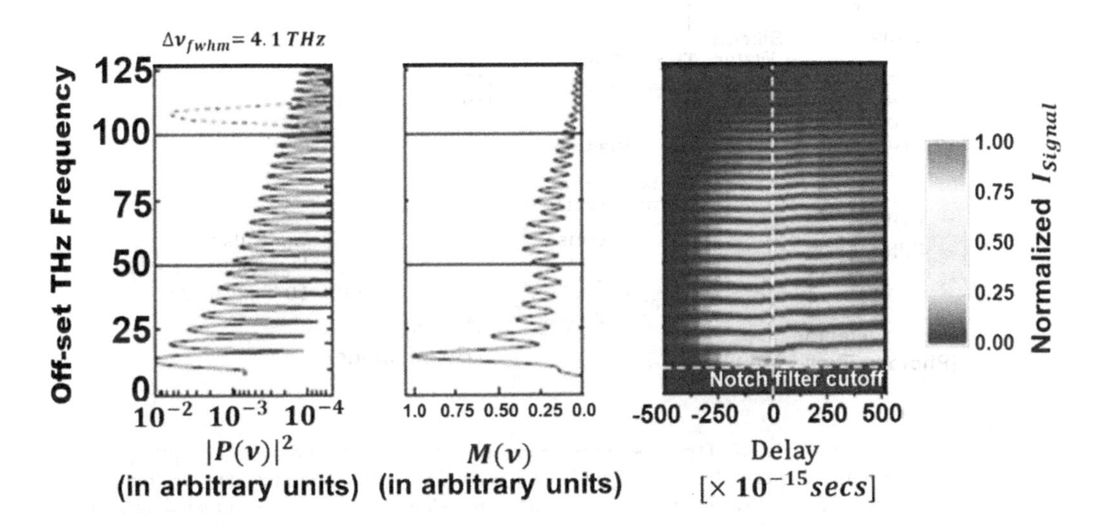

FIGURE 5.16
THz spectrum obtained using the SFG detection technique. The rightmost part of the figure is a plot of the signal intensity (normalized) with respect to pulse delay τ, and the left side shows the off-set THz frequency spectrum including the optical spectrum (dotted line in the leftmost part).

to obtain a matrix equation from which the THz spectral information could be derived by multiplying with a suitable Kernel function. Hence

$$P_i^{(2)}(t) \propto \varepsilon_0 \chi_{ijk}^{(2)} E_j^{\omega_0}(t) \varepsilon_k(t-\tau) \tag{5.3}$$

where $\varepsilon_k(t)$ is the THz time-domain signal and τ is the delay between the pulses.

5.3.4 Terahertz Detection Using ABCD

This ABCD method is just as useful as the previously described technique, and both are commonly employed in THz-TDS and ultrafast spectroscopy. THz-ABCD was put forward and developed from scratch by Karpowicz et al. (2008) and a few others (Dai et al. 2006; Lu et al. 2008; Ho et al. 2010; Lu and Zhang 2011; Lü et al. 2012). The technique is just the reverse of the FWR process. While two photons at the fundamental frequency ω_0 and one photon at second harmonic $2\omega_0$ are mixed together in a third-order nonlinear process to give rise to a THz pulse in FWR, two photons at ω_0 and a THz pulse mix together to give a second-harmonic pulse at $2\omega_0$ in ABCD. The spectrum of this second-harmonic pulse is analysed to obtain the spectrum of the THz pulse. Third-order nonlinearity in air plasma is the underlying process for both. The experimental setup is shown in Figure 5.17 (Dai et al. 2006).

The SHG in air can actually take place in two ways: one could simply apply a DC bias at the plasma where the fundamental beam is focussed, in which case the method is referred to as the electric field–induced second-harmonic (EFISH) technique, or one

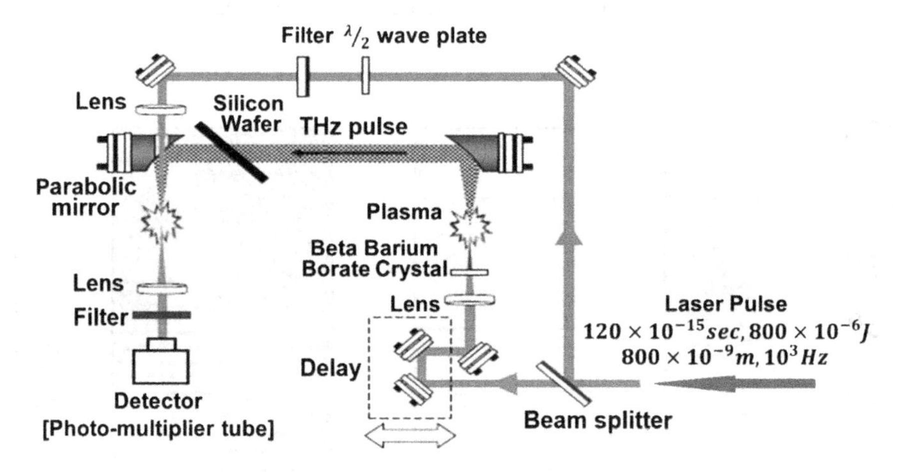

FIGURE 5.17
Experimental setup for the THz-ABCD process. At the first air-plasma point, the mixture of the pump beam (ω) with its second harmonic results in the generation of a THz wave. THz beam is collimated by the first parabolic mirror. The Si-filter/wafer is used to remove residual 400- and 800-nm pulses. The collimated THz beam is focussed by the second parabolic mirror. The $\frac{\lambda}{2}$ plate serves to change the polarization of the pump beam. The detection of the THz wave is achieved through the measurement of time-resolved second-harmonic signals generated by the mixture of the THz field and the pump beam (ω) at the second plasma point.

could use the THz pulse instead of this bias to obtain the same, in which case it is called the THz field-induced second-harmonic (TFISH) technique. Since the detector is simply a power detector capable of measuring only the area-integrated intensity, the EFISH and TFISH signals are used together to obtain information about the pulse phase and intensity carrier envelope.

The time-domain expressions for the third-order polarizations are simply given as

$$P^{(3)}(t) \propto \varepsilon_0 \chi^{(3)} E^{\omega_0}(t) E^{\omega_0}(t) E_{DC}(t-\tau) \ldots\ldots \text{EFISH}$$

$$P^{(3)}(t) \propto \varepsilon_0 \chi^{(3)} E^{\omega_0}(t) E^{\omega_0}(t) \hat{\varepsilon}(t-\tau) \ldots\ldots \text{TFISH}$$

The nonlinear polarizations as described above act as source terms in Maxwell's wave equation, and using mathematical tools like slow varying amplitude approximation (SVAA) and Fourier transform, the frequency-domain electric field is derived as

$$A_y(z,\Omega) = \frac{-3i(2\omega_0+\Omega)z}{16\pi^2 cn(2\omega_0+\Omega)} \operatorname{sinc}\left(\frac{\Delta kz}{2}\right) e^{-i\frac{\Delta kz}{2}} \int_{\infty}^{-\infty}\int_{\infty}^{-\infty} d\Omega' d\Omega'' \chi_{yxxx}^{-(3)} E_{0x}(\Omega') E_{0x}(\Omega'') \widehat{\varepsilon_x}(\Omega-\Omega'-\Omega'') \tag{5.4}$$

where Ω represents the broadband THz frequency, and the term $(2\omega_0+\Omega)$ implies that the power spectrum is centred around the second-harmonic frequency. Also, $\widehat{\varepsilon_x}$ may represent the DC signal alone or the THz signal (hence EFISH and TFISH), and Δk is the phase mismatch between the input fundamental and output second harmonic. In the event of a perfect phase match, the field amplitude grows linearly with pulse propagation distance, which is due to the SVAA.

The following equation may give the reader an idea about the mathematical form of the second harmonic power spectrum.

$$\Delta S(\tau) = (\chi_{yxxx}^{-(3)})^2 \frac{9E_{DC} z^2 \operatorname{sinc}^2\left(\frac{\Delta kz}{2}\right)}{64\pi^3 c^2} \int_{\infty}^{-\infty} d\Omega \frac{(2\omega_0+\Omega)^2}{n^2(2\omega_0+\Omega)} G(\Omega) \left[\frac{1}{\pi}\int_{\infty}^{-\infty} d\Omega_1 (\widehat{\varepsilon_x}(\Omega_1)) G(\Omega-\Omega_1) e^{i\Omega_1\tau}\right] \tag{5.5}$$

where $G(\Omega-\Omega_1) = \int_{-\infty}^{\infty} d\Omega_2 E_{0x}(\Omega_2) E_{0x}(\Omega-\Omega_1-\Omega_2)$ and

$$G(\Omega) = \int_{-\infty}^{\infty} d\Omega_2 E_{0x}(\Omega_2) E_{0x}(\Omega-\Omega_2)$$

Eq. (5.5) is plotted in MATLAB and is depicted in Figure 5.18. It can be clearly seen from the figure that the spectrum is sharply centred around the second-harmonic frequency. On setting the frequency offset to zero, one finds that the spectrum covers the entire THz gap and more.

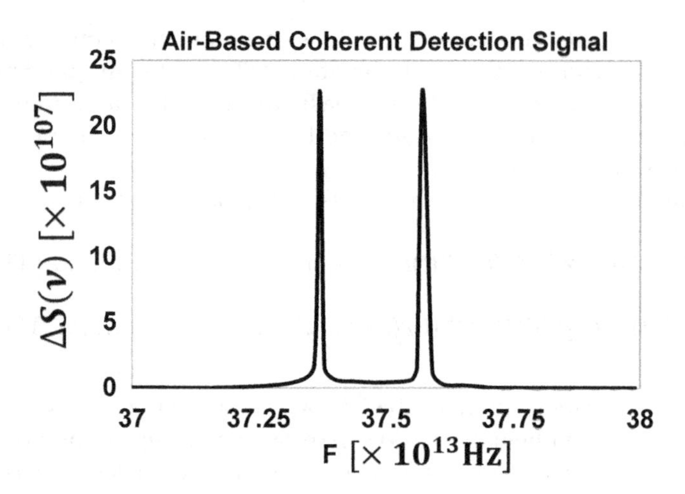

FIGURE 5.18
Sample ABCD spectrum in the frequency domain.

The time-domain signals for second-harmonic waveforms for different pump intensities and the frequency-domain spectrum for power corresponding to 920 TW/cm^2 is shown in Figure 5.19. Here, two dips at 0.5×10^{12} Hz and 3.25×10^{12} Hz are observed. For different probe intensities, these dips can be shifted and also disappear (Dai et al. 2006).

Since two photons of the fundamental pulse were used up per photon of the second harmonic created, the intensities should have obeyed a quadratic relation, while those with the THz pulse must have obeyed a linear relation. This was also confirmed by the same team (Dai et al. 2006) through Figure 5.20. In this figure, the second-harmonic intensity, $I_{2\omega}$, is found to obey a quadratic relation with the probe intensity I_ω for the probe intensity below 5×10^{14} W/cm^2 only. On the other hand, the second-harmonic intensity, $I_{2\omega}$, shows a linear dependence on the THz intensity I_{THz}, where I_ω was kept fixed at 1.8×10^{14} W/cm^2.

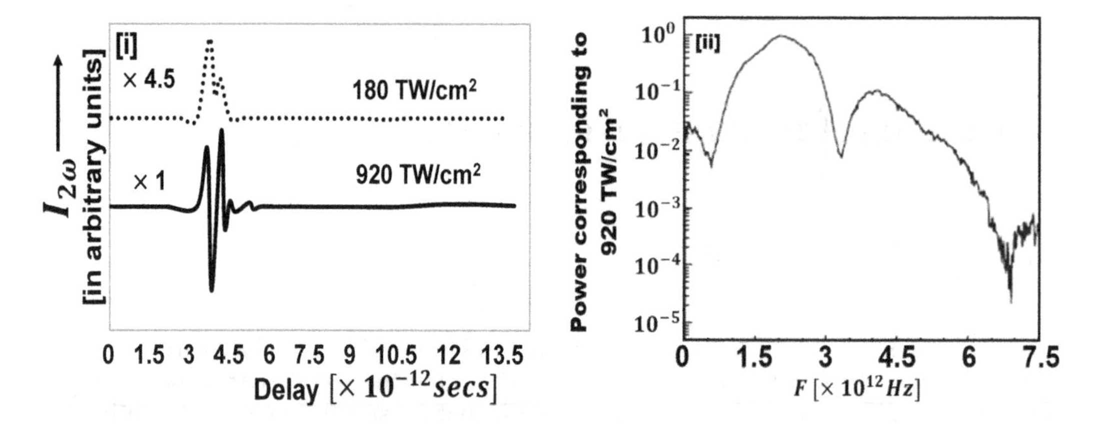

FIGURE 5.19
[i] Time-domain signals for second-harmonic waveforms for different pump intensities and [ii] frequency-domain spectrum for power corresponding to 920 TW/cm^2.

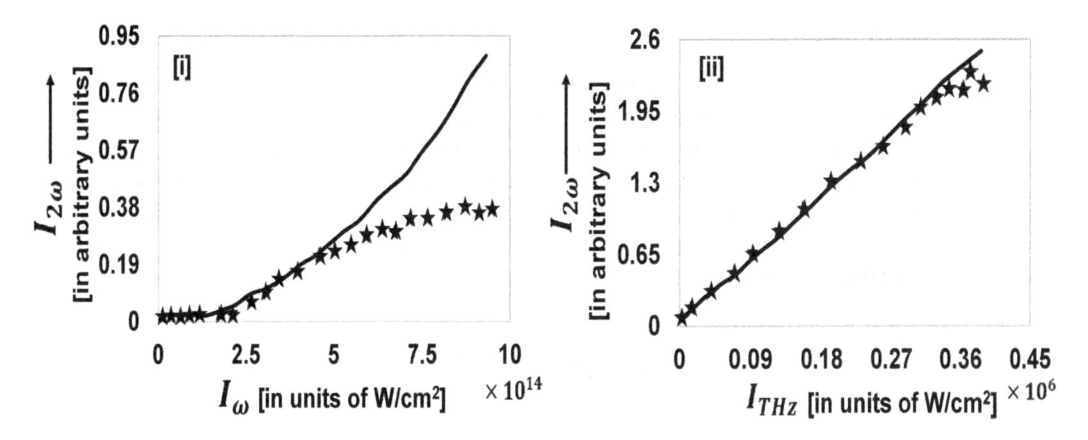

FIGURE 5.20
[i] Behaviour of second-harmonic intensity, $I_{2\omega}$, as a function of the probe intensity I_{ω} (starred points). Solid line is the quadratic fit to the data. [ii] Behaviour of second-harmonic intensity, $I_{2\omega}$, as a function of the THz intensity I_{THz} (starred points). The solid line is the linear fit to the data (starred points).

5.4 Absorption Processes of Dielectrics in the THz Region

The accessible THz frequency gap also has a significant impact on the dielectric properties (especially optical properties) of different materials that can be manipulated and hence, be utilized in numerous applications. In the THz gap, on the one hand, the low-frequency effects (below the microwave band) fall off with an increment in frequency; on the other, high-frequency effects are reduced as the frequency decreases. Here, we perceive mainly three microscopic absorption processes of materials dominating in the THz gap, namely the Drude model, the Debye model, and the lattice vibration model.

5.4.1 Drude Model

The transportation of the free charge carriers, i.e., the electrical conductivity (σ) of a material as a function of frequency (ω) is given as

$$\sigma(\omega) = \frac{\sigma_0}{1 - i\omega\tau} \tag{5.6}$$

where $\sigma_0 = \dfrac{n_e e^2 \tau}{m_e}$ is the static conductivity with m_e as the mass of electrons, n_e as the electron density, e as the electronic charge, and τ as the relaxation time. The electric permittivity of the material using the Drude model is given as

$$\varepsilon(\omega) = \varepsilon_0 + i\frac{\sigma(\omega)}{\omega} = \varepsilon_0 + \frac{i\sigma_0}{\omega(1 - i\omega\tau)} \tag{5.7}$$

Among the different parameters in Eq. (5.7), the relaxation time seems to be controlling those properties of the material, which are dependent on the frequency. For an intrinsic semiconductor, this relaxation timescale is of the order of picoseconds, which is a few orders higher than other dielectric solids. The inverse of this timescale falls in the THz frequency range for dielectric medium.

5.4.2 Debye Model

Second, the delayed response of dielectric materials on account of the electric field, termed as dielectric relaxation or Debye relaxation (τ_D), is due to disorganized thermal fluctuations. These random fluctuations delay the dipole moment reorientations in the medium. The dielectric material system strongly affects the relaxation time, ranging from microseconds to nanoseconds at room temperature. Hence, the expression for dielectric relaxation is directly linked with the frequency-dependent electric permittivity in terms of static permittivity $\varepsilon(0)$ and high-frequency permittivity $\varepsilon(\infty)$, given by

$$\varepsilon(\omega)=\varepsilon(0)+\frac{\varepsilon(0)-\varepsilon(\infty)}{(1-i\omega\tau_D)} \tag{5.8}$$

5.4.3 Lattice Vibration Model

Third, using the harmonic oscillator model, we can approximate the dielectric response of lattice vibrations in the medium (with oscillator length L) as

$$\varepsilon(\omega)=\varepsilon_L(0)+\frac{L}{(\omega_L^2-\omega^2-i\omega\gamma_L)} \tag{5.9}$$

where $\varepsilon_L(0)$, γ_L, and ω_L stand for the static permittivity, damping constant, and resonance frequency of lattice variations, respectively. Generally, the optical phonon resonance of a dielectric solid lies in the above 10 THz categories.

Finally, we sum up the effects of these three microscopic absorption processes in the form of refractive index $n(\omega)$ and the absorption coefficient $\alpha(\omega)$ as

$$n(\omega)=\mathbb{R}\left[\sqrt{\varepsilon_r(\omega)}\right] \tag{5.10}$$

$$\alpha(\omega)=\frac{2\omega}{c}\Im\left[\sqrt{\varepsilon_r(\omega)}\right] \tag{5.11}$$

This frequency-dependent behaviour of refractive index n and absorption coefficient α is also plotted in Figure 5.21 for a comparative study of these three mechanisms.

Figure 5.21 shows that, in the case of lattice vibration model, the refractive index of the dielectric material first remains unaltered until around the 5 THz frequency and then it increases exponentially with the frequency and shows the optical phonon resonance phenomenon around the 10 THz frequency. However, the absorption coefficient shows a linear behaviour with the THz frequency and it continuously increases with the frequency. As per the Drude model, the refractive index remains unaltered until around the 3 THz frequency, consistent with the lattice vibration model but for smaller values of THz frequency, and then it starts increasing slowly. Here the absorption coefficient remains constant until around the 5 THz frequency, and then it starts decreasing at a faster rate. It means the refractive index and the absorption coefficient show almost the opposite behaviour with the THz frequency. Their opposite behaviour is also noticed in view of the Debye model, where the refractive index first decreases until around 0.4 THz and then becomes almost constant with the THz frequency.

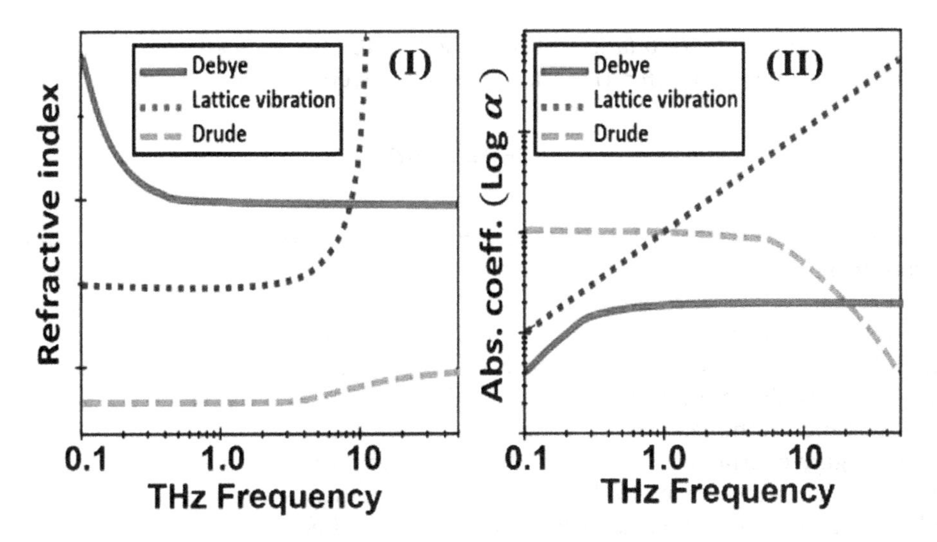

FIGURE 5.21
Behaviour of (I) refractive index and (II) absorption coefficient in terms of THz frequency for a dielectric medium, showing a comparative study of Drude, Debye, and lattice vibration models.

5.5 Artificial Materials at Terahertz Frequencies

Here metamaterials will be introduced as the artificial materials and their response to the THz waves / frequencies will be discussed. Since the plasmons are excited at the metal-dielectric interfaces, the metamaterials have a role in the excitation and use of plasmons. In view of this, plasmonics will also be discussed as a separate topic.

5.5.1 Metamaterials

Metamaterials refer to the man-made composites structured with an enhanced electromagnetic property that is unachievable by the naturally occurring materials. Electromagnetic metamaterials are characterized by effective permittivity and permeability of the medium. For example, if the structural elements of a metamaterial are smaller than the wavelength of incident laser, then this metamaterial will act as a homogenous medium for the laser. About two decades ago, a shred of experimental evidence demonstrating that these metamaterials had a negative refractive index brought attention to THz metamaterials (Smith et al. 2000; Yen et al. 2004).

To have a clearer picture of metamaterials and their manipulative behaviour towards the THz waves, we assume an electromagnetic wave interacting with an isotropic medium whose refractive index $n(\omega)$ is usually defined as

$$n^2(\omega) = \varepsilon_r(\omega)\,\mu_r(\omega) \tag{5.12}$$

This clearly shows that the value of the refractive index will be either $+\sqrt{\varepsilon_r(\omega)\,\mu_r(\omega)}$ or $-\sqrt{\varepsilon_r(\omega)\,\mu_r(\omega)}$ depending on the other parameters defined as $\varepsilon_r(\omega) = \varepsilon(\omega)/\varepsilon_0$ and $\mu_r(\omega) = \mu(\omega)/\mu_0$. The electromagnetic wave is assumed to be incident on the medium with an angle of incidence θ (angle of refraction be θ_r). Using Snell's law, refraction at an interface between vacuum ($n = 1$) to the medium is given by

$$\sin\theta = n\sin\theta_r \tag{5.13}$$

If the boundary conditions satisfy at the interface and warrant $\theta_r > 0$, then the refractive index is also positive, which happens mostly with the naturally occurring materials. Such media are categorized as right-handed material (RHM) as the directions of the electric and magnetic field, and the wave vector follow the right-hand rule. However, if either $\varepsilon_r(\omega)$ or $\mu_r(\omega)$ is negative after satisfying the boundary conditions, then the refractive index is negative and such materials are known as left-handed material (LHM). The behaviour of these two types of the materials is shown in Figure 5.22.

These negative refractive index materials lead to the perfect focusing of electromagnetic waves, which can be utilized for high-resolution imaging. Acting as a perfect lens, the slab of LHM can focus both the near-field as well as the far-field components of a point source as, unlike RHMs, the evanescent waves grow exponentially in metamaterials. In addition to this, the negative refractive index of metamaterials countereffects other properties of RHMs such as the phase velocity, Doppler shift, and the direction of Cherenkov radiation.

The geometry of THz metamaterials modulator / switch and its equivalent circuit are shown in Figure 5.23(a). The elements of metamaterials are linked together through metal wires so that they can deliver as a Schottky (metallic) gate. Near the split gaps, the substrate charge carrier density is controlled by the applied voltage between the ohmic and Schottky contacts [Figure 5.23(b)]. Behaviour of active THz metamaterial devices for different gate voltages is depicted in Figure 5.23(c). Here, the connecting wires and polarization

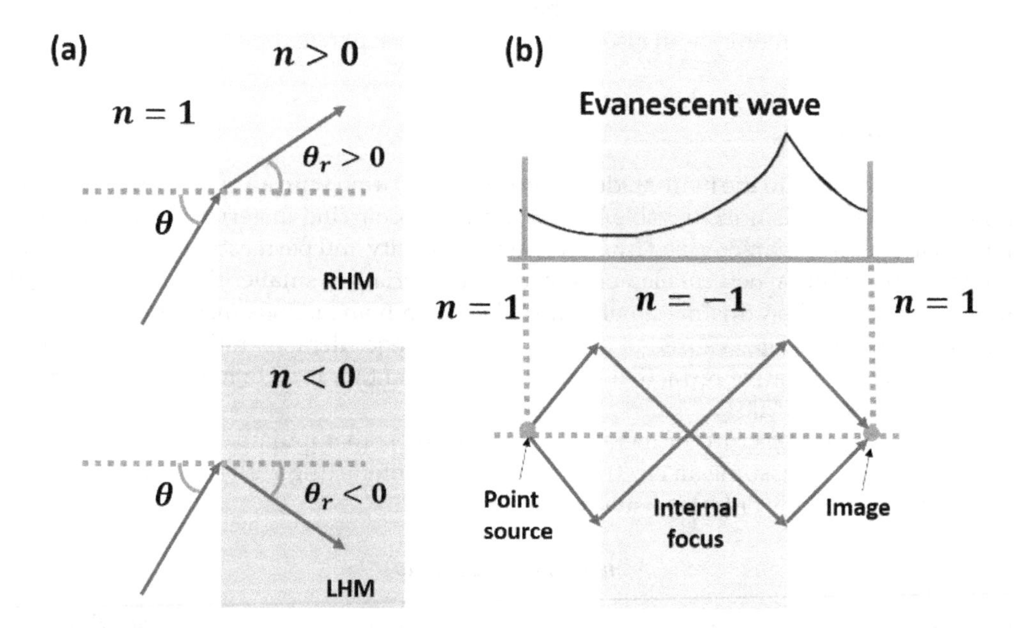

FIGURE 5.22
(a) Refraction by materials with different refractive indexes i.e., $n > 0$ and $n < 0$. (b) Focusing of the electromagnetic wave by metamaterials.

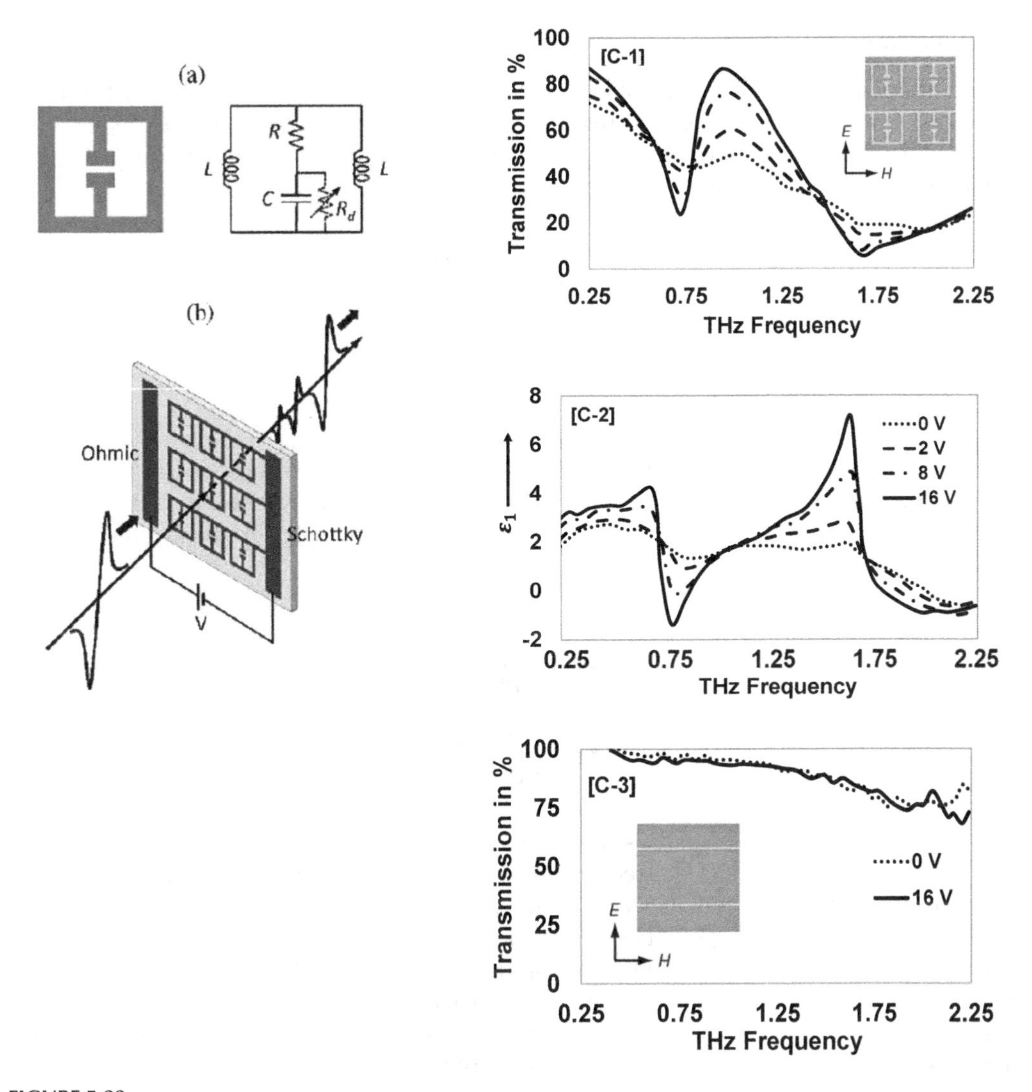

FIGURE 5.23
(a) Geometry of THz metamaterials modulator/switch and its equivalent circuit, (b) pattern of metamaterials elements, and (c) comparative study of THz transmission spectra with and without active metamaterial elements. The legends of (c-2) also correspond to (c-1).

of the THz electric field are perpendicular to each other. Figure 5.23(c-1) shows the percentage transmission intensity of the THz radiations as a function of THz frequency for different gate voltages. The analogous permittivity is depicted in Figure 5.23(c-2). Figure 5.23(c-3) illustrates the percentage transmission intensity of the THz radiation without metamaterials as a function of THz frequency for gate voltages of 0 and 16 V. The polarization configuration of magnetic and electric fields is shown in the insets. A comparison of Figs. 5.23 (c-1) and (c-3) reveals that the THz radiation is transmitted less when the active metamaterial elements are used and there are peaks and dips at particular values of THz frequency. These peaks become prominent at higher values of the applied gate bias voltage (Chen et al. 2006). The modified electric response of the THz device concerning gate bias is also evident from Figure 5.23 (c-2).

5.5.2 Plasmonics

As mentioned earlier also, the collective oscillations of the free electrons at the metal-dielectric interfaces are termed as surface plasmons. Assuming surface plasmon waves to propagate in the x-direction, their dispersion relation for the metal-air interface at a given frequency ω can be written as

$$k_d^2 = -\frac{1}{1+\varepsilon_m}\frac{\omega^2}{c^2} \tag{5.14}$$

$$k_m^2 = -\frac{\varepsilon_m^2}{1+\varepsilon_m}\frac{\omega^2}{c^2} \tag{5.15}$$

The dielectric constant of the metal is denoted by ε_m, the wave number in the metal is k_m, and k_d is the wave number in air medium (as a dielectric). Since the metal dielectric constant is imaginary ($|\varepsilon_m| \gg 1$) in the THz regime, the attenuation lengths for dielectric, ς_d, and metal, ς_m, are related to the free-space wavelength as

$$\varsigma_d = \frac{1}{\Re[k_d]} \gg \lambda_0 \tag{5.16}$$

$$\varsigma_m = \frac{1}{\Im[k_m]} \ll \lambda_0 \tag{5.17}$$

Eq. (5.16) depicts the extension of surface modes with the distances of many wavelengths into a dielectric medium at THz frequencies. For example, in metals the value of the dielectric constant is $|\varepsilon_m| \sim 10^6$, which gives rise to $\varsigma_d \sim 10$ cm in the THz range. The experiments related to the propagation of THz surface waves on a metal surface have been investigated already (O'Hara et al. 2005; Jeon and Grischkowsky 2006). Moreover, by using a periodically corrugated metallic cone, the surface plasmons can be guided as well as super-focussed at the tip of the cone. In Figure 5.24, the contour plot represents the field amplitude on a logarithmic scale of two orders of magnitude (Maier et al. 2006).

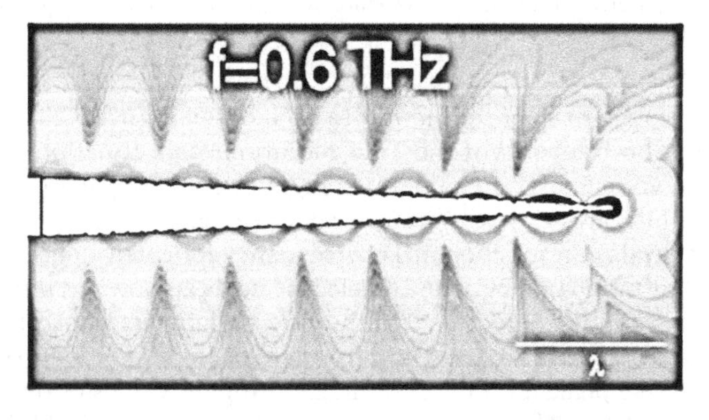

FIGURE 5.24
Focusing of surface plasmons in the corrugated cone.

5.6 Applications of THz Radiation for Detection

The unique non-ionizing ability of THz radiation distinguishes it from other high-energy radiations like X-rays. This property allows numerous applications in the physical world. In the field of medical science, THz radiation replaces the X-ray as it does not harm the healthy tissues of the human body while treating cancer. This radiation is also absorbed by the water molecules present in the human body and extinguishes itself, which makes it harmless. THz radiation successfully differentiates the histopathological samples by transmission images. With the ability of dielectrics penetration, THz radiation can detect and identify the threats even under clothing and through liquid (up to some extent) and solid materials. Also, many substances have spectroscopic signatures in the THz range that can be detected by this radiation. Today, the potential economic value of the THz marketplace is visible in communication, security system, food, and drink analysis.

5.7 Conclusions

The momentous applications of THz radiation have arisen because of its unique placement between IR and microwave bands, which in turn, allows for the high-resolution characterization without the risk of ionization. Three main methods were discussed that produce THz radiation through laser-plasma interaction. The first method is THz generation through laser wakefield oscillations in plasma, which relies on the wakefields produced by high-intensity laser pulses. THz emission from plasma with perturbed density profiles accounts for the second method of plasma-based THz generation. This method is mediated by the ponderomotive force of intense lasers. The third method is THz generation through nonlinear effects in plasma. In this method, instead of wakefields and ponderomotive forces, higher-order nonlinearities of plasma are exploited for THz radiation generation. The detection of THz radiation is sometimes even more cumbersome but more important than generating them because any inaccuracy in detection mechanisms may lead one to misjudge the validity of the generation process. Therefore, the phenomenon and working of various detection techniques such as photoconductive sampling, electro-optic sampling, SFG, and THz-ABCD were discussed in detail.

5.8 Selected Problems and Solutions

PROBLEM 5.1

Consider a Gaussian laser beam with angular frequency $\omega_1 = 2.4 \times 10^{14}$ rad/s and wavelength 7.85 μm is incident on an underdense plasma with an electric field $E_0 = 5.0 \times 10^{10}\,\mathrm{Vm}^{-1}$. Calculate the intensity of the incident laser beam and check whether the motion of electrons would be relativistic or not.

SOLUTION

The laser's intensity is determined by $I = \frac{1}{2} c\varepsilon_0 E_0^2$, which gives rise to

$$I = \frac{3\times10^{8}\times8.85\times10^{-12}\times\left(5\times10^{10}\right)^{2}}{2} = 3.31\times10^{14}\,\text{W/cm}^2$$

The laser strength parameter $a_0 = \frac{eE_0}{m_e c\omega_0}$ determines the relativistic (if >1) or nonrelativistic (if ≤1) nature / motion of the electrons. Putting in the respective values we find the magnitude of a_0 to be less than 1. This means the motion of electrons would be nonrelativistic under the influence of the said incident laser field.

PROBLEM 5.2

A laser beam imparts an oscillatory velocity to the electrons when it penetrates into the plasma having a density $1.9\times10^{18}\,\text{m}^{-3}$. The total nonlinear velocity attained by the plasma electrons is $v_T = 4.5\times10^{6}\,\text{m/s}$ under the action of various nonlinear forces and density perturbations. What would be the total current induced by the laser beam inside the plasma that generates the THz radiation?

SOLUTION

The total current induced by the laser beam inside the plasma is $J_T = nev_T$, where n is the plasma density and e is the electronic charge.

$$J_T = 1.9\times10^{18}\times1.6\times10^{-19}\times4.5\times10^{6}$$

$$J_T = 1.4\times10^{8}\ \text{Am}^{-2}$$

Suggested Reading Material

Ahi, K., & Anwar, M. (2016). Advanced terahertz techniques for quality control and counterfeit detection. In *Terahertz Physics, Devices, and Systems X: Advanced Applications in Industry and Defense*, Vol. 9856, International Society for Optics and Photonics, p. 98560G.

Baiz, C. R., & Kubarych, K. J. (2011). Ultrabroadband detection of a mid-IR continuum by chirped-pulse upconversion. *Optics Letters, 36*(2), 187–189.

Bandurin, D. A., Svintsov, D., Gayduchenko, I., … & Watanabe, K. (2018). Resonant terahertz detection using graphene plasmons. *Nature Communications, 9*(1), 1–8.

Beard, M. C., Turner, G. M., & Schmuttenmaer, C. A. (2002). Measuring intramolecular charge transfer via coherent generation of THz radiation. *The Journal of Physical Chemistry A, 106*(6), 878–883.

Blanchard, F., Sharma, G., Ropagnol, X., Razzari, L., Morandotti, R., & Ozaki, T. (2009). Improved terahertz two-color plasma sources pumped by high intensity laser beam. *Optics Express, 17*(8), 6044–6052.

Blank, V., Thomson, M. D., & Roskos, H. G. (2013). Spatio-spectral characteristics of ultra-broadband THz emission from two-colour photoexcited gas plasmas and their impact for nonlinear spectroscopy. *New Journal of Physics, 15*(7), 75023.

Byrd, J. M., Hao, Z., Martin, M. C., ... Zolotorev, M. S. (2006). Laser seeding of the storage-ring microbunching instability for high-power coherent terahertz radiation. *Physical Review Letters, 97*(7), 74802.

Bystrov, A. M., Vvedenskii, N. V., & Gildenburg, V. B. (2005). Generation of terahertz radiation upon the optical breakdown of a gas. *Journal of Experimental and Theoretical Physics Letters, 82*(12), 753–757.

Castilla, S., Terrés, B., Autore, M., ... & Vitiello, M. S. (2019). Fast and sensitive terahertz detection using an antenna-integrated graphene pn junction. *Nano Letters, 19*(5), 2765–2773.

Chen, H. T., Padilla, W. J., Zide, J. M. O., Gossard, A. C., Taylor, A. J., & Averitt, R. D. (2006). Active terahertz metamaterial devices. *Nature, 444*(7119), 597.

Cheon, H., Paik, J. H., Choi, M., Yang, H. J., & Son, J. H. (2019). Detection and manipulation of methylation in blood cancer DNA using terahertz radiation. *Scientific Reports, 9*(1), 1–10.

Cook, D. J., & Hochstrasser, R. M. (2000). Intense terahertz pulses by four-wave rectification in air. *Optics Letters, 25*(16), 1210–1212.

Dai, J., Xie, X., & Zhang, X.-C. (2006). Detection of broadband terahertz waves with a laser-induced plasma in gases. *Physical Review Letters, 97*(10), 103903.

Dobroiu, A., Otani, C., & Kawase, K. (2006). Terahertz-wave sources and imaging applications. *Measurement Science and Technology, 17*(11), R161.

Dyakonov, M., & Shur, M. (1993). Shallow water analogy for a ballistic field effect transistor: New mechanism of plasma wave generation by dc current. *Physical Review Letters, 71*(15), 2465.

Dyakonov, M. I., & Shur, M. S. (1996). Plasma wave electronics: novel terahertz devices using two dimensional electron fluid. *IEEE Transactions on Electron Devices, 43*(10), 1640.

Dyakonova, N., Teppe, F., Łusakowski, J., ... Cappy, A. (2005). Magnetic field effect on the terahertz emission from nanometer InGaAs/AlInAs high electron mobility transistors. *Journal of Applied Physics, 97*(11), 114313.

El Fatimy, A., Dyakonova, N., Meziani, Y., ... Skierbiszewski, C. (2010). AlGaN/GaN high electron mobility transistors as a voltage-tunable room temperature terahertz sources. *Journal of Applied Physics, 107*(2), 024504.

Faure, J., Van Tilborg, J., Kaindl, R. A., & Leemans, W. P. (2004). Modelling laser-based table-top THz sources: Optical rectification, propagation and electro-optic sampling. *Optical and Quantum Electronics, 36*(8), 681–697.

Gaal, P., Reimann, K., Woerner, M., Elsaesser, T., Hey, R., & Ploog, K. H. (2006). Nonlinear terahertz response of n-type GaAs. *Physical Review Letters, 96*(18), 187402.

Hamster, H., Sullivan, A., Gordon, S., & Falcone, R. W. (1994). Short-pulse terahertz radiation from high-intensity-laser-produced plasmas. *Physical Review E, 49*(1), 671.

Hamster, H., Sullivan, A., Gordon, S., White, W., & Falcone, R. W. (1993). Subpicosecond, electromagnetic pulses from intense laser-plasma interaction. *Physical Review Letters, 71*(17), 2725.

Hangyo, M., Tani, M., & Nagashima, T. (2005). Terahertz time-domain spectroscopy of solids: A review. *International Journal of Infrared and Millimeter Waves, 26*(12), 1661–1690.

Ho, I. C., Guo, X., & Zhang, X. C. (2010). Design and performance of reflective terahertz air-biased-coherent-detection for time-domain spectroscopy. *Optics Express, 18*(3), 2872–2883.

Ishigaki, K., Shiraishi, M., Suzuki, S., Asada, M., Nishiyama, N., & Arai, S. (2012). Direct intensity modulation and wireless data transmission characteristics of terahertz-oscillating resonant tunnelling diodes. *Electronics Letters, 48*(10), 582–583.

Jeon, T. I., & Grischkowsky, D. (2006). THz Zenneck surface wave (THz surface plasmon) propagation on a metal sheet. *Applied Physics Letters, 88*(6), 061113.

Jiang, Z., & Zhang, X.-C. (1998). Single-shot spatiotemporal terahertz field imaging. *Optics Letters, 23*(14), 1114–1116.

Karpowicz, N., Dai, J., Lu, X., ... Price-Gallagher, M. (2008). Coherent heterodyne time-domain spectrometry covering the entire "terahertz gap." *Applied Physics Letters, 92*(1), 11131.

Knap, W., Lusakowski, J., Parenty, T., … Shur, M. S. (2004). Terahertz emission by plasma waves in 60 nm gate high electron mobility transistors. *Applied Physics Letters*, 84(13), 2331.

Kress, M., Löffler, T., Eden, S., Thomson, M., & Roskos, H. G. (2004). Terahertz-pulse generation by photoionization of air with laser pulses composed of both fundamental and second-harmonic waves. *Optics Letters, 29*(10), 1120–1122.

Li, N., Bai, Y., Miao, T., Liu, P., Li, R., & Xu, Z. (2016). Revealing plasma oscillation in THz spectrum from laser plasma of molecular jet. *Optics Express, 24*(20), 23009–23017.

Liu, K., Xu, J., Yuan, T., & Zhang, X.-C. (2006). Terahertz radiation from InAs induced by carrier diffusion and drift. *Physical Review B, 73*(15), 155330.

Löffler, T., Hahn, T., Thomson, M., Jacob, F., & Roskos, H. G. (2005). Large-area electro-optic ZnTe terahertz emitters. *Optics Express, 13*(14), 5353–5362.

Löffler, T., Jacob, F., & Roskos, H. G. (2000). Generation of terahertz pulses by photoionization of electrically biased air. *Applied Physics Letters, 77*(3), 453–455.

Löffler, T., & Roskos, H. G. (2002). Gas-pressure dependence of terahertz-pulse generation in a laser-generated nitrogen plasma. *Journal of Applied Physics*, 91(5), 2611–2614.

Lu, X., Karpowicz, N., Chen, Y., & Zhang, X.-C. (2008). Systematic study of broadband terahertz gas sensor. *Applied Physics Letters, 93*(26), 261106.

Lu, X., & Zhang, X.-C. (2011). Balanced terahertz wave air-biased-coherent-detection. *Applied Physics Letters, 98*(15), 151111.

Lü, Z., Zhang, D., Meng, C., … Yuan, J. (2012). Polarization-sensitive air-biased-coherent-detection for terahertz wave. *Applied Physics Letters, 101*(8), 81119.

Maier, S. A., Andrews, S. R., Martín-Moreno, L., & García-Vidal, F. J. (2006). Terahertz surface plasmon-polariton propagation and focusing on periodically corrugated metal wires. *Physical Review Letters, 97*(17), 176805.

Malik, A. K., Malik, H. K., & Kawata, S. (2010). Investigations on terahertz radiation generated by two superposed femtosecond laser pulses. *Journal of Applied Physics*, 107(11), 113105.

Malik, A. K., Malik, H. K., & Nishida, Y. (2011). Terahertz radiation generation by beating of two spatial-gaussian lasers. *Physics Letters A, 375*(8), 1191–1194.

Malik, A. K., Malik, H. K., & Stroth, U. (2012). Terahertz radiation generation by beating of two spatial-gaussian lasers in the presence of a static magnetic field. *Physical Review E, 85*(1), 16401.

Malik, H. K. (2020). Generalized treatment of skew-cosh-gaussian lasers for bifocal terahertz radiation. *Physics Letters A*, 384(15) 126304.

Murphy, T. E., Jadidi, M. M., Mittendorff, M., Sushkov, A. B., Drew, H. D., & Fuhrer, M. S. (2018). Terahertz detection in 2D materials. In *Quantum Sensing and Nano Electronics and Photonics XV*, Vol. 10540, International Society for Optics and Photonics, p. 105401X.

O'Hara, J. F., Averitt, R. D., & Taylor, A. J. (2005). Prism coupling to terahertz surface plasmon polaritons. *Optics Express, 13*(16), 6117.

Roskos, H. G., Thomson, M. D., Kreß, M., & Löffler, T. (2007). Broadband THz emission from gas plasmas induced by femtosecond optical pulses: From fundamentals to applications. *Laser & Photonics Reviews, 1*(4), 349–368.

Segschneider, G., Jacob, F., Löffler, T., … Döhler, G. (2002). Free-carrier dynamics in low-temperature-grown GaAs at high excitation densities investigated by time-domain terahertz spectroscopy. *Physical Review B, 65*(12), 125205.

Sharma, D., Singh, D., & Malik, H. K. (2020). Shape-dependent terahertz radiation generation through nanoparticles. *Plasmonics, 15*(1), 177–187.

Sheng, Z.-M., Mima, K., Zhang, J., & Sanuki, H. (2005). Emission of electromagnetic pulses from laser wakefields through linear mode conversion. *Physical Review Letters, 94*(9), 95003.

Singh, D., & Malik, H. K. (2015). Enhancement of terahertz emission in magnetized collisional plasma. *Plasma Sources Science and Technology, 24*(4), 45001.

Singh, D., & Malik, H. K. (2020). Terahertz emission by multiple resonances under external periodic electrostatic field. *Physical Review E, 101*(4), 043207.

Sizov, F. (2018). Terahertz radiation detectors: the state-of-the-art. *Semiconductor Science and Technology, 33*(12), 123001.

Smith, D. R., Padilla, W. J., Vier, D. C., Nemat-Nasser, S. C., & Schultz, S. (2000). Composite medium with simultaneously negative permeability and permittivity. *Physical Review Letters, 84*(18), 4184.

Sun, Y., Sy, M. Y., Wang, Y.-X. J., Ahuja, A. T., Zhang, Y.-T., & Pickwell-MacPherson, E. (2011). A promising diagnostic method: Terahertz pulsed imaging and spectroscopy. *World Journal of Radiology, 3*(3), 55.

Yen, T. J., Padilla, W. J., Fang, N., ... Zhang, X. (2004). Terahertz magnetic response from artificial materials. *Science, 303*(5663), 1494.

Yoshii, J., Lai, C. H., Katsouleas, T., Joshi, C., & Mori, W. B. (1997). Radiation from cerenkov wakes in a magnetized plasma. *Physical Review Letters, 79*(21), 4194.

Yugami, N., Higashiguchi, T., Gao, H., ... Katsouleas, T. (2002). Experimental observation of radiation from Cherenkov wakes in a magnetized plasma. *Physical Review Letters, 89*(6), 65003.

6

Plasma-Based Particle Acceleration Technology

6.1 Introduction

A particle accelerator is a machine, adopted to enhance the energies and speeds of the charged particles, like protons, electrons, ions, and so forth, to a great extent and also to steer them. Two fundamental classes of particle accelerators are electrostatic and electrodynamic (electromagnetic) accelerators. As the name suggests, in electrostatic accelerators, charged particles are accelerated through the static electric field. Van de Graaff and Cockcroft Walton generators fall under this category. In such accelerators, an accelerating voltage decides the maximum gain of energies for the charged particles. Also, in these cases, the acceleration of the charged particles is limited due to the limitation of electric breakdown. Therefore, accelerators based on the dynamic fields are mostly adopted compared with that of static fields to achieve higher accelerating energies of the charged particles. Such accelerators are designated electromagnetic or electrodynamic accelerators. Here, acceleration and steering of the charged particles are attained either by oscillating radio-frequency (RF) fields or magnetic induction. The most common exmaples of this category are betatrons, cyclotrons, the linear particle accelerator (linac), synchrotrons, and synchrocyclotrons. In these particle accelerators, the maximum energy gain for the charged particles is not limited to the accelerating field strength because of multiple movement of particles through the similar accelerating field.

The high energetic beams of particles are beneficial in both applied and fundamental research. Around 30,000 accelerators have been estimated worldwide. These accelerators are used for radiotherapy, industrial research and processing, ion implantation, and low-energy and biomedical applications. In high-energy physics or particle physics, the broad knowledge of physical laws that carry out energy, matter, time, and space has been achieved through particle accelerators. The heavier particles accelerate through ion beam accelerators, which find imperative applications in the manufacturing of chips in semiconductor industries, and in the surface hardening of those materials that adopt an artificial joint. In many clinics and hospitals, accelerator-based therapy and diagnoses have been given to patients. For example, through the use of external beams, radiation therapy has become an exceedingly efficient approach for the treatment of cancer patients. Apart from these, particle accelerators also play an effective role in national security.

Graphically, the improvement in accelerator performances over time is portrayed in Figure 6.1. In Figure 6.1(a), constituent centre-of-mass (CM) energy or constituent collision energy (in the units of TeV (10^{12} eV)) is depicted as a function of time (in terms of years). The maximum energy of the particles in several particle accelerators as a function of time (in terms of years) is depicted in Figure 6.1(b), which is also called a Livingston plot. There is an exponential increment in the CM energy and maximum gained energy over the course of several decades, which can be credited to the development of different accelerator technologies (Seryi 2015). However, it is clear that with time, there occurs a saturation of the peak energies achievable with any single technology, and it is at such times that new

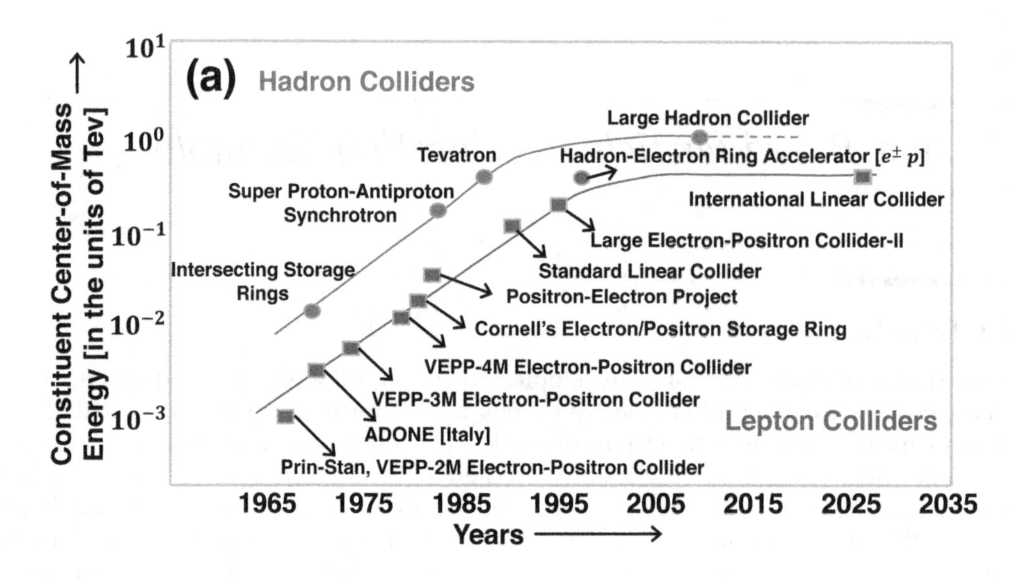

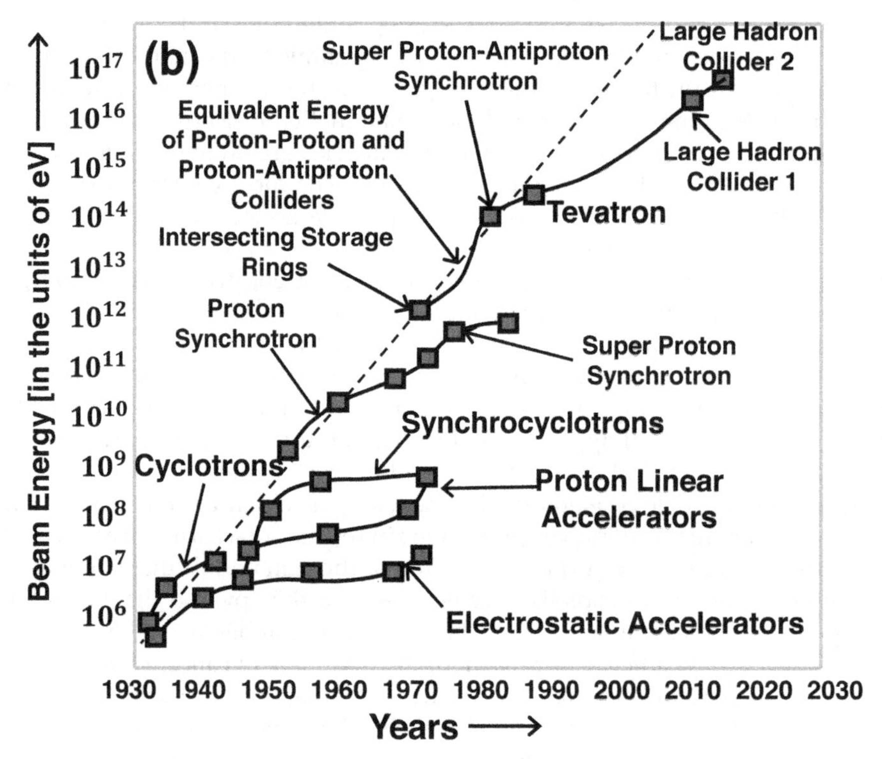

FIGURE 6.1
(a) Behaviour of constituent centre-of-mass energy as a function of time. (b) Livingston plot of the evolution of accelerators.

methods are adopted. Researchers are convinced that conventional RF accelerators have reached their peak performance, and new ideas need to be explored to push beyond the present energy frontier.

The performance of an accelerator is evaluated in terms of the acceleration gradient, which is nothing but the energy gained per unit length. Because of the material breakdown and discharges, conventional accelerators have a gradient of around 100 MeV/m (Joshi and Katsouleas 2003), which means for high-end operations [like the TeV beams in the Large Hadron Collider (LHC)], the accelerators need to be more than 10 km long. Thus, the size (and expense) of such conventional particle accelerators keep rising and have started to overshadow the potential benefits in terms of beam energy.

Because of the requirement of longer linear accelerators to produce a beam of required energy, a transition to circular accelerators from the linear accelerators was made to increase the energy of the particle beam. In this method, a multiple movement of the particle beam through the similar accelerating field is achieved. This results in the considerable reduction of a number of components required and thus a reduction in the amount of components. However, charged particles start to lose energy as they come under the effect of the magnetic field because of synchrotron radiations. The magnitude of power loss is given by $W_{\text{loss}} \propto \frac{\gamma^4}{\rho^2}$, where ρ and γ, respectively, are the deflection radius and the particle's Lorentz factor. For the maintainance of the power loss to be constant, one must quadruple the circumfrance to ~108 km to double the output energy of the particle beam in a large electron-positron (LEP) collider. Otherwise, there is a significant immense energy loss (approximately 16 times more) of the order of gigawatts. Therefore, we need to move back to linacs. To achieve the required beam energy for future processings, a linear accelerator of several tens of kilometers is needed. One of the effective ways to overcome the limitation of higher-length accelerators is to enhance the acceleration gradient. Consequently, particles will gain a considerable amount of energy within a shorter distance.

Research in particle acceleration thus needs an immediate revolution, and plasma-based accelerators are turning out to be promising for physics, at both the highest energies (100 GeV and up) for large-scale applications, and in the range 100 MeV to 1 GeV, which is used in materials science, biology, nuclear science, and medicine. This plasma-based approach offers the possibility to vastly increase acceleration gradients (about a 1000 times more than conventional accelerators), while decreasing the size and cost of the machines, to the extent that the next generation of powerful accelerators may just occupy a small room, and become so-called 'tabletop' accelerators (Joshi and Katsouleas 2003).

Plasma-based accelerators are based on the acceleration of charged particles, like electrons and positrons, in 'wakefields' produced by the interaction of the plasma with extremely high intensity lasers, or with driving charged particle beams. The former approach is called laser wakefield acceleration (LWFA), whereas the latter approach is a beam-driven plasma wakefield accelerator (PWFA). Before the advent of very high intensity lasers that made LWFA possible, researchers relied on two subcategories, plasma beat-wave acceleration (PBWA) and self-modulated LWFA (SM-LWFA). In this chapter, we shall be taking a closer look at the concepts behind each of these techniques. Along with the attractive prospects of high energies at low cost and small sizes, plasma-based accelerators can be tailored to provide focussed beams with low-energy spreads and beam emittances. There are also plans of improving the present linear accelerator facilities by using plasma 'afterburners' (Joshi and Katsouleas 2003). Different kinds of plasma-based particle acceleration techniques, i.e. (I) LWFA, (II) PWFA, (III) PBWA, and (IV) SM-LWFA are portrayed in Figure 6.2, and a detailed description of these techniques is given in Section 6.4.

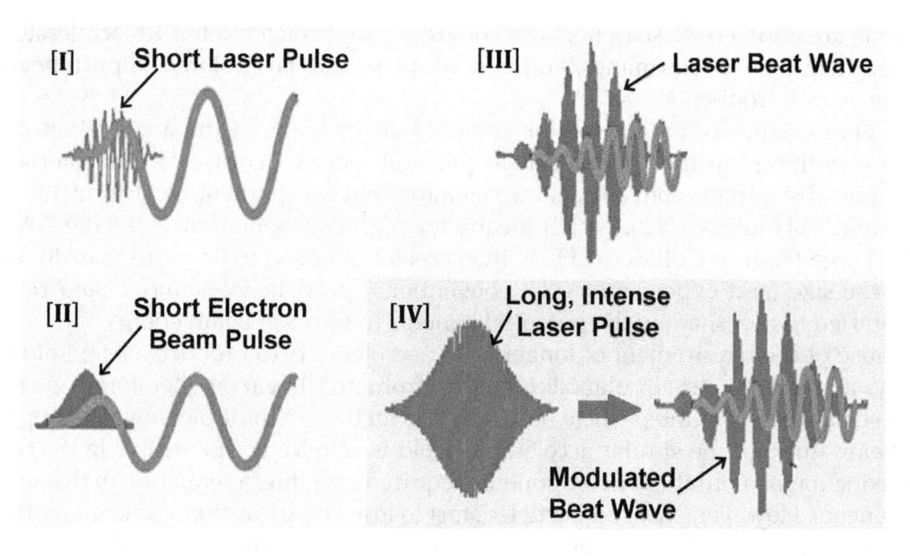

FIGURE 6.2
Different kinds of plasma-based particle acceleration techniques: (I) LWFA, (II) PWFA, (III) PBWA, and (IV) SM-LWFA.

6.2 Theory of Plasma-Based Accelerators

The basic process of plasma-based accelerators was initially described by the late John Dawson along with his collaborator Toshi Tajima (Tajima and Dawson 1979). As stated earlier, the fundamental benefit of such accelerators is their ability to favour ultrahigh acceleration gradients without the risk of material breakdown, since the plasma is already an ionized medium. The electron density (n) of the plasma determines the collective electric field (E) as per the relation $E \propto \sqrt{n}$, n in m^{-3}. For example, an electron plasma density of the order of 10^{18} cm^{-3} is able to sustain the field up to 100 GeV/m, which is 1000 times stronger than conventional RF linear acceleration gradients.

Irrespective of whether the source driving the wakefields is a high-intensity laser pulse or energetic charged particle beams, the basic phenomenon behind the process remains the same: the interaction of the beams with plasma leads to a significant force that pushes the electrons and ions away from each other creating a charge separation. This force is the well-known nonlinear ponderomotive force in the case of intense laser pulses and repulsive Coulomb force in the case of charged particle beams. The charge separation oscillates for a few cycles at the plasma frequency, given by $\omega_p = \sqrt{\frac{n_e e^2}{m_e \varepsilon_0}}$, which generates huge longitudinal and transverse acceleration gradients, and travels along the plasma trailing behind the driving source at a speed equal to the group velocity of the source or the beam velocity. Any charged particles placed at one end of this charge separation wakefield, pre-accelerated to a velocity close to that of the wakefield, will be accelerated to high energies by the longitudinal electric fields.

6.3 Wakefield Acceleration Using Intense Laser Systems

The concept of charged particle acceleration using wakefield generated by intense lasers in plasma was introduced by Tajima and Dawson (1979) in their pioneering work. Unlike many modern methods, which require injection of external charges pre-accelerated to wakefield velocities, they described a simple method wherein an energetic laser pulse moving through an underdense plasma ($\omega_p < \omega_0$, ω_p and ω_0 are the plasma frequency and laser frequency, respectively) pushes electrons longitudinally through a nonlinear ponderomotive force, and creates a charge separation that oscillates at ω_p. This separation traps plasma electrons inside it, which execute trapping oscillations and accelerate to large energies. Moreover, since the acceleration gradients are very high, the electron energies become relativistic, and further acceleration largely leads to gains in mass, keeping the particles in phase with the trapped oscillations. This allows for energy gain over longer time periods.

Using both analytical methods and simulations, Tajima and Dawson (1979) were able to predict the dependence of peak electron energies on plasma frequency and laser frequency through

$$\gamma^{\max} = \frac{2\omega^2}{\omega_p{}^2} \tag{6.1}$$

where $\gamma = 1\Big/\sqrt{1-\frac{v_e^2}{c^2}}$ together with v_e as the electron velocity. They measured the maximum electron energy for different photon frequencies $\left(\frac{\omega}{\omega_p}\right)$ and presented a comparison between the results of theoretical prediction (Eq. 6.1) and simulations in Figure 6.3. Simulation results are found to follow Eq. (6.1) closely.

They deduced that the wake generation is the most efficient when laser pulse length (L) at full-width at half-maximum (FWHM) was equal to one-half of the plasma wavelength (λ) defined as $\lambda = \frac{2\pi c}{\omega_p}$, i.e. $L = c\tau_L = \frac{\pi c}{\omega_p}$, with τ_L being the laser FWHM pulse width (in time). The dependence of electron acceleration on the pulse width of the laser was later studied by various groups, and it led to the development of SM-LWFA and to the use of the 'bubble' regime when describing conventional LWFA. Both these concepts will be discussed in more detail later.

6.4 Plasma-Based Particle Acceleration Techniques

We will discuss the basic four types of plasma-based particle acceleration techniques, which are called PBWA, SM-LFWA, LWFA, and PWFA.

6.4.1 Plasma Beat-Wave Accelerator (PBWA)

The production of immense amplitude and relativistic plasma waves are of great interest because of their potential to accelerate the particles with ultrahigh gradients. The plasma

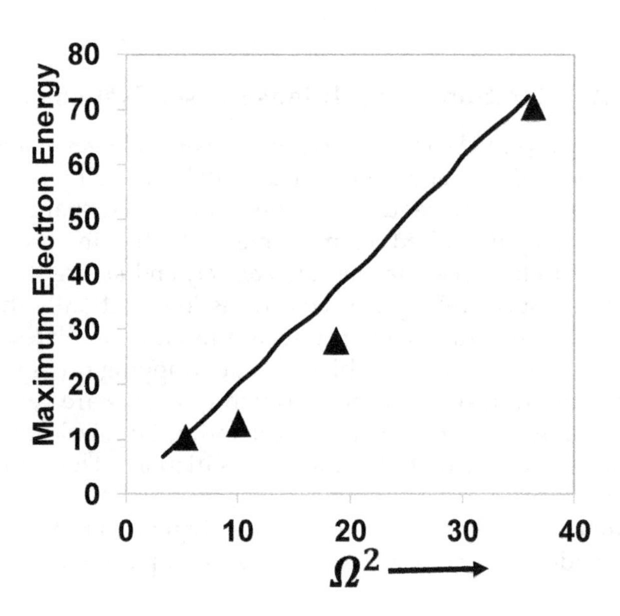

FIGURE 6.3

Maximum electron energy (γ^{max}) versus frequency ratio $\Omega^2=\left(\frac{\omega_o}{\omega_p}\right)^2$. The solid line corresponds to theoretical prediction (Eq. 6.1), whereas the triangles correspond to simulations.

waves can be excited by ultrashort intense laser pulses or beating of two frequency lasers, so that these waves propagate with the phase velocity close to speed of light. To achieve this, mainly two mechanisms, LWFA and PBWA, are used. The latter one is explained in this section. As mentioned above, plasma wave propagating with a velocity close to the speed of light results in the generation of the concept of charge separation, which in turn produces a longitudinal electrostatic field. Such longitudinal electrostatic fields are responsible for the acceleration in PBWAs. The plasma wave would be produced if the frequency difference of two collinear laser pulses is in such a way that the plasma frequency matches with the beat frequency. Initially Tajima and Dawson (1979) proposed the concept of PBWA, in which two long pulse (10^{-9} sec) laser beams were spatially overlapped in the plasma, with their frequency difference equivalent to the plasma frequency, i.e. $\Delta\omega \sim \omega_p$. This allowed them to produce intensity high enough to resonantly excite plasma oscillations, while simultaneously fulfilling the restriction on the pulse width. The concept of PBWA was further investigated by many groups, but the best experimental results in that time were reported by Clayton et al. (1993, 1994) at University of California, Los Angeles (UCLA), where a sequence of PBWA experiments adopting two lines of CO_2 laser in the plasma of density ~10^{15} cm^{-3} were conducted (Clayton et al. 1993, 1994; Marsh 1994; Esarey et al. 1996). Clayton et al. (1993) was the first group to demonstrate the ultrahigh acceleration gradient of externally inserted electrons in relativistic wakefields excited by the beating of two laser beams. The experimental setup used by this team is shown in Figure 6.4.

The experimental setup used was quite simple (Clayton et al. 1993). The laser source was a CO_2 system producing two frequencies: 10.59 μm with 60 ± 10 J energy and 10.29 μm with 10 ± 5 J energy, in the form of 300 ps pulses. Because the beat frequency is equal to the plasma frequency in this type of accelerator, the resonance density was obtained as 8.6 × 10^{15} cm^{-3}. The beams were focussed onto a diffraction limited spot of 300 μm diameter, using f/11.5 off-axis parabolic mirrors. The vacuum chamber was filled with H_2

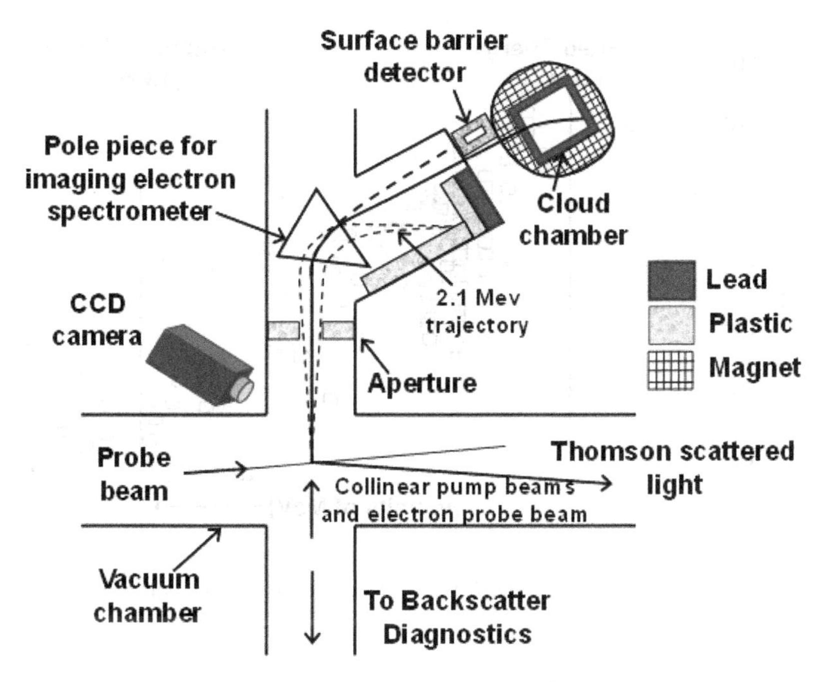

FIGURE 6.4
Experimental arrangement used by UCLA team. The optical diagnostics were made to study the wakefield characteristics, while the CCD (charge coupled device), surface barrier detector (SBD) and cloud chamber images probed the plasma, electron energy, and trajectory, respectively.

at pressures ranging from 110 to 200 mTorr (a range of pressures was taken so the necessity of resonance beat frequencies could be proven). The CCD (charge coupled device) captured an image of the plasma, which was 20 mm in length along the laser beam axis with a fully ionized core 10 mm in length. The external electron source was a 9.3×10^9 Hz RF linear accelerator that produced 20 ps long electron bunches of 2.1 MeV energy, with a 5% FWHM energy spread. The team captured images of electron trajectories in the cloud chamber and compared them with theoretical trajectories for 5.1 MeV electrons. The entire experiment was conducted for a range of pressures. Also, using optical diagnostic tools like Thomson scattering and Raman scattering, they proved the correlation between the electron energy and plasma wave amplitude. The group was able to demonstrate acceleration of externally injected electrons using PBWA, and maximum energy of electrons of 9.1 MeV (detection limit) was observed. They also observed that no signal was detected when a single frequency laser was used, or when no external electrons were injected. This means that only through beating of two waves could a wakefield be generated, and no self-trapping was possible in this approach. Finally, the peak in electron signal at 140 mTorr implied that resonance is necessary for the efficient acceleration (for more details, please see Clayton et al. 1993).

By extending the range of their detection apparatus, the same group was able to measure final electron energies of up to 30 MeV, for the same injection energy of 2 MeV (Clayton et al. 1994). In the same article, they also made more intensive optical diagnostics of the plasma wave structure using both Thomson scattering and study of electron beam perturbation, and found that although the actual wave was long-lived (diffuse scattering for 0.5–1 ns after Thomson scattering terminates and full beam deflection for 5 ns), the actual wave coherence (given by the Thomson scattering signal) lasts only for about 100 ps, which

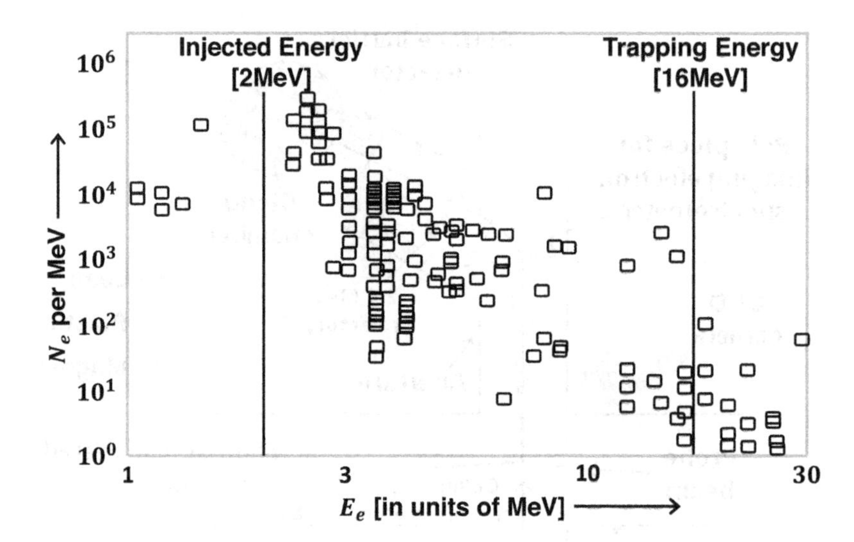

FIGURE 6.5
Electron number (N_e) (signal) versus energy (E_e) from electron spectrometer. All electrons above the energy ~16 MeV are trapped in the plasma wave.

means the electron acceleration has a rather short window. As a result, the group was able to prove that the latest gain of 30 MeV detected was due to trapping of a few injected electrons by the plasma wave, which was a theoretical requirement for efficient energy transfer (Marsh 1994). This result was arrived at by calculating the Lorentz factor of the plasma wave ($\gamma_{ph} \sim 33$, which corresponded to an electron energy of 16 MeV) and studying the electron spectrometer readings (Figure 6.5), which showed a sizeable proportion of the accelerated electrons above this energy. This was only possible in the presence of trapping.

Many other groups built on the original proposal of Tajima and Dawson (1979) by developing theoretical and computational models analyzing various aspects of PBWA. Some groups conducted more experiments on the same, using a similar approach followed by the UCLA team. Katsouleas and Dawson (1983) suggested application of a magnetic field perpendicular to the direction of plasma oscillations (surfatron) to add a component of longitudinal force to the electrons, so the issue of phase detuning between plasma oscillations and accelerated electrons could be removed. Tang et al. (1984) analyzed the process in more detail and arrived at the conclusion that the plasma wave amplitude is maximum for a slight mismatch between the beat frequency and plasma frequency (Figure 6.6). They observed that the relativistic effects on the motion of electrons is considerable when they enhanced the laser power. Such relativistic effects resulted in the maximization of accelerating field at a particular laser beat frequency. This laser beat frequency is less than the effective plasma frequency, given as $f_{\text{eff}} = 1 - \frac{1}{2}\left(\frac{9\varepsilon}{8}\right)^{\frac{2}{3}}$, where ε is the laser power parameter.

The comparative study of numerical and analytical results for the behaviour of peak amplitude of normalized accelerating electric field as a function of normalized laser beat frequency with different laser power parameters (ε) is depicted in Figure 6.6. From this figure, it is clearly seen that with increasing laser power the magnitude of the peak amplitude of the accelerating field is enhanced and the maximum is achieved for a slightly higher mismatch between the beat frequency and plasma frequency. When $f = f_{\text{eff}}$, there is a discontinuity in the amplitude and phase velocities correlated with the peak accelerating

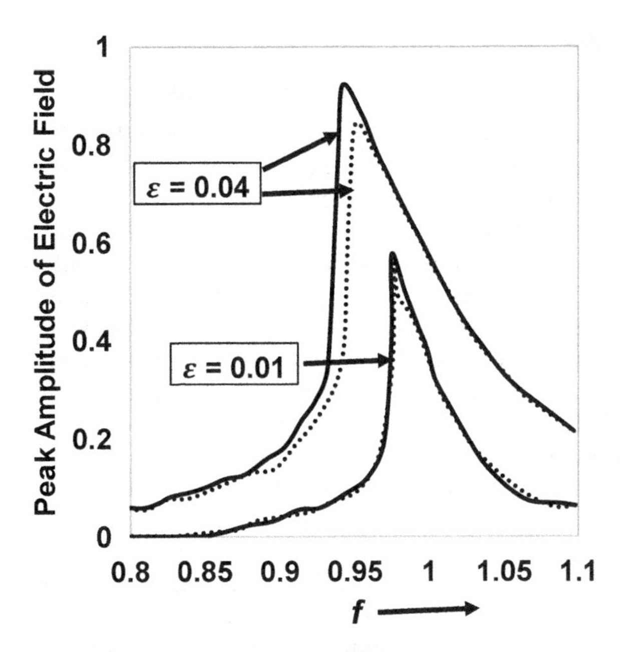

FIGURE 6.6
Peak amplitude of the accelerating field (normalized) as a function of normalized laser beat frequency (f) for different laser power parameters (ε) with numerical (dotted line) and analytical (solid line) calculations. Here, f is the frequency mismatch parameter and is given by $\frac{\Delta\omega}{\omega_p} = \frac{\omega_1 - \omega_2}{\omega_p} = \frac{\text{laser beat} - \text{wave frequency}}{\text{plasma frequency}}$.

electric field. For $f > f_{\text{eff}}$, phase velocities correlated with the peak accelerating electric field are smaller than the speed of light, whereas, for $f < f_{\text{eff}}$, phase velocities correlated with the peak accelerating electric field are greater than the speed of light.

Today, PBWAs are adopted to produce the beams of ultralow emittance using two-colour laser ionization (Schroeder et al. 2018). In this direction, Mahmoud et al. (2018) have conducted the acceleration of electrons in plasma using the beating of beams of two intense cross-focussed hollow Gaussian lasers. In collisional dominated relativistic plasma, the concept of plasma excitation from the beating of two lasers has been conducted by Kaur and Gupta (2018).

6.4.2 Self-Modulated Laser Wakefield Acceleration (SM-LFWA)

It was mentioned before that when Tajima and Dawson (1979) presented their idea of LFWA, they emphasized that the spatial width of the laser pulse should not be more than one-half of the plasma wave period for the efficient wakefield generation. This put a constraint on the plasma density with respect to the frequency of the incident laser being used (the condition of underdense plasma had to be obeyed).

In the above section, the concept of PBWA as an alternative to single laser-driven plasma accelerators was discussed in some detail. The objective was to produce a pulse short enough and strong enough to resonantly drive the plasma wave using wave beating, since producing intense pulses using chirped pulse amplification (CPA) was yet to be extended to the optical range. While examining the effect of the plasma wake on the laser pulse profile (Andreev et al. 1992), researchers stumbled across another method to achieve

LWFA using the pulses available at the time, which involved using optical pulses that satisfied the condition $c\tau_L > \frac{2\pi c}{\omega_p}$, τ_L being the laser pulse width, and for relativistic self-guiding, power slightly greater than the critical power is required $(P \geq P_c)$. Physically, this meant using sub-picosecond pulses with high intensities (~10^{18} W/cm^2) but with high plasma densities (Esarey et al. 1996), and in such a regime, the laser pulse underwent self-modulation, which resulted in two phenomena: the leading and trailing edges of the pulse steepened, shortening the pulses (Andreev et al. 1992), and the self-guiding pulse compensated for diffraction, causing the beam to remain in focus for distances greatly exceeding the Rayleigh length. This extended focussing allowed the beam to interact longer with the plasma, allowing for power buildup, which caused the beam to split into multiple beamlets, each of which drove the plasma resonantly. These highly anharmonic Langmuir waves led to enhanced acceleration gradients (Schmid 2011). The pulse shortening is the reason why it was recommended to use pulses that are longer than several plasma wavelengths. Many groups analyzed the concept and reported advantages over conventional LWFA.

The process of beam splitting by the plasma was first reported by Esarey et al. (1989), who were examining the relativistic optical guiding of laser beams in plasma beat-wave accelerators. They arrived at the conclusion that when the plasma wave amplitude was large enough to obey the inequality, the relative density perturbation follows the relation

$$\frac{\delta n_0}{n_0} \geq \left(\frac{1+\varepsilon^2}{2\pi}\right)\frac{\lambda_{p0}^2}{r_p^2} \tag{6.2}$$

where $\varepsilon^2 = \frac{2r_p^2}{r_s^2}$ together with r_p and r_s as the plasma wave spot size and beam spot size, respectively, and λ_{p0} is the plasma wavelength in the ambient density n_0; here the beam focussing was enhanced. This led to splitting of the beam into multiple beamlets of Rayleigh length $\leq \frac{\lambda_p}{2}$, and those beamlets occurring around the density minima of the original plasma wave remained optically guided (experienced enhanced focusing) while the ones around the density maxima experienced enhanced diffraction (Figure 6.7). Detailed numerical simulations demonstrating the self-modulation of laser pulses can also be found in the work by Andreev et al. (1992). The effect of the Raman instability on the propagation of short laser pulses was discussed by Antonsen and Mora (1992). Their

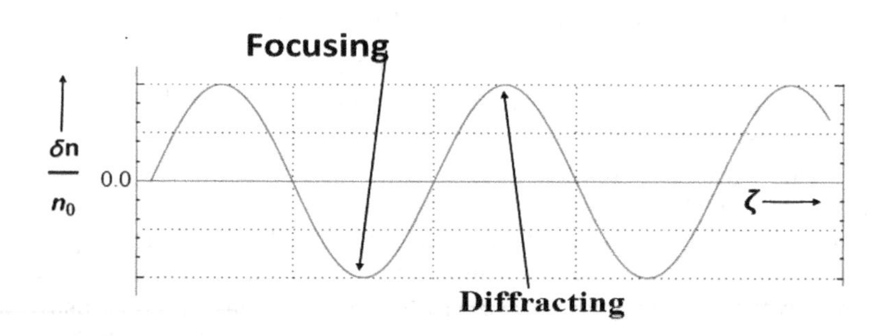

FIGURE 6.7
Schematic diagram showing the position of optically guided and increasingly diffracted beamlets. The plot is between the relative density perturbation and generalized spatial coordinate for the laser.

results showed that although self-guiding pulses did take place in the plasmas, there was a limit to how long the pulses could be, and stable propagation for over many Rayleigh lengths could only be expected for pulses that were not too long. Because the interaction time had a direct effect on the plasma wakefield strength, the pulse width needed to be taken into account when using the SM-LWFA approach. Also, they predicted that the plasma wake generated due to small angle scattering was coherent, although its applicability to particle acceleration was not investigated by them.

For better visualization of the pulse propagation characteristics, we refer to the detailed particle-in-cell (PIC) simulations made by Bulanov et al. (1995), which described, not only the self-focussing of intense, short, and ultrarelativistic pulses in plasmas, but also described the structure of the generated wakefield, the electric field distribution, and the induced focusing due to the interaction among plasma waves and laser pulses. They predicted that pulses shorter than the plasma wavelength, but wider than it (in terms of spatial focusing) excite a wakefield with a regular electric field that can be adopted to accelerate the electrons and ions.

The first group to propose a theoretical model outlining the process of SM-LWFA as a whole was that of Krall et al. (1993) at the Beam Physics Branch of the Naval Research Laboratory in Washington, DC. They considered an incident Gaussian laser pulse with $\lambda_0 = 1\ \mu$m (laser wavelength), $a_0 = 0.70$ (laser strength parameter), $r_0 = 31\ \mu$m (minimum spot size in vacuum), which implies a Rayleigh length $Z_R = 0.3$ cm, and a spatial width $L = 45\ \mu$m ($\tau = 150$ fs). The peak laser power P was thus 10 TW. The team studied the laser modulation, and its effect on particle acceleration, by making a comparison between LWFA and SM-LWFA. Using the initial constraint of $\lambda_0 = 2L$ by Tajima and Dawson (1979), the plasma density for the former method was set at $n_0 = 1.4 \times 10^{17}\,\text{cm}^{-3}$, while the same for the latter method was set at $2.8 \times 10^{18}\,\text{cm}^{-3}$. This implied that $P \approx 1.5\ P_c$ (P_c is critical power) for the self-modulation case. Under conventional LWFA, the laser power was much less than the critical power, hence relativistic guiding effects were absent, but under SM-LWFA, the over-dense plasma caused self-focussing over multiple Rayleigh wavelengths, leading to prolonged interaction and enhanced laser power, and all the beamlets that split from the main beam drove the plasma wave resonantly. Moreover, according to the original predictions, $E \sim \sqrt{n_0}$, a higher plasma density automatically meant greater wakefield strengths. Taking all these factors into consideration, the team expected to find enhancement in the acceleration gradients, and this was indeed observed in their simulations (Figures 6.8–6.10).

Figures 6.8–6.10 show three imperative physical impacts, which are stated as follows. The first is a wakefield of extremely high amplitude, which will be excited by self-modulated laser pulse because $\lambda_P < L$ (λ_P and L, respectively, are plasma wavelength and FWHM length) and $P_c \le P$ for the self-modulated case. Second, there is an increment in the laser intensity and an enhancement in the length of acceleration due to the focussing and optical guiding over various Z_R of the sections of longer laser pulse with $P_c \le P$. Third, we obtained that $\Delta\gamma_{max} = 170$ MeV.

The first measurement of acceleration of the electrons from the mechanism of the SM-LWFA was reported by Nakajima et al. (1995). The same group also analyzed the simulation results for the experiments on laser plasma interactions in nonlinear region. They used an Nd:glass laser system to focus a 3 TW pulse with pulse duration of ~1 ps into a vacuum chamber filled with He, which produced a fully ionized plasma about 20 mm on either side of the focus. The electrons for acceleration were generated by using an Al solid target, which was irradiated by a laser pulse of 200 ps. At the exit of the interaction chamber, the dipole field of magnetic spectrometer analyzed the momentum of the electrons. The measured momentum spectra reveal the maximum energy gain for the electron and

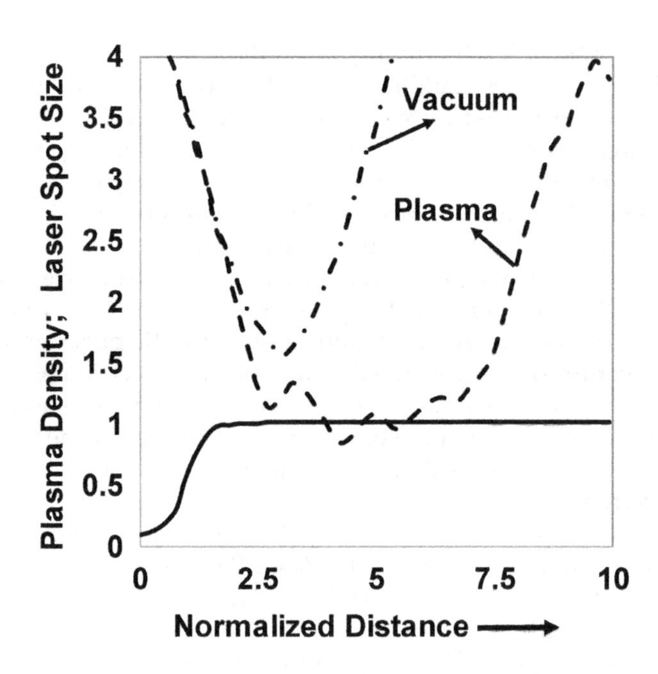

FIGURE 6.8

Normalized plasma density $\left(\frac{n_P}{n_0}\right)$ and normalized laser spot size $\left(\frac{r_s}{\lambda_P}\right)$ as a function of normalized distance $\left(\frac{c\tau}{Z_R}\right)$. Here, simulations are started at $\tau = 0$ which corresponds to the laser pulse outside the plasma and are continued up to $c\tau = 10Z_R = 0.03$ m. The plasma density reaches to full density at around $\frac{c\tau}{Z_R} = 2$, which is shown by the solid line, where Z_R corresponds to Rayleigh length or diffraction length. In the absence of self-modulation, the pulse would be expected to focus or attaining minimum spot size at $\frac{c\tau}{Z_R} = 3$ (shown by the dip in the dotted-dashed curve marked 'vacuum') but in the presence of relativistic self-guiding, the beam splits into beamlets that focus at different points (dotted curve marked 'plasma').

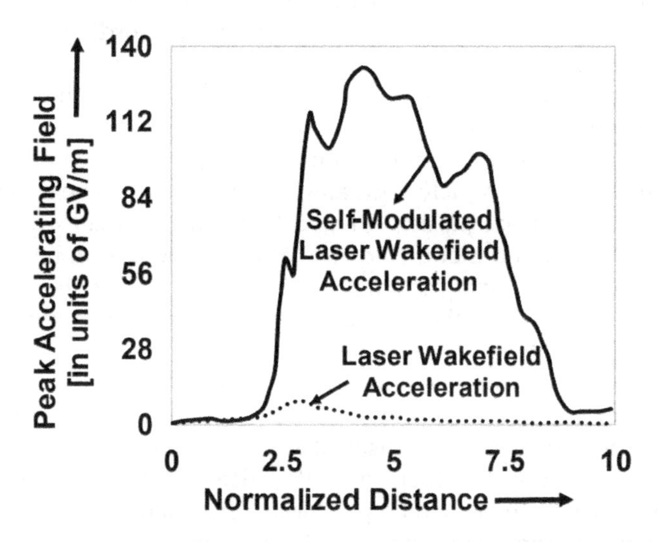

FIGURE 6.9

Peak accelerating field (E_z) as a function of normalized distance $\left(\frac{c\tau}{Z_R}\right)$ for conventional LWFA with plasma density of 1.4×10^{17} cm^{-3} and SM-LWFA with plasma density of 2.8×10^{18} cm^{-3}. Here, simulations are started at $\tau = 0$, which corresponds to the laser pulse outside the plasma and are continued up to $c\tau = 10Z_R = 0.03$ m.

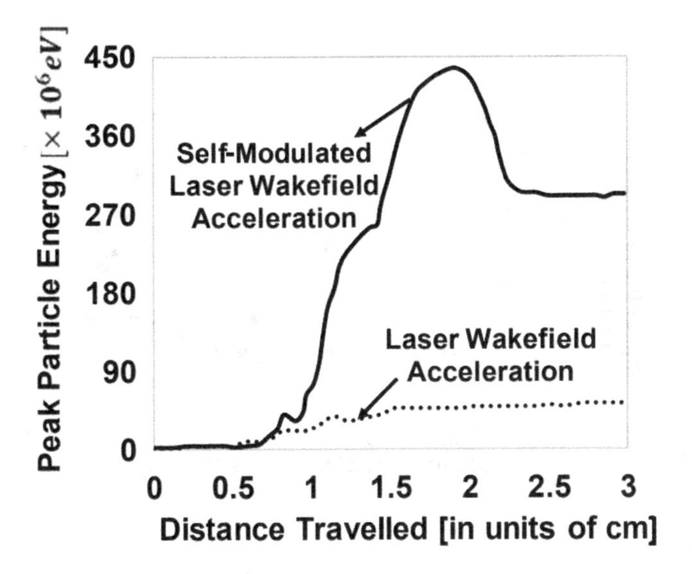

FIGURE 6.10
Peak particle energy versus distance travelled ($c\tau$) for conventional LWFA with plasma density of 1.4×10^{17} cm^{-3} and SM-LWFA with plasma density of 2.8×10^{18} cm^{-3}.

is depicted in Figure 6.11. From this figure, it is clear that for low plasma densities the linearized theory is in good agreement with the experimental data. However, the prediction of linearized theory failed in the cases of plasma densities greater than 10^{22} m^{-3}. For higher plasma densities (>10^{22} m^{-3}), the experimental results agree more with the nonlinear model. This suggests that because of the mismatching between laser pulse duration and plasma wave period, the excitation of the wakefield decreases, which is preventing

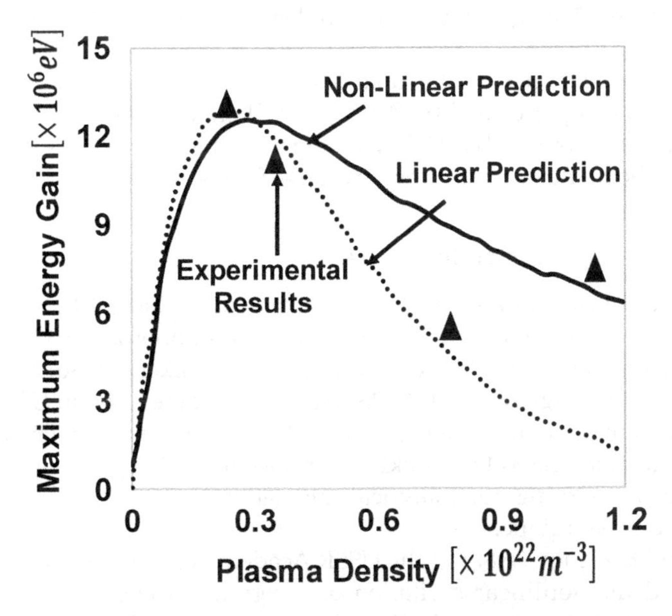

FIGURE 6.11
Comparison of experimental results of maximum energy gain of the injected electrons with linear and nonlinear predictions with the variation of plasma density. The experimental results agree more with nonlinear models for higher densities, implying self-modulation.

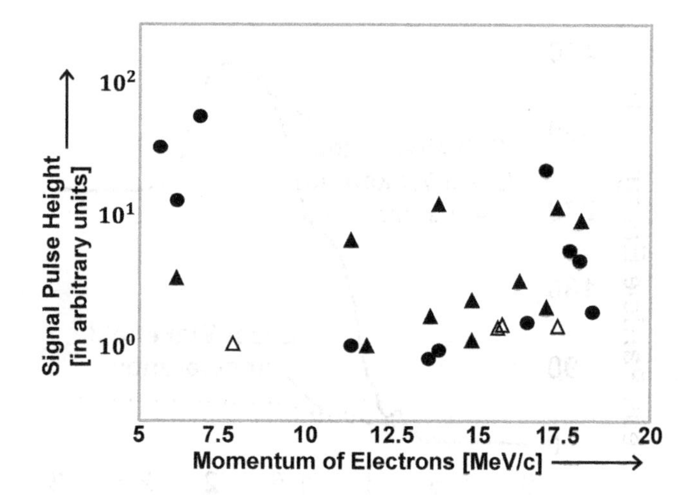

FIGURE 6.12
Signal pulse height as a function of momentum of electrons (case of He gas jet) at 7.8 atm pressure. Hollow triangles represent 3.7 TW (1 ps) with no electron injected (200 ps), solid triangles represent 3.7 TW (1 ps) with 15 J (200 ps), and solid dots represent 2.8 TW with 14 J (200 ps).

the nonlinear behaviour of plasma waves. Highly nonlinear effects caused the excitation of plasma waves in a more efficient way at higher densities.

When electrons were injected, a considerable number of signals were observed as portrayed in Figure 6.12 (Nakajima et al. 1995). The levels of observed signals, when no electrons were injected, were as minuscule as background signals. Around a hundred of the injected electrons were trapped in the plasma and were accelerated with the momenta higher than 5 MeV/c. The maximum momentum of 18 ± 0.8 MeV/c was observed for the accelerated electrons. This group was able to demonstrate a peak accelerating gradient of 30 GeV/m.

The theoretical and experimental analysis of betatron X-ray radiation in the region of SM-LWFA has been performed by Albert and Thomas (2016). The generation of beams of relativistic electron from the SM-LWFA are today adopted to enhance the characteristics of X-rays like high photon flux and wide energy range from keV to several MeV (King et al. 2019).

6.4.3 Laser Wakefield Accelerator (LWFA)

The intense laser pulse can also drive the wakefield in plasma. In such cases, with the gradient of laser pulse intensity, there is an associated ponderomotive force, which results in the displacement of plasma electrons and drives the wakefield. Such plasma-based particle accelerators are designated as LWFAs. Almost a decade after an LWFA was proposed by Dawson and Tajima, and around the same time as the use of CPA to power up optical pulses was demonstrated by Strickland and Mourou (1985), researchers revisited the concept and began carrying out analytical calculations and numerical simulations on the many facets of this technique.

Gorbunov and Kirsanov (1987), at the USSR Academy of Sciences in the erstwhile Soviet Union, analyzed the nonlinear excitation of Langmuir waves by short electromagnetic pulses. They concluded that the injection of a charged particle into a wake of sufficiently high amplitude, at a velocity close to that of the plasma wave group velocity v_E, could result in the acceleration of the charged particles to high energies, if the injection happened at a time and point when the electric field was directed along the wave motion. They derived an approximate expression for the electric field of the wakefield, given as

$$E_L \sim \frac{4\pi n_0 e L v_E^2}{4c^2} \tag{6.3}$$

where v_E refers to the magnitude of electron's velocity in oscillatory motion in the transverse field. For a pulse of duration 10^{-13} s and wavelength 10 μm, in a plasma of density, $n_0 = 10^{17}$ cm^{-3} the field came out to be around 8×10^6 V/cm, which was comparable to the experimentally obtained fields of (3–10)$\times 10^6$ V/cm in PBWA. However, a good adjustment between plasma frequency and beat frequency is required in PBWA, and because PBWA relied on resonance, which was hard to ensure in those times when the plasma was highly inhomogeneous, LWFA was a more elegant and practical alternative, as its operation was independent of resonance. Theoretical calculations concerning electron acceleration with LWFA were also made by Sprangle et al. (1988) at the Naval Research Laboratory, and they reiterated the importance of relativistic self-guiding for prolonged laser-plasma interaction. For their calculations, the group considered a short laser pulse with $\tau \sim 2\pi\omega_p^{-1}$ and power just above the critical laser power so that relativistic guiding could take place, but not so much that pulse filamentation and stimulated Raman forward scattering (SRS-F) occurred. Through the results of theoretical calculations and simulations, this group predicted maximum acceleration gradients of about 2.6 GeV/m (Figure 6.13).

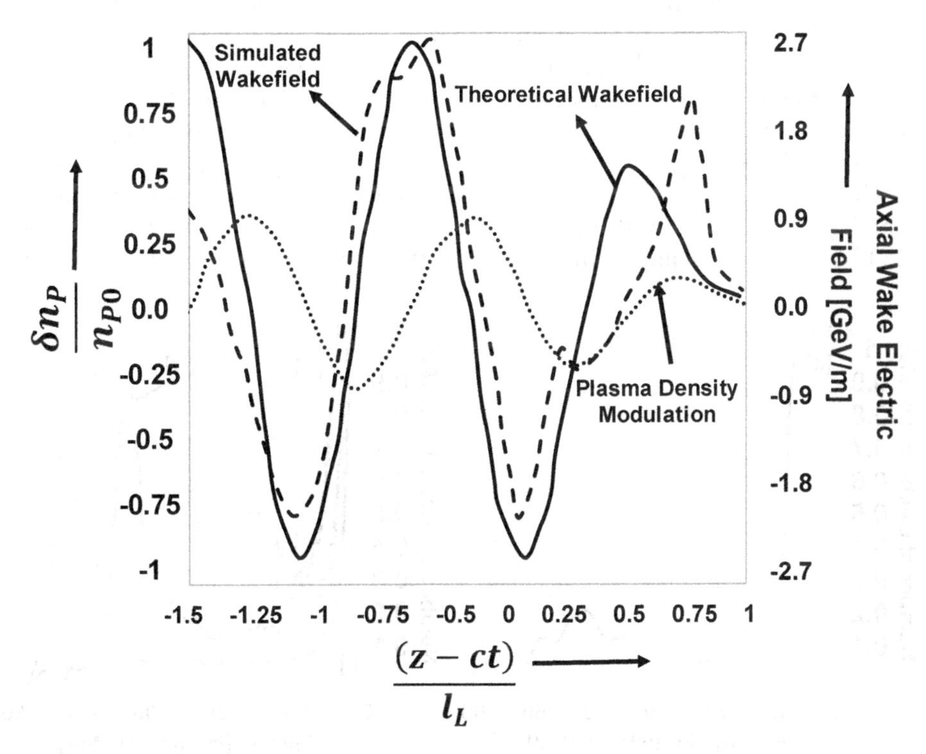

FIGURE 6.13

Axial wakefield (E_Z) and density wake $\left(\frac{\delta n_P}{n_{P0}}\right)$ for the theoretical calculations are represented by a solid and dotted line, respectively. The dashed line refers to the axial wakefield determined from simulations (FRIEZER–particle code).

Present theoretical work in LWFA greatly revolves around the bubble regime of laser-plasma interaction, which was hinted at before when discussing the constraint on laser pulse-width. The bubble regime was introduced by Pukhov and Meyer-ter-Vehn (2002) and further studied by Kostyukov et al. (2004). Under normal LWFA, the laser pulse pushes the electrons out of its way due to the ponderomotive force, but these remain trapped in the plasma potential, and execute simple harmonic oscillations at the plasma frequency. The oscillations in the longitudinal direction lead to acceleration either though potential trapping (leading to maximum gain in energy) or abrupt exit from the plasma the moment the field reverses direction. However, the bubble regime works under the presumption that the laser radiation pressure (read ponderomotive force) is so strong that the wave breaks after the very first oscillation, and the charge separation field is insufficient to cause electron oscillation in any direction. The laser pulse is able to propagate many Rayleigh lengths through the plasma without significant spreading, and a cavity completely devoid of cold plasma electrons trails it at the same velocity. Since the cavity or bubble is composed of only ions, bulk electrons near the end of the bubble are accelerated along the axis; thus, a dense bunch of relativistic electrons with a monoenergetic spectrum is generated. In a way, the bubble regime of LWFA is similar to what Tajima and Dawson (1979) first proposed, not only because both methods operate without external injection of charged particles, but because peak fields are observed when $c\tau_L \sim \frac{\lambda_p}{2}$.

Figure 6.14 shows the results of the PIC simulations made by Pukhov and Meyer-ter-Vehn (2002) concerning energy spectrum of accelerated electrons. In Figure 6.14(a), a small peak at the electron energy of about 45 MeV is portrayed in a plateau-like structure. They observed that total 10^9 relativistic electrons have energies greater than 5 MeV. The emittance is found better as compared with that of conventional accelerators.

In Figure 6.14(b), a plateau-like spectrum, similar to that detected in Figure 6.14(a), still appears after 350 laser cycles. However, a significant change in the spectrum is observed after 450 laser cycles, as there is a development of peak, which grows with time and shifted towards higher energies with increasing laser cycles. Between the energy interval of 300–360 MeV, there are approximately 3.5×10^{10} electrons after 750 laser cycles. Many

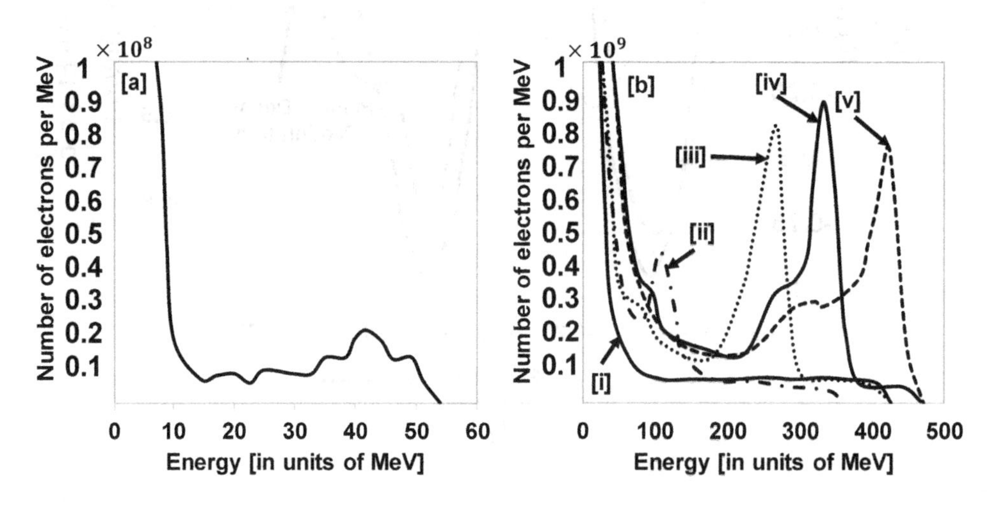

FIGURE 6.14
Electron energy spectrum simulations. (a) Final energy spectrum in the case of propagation of a laser pulse of 6.6 fs, 20 mJ in plasma. (b) Time evolution of energy spectrum in the case of a 12-J, 33-fs pulse with laser cycles of [i] 350, [ii] 450, [iii] 550, [iv] 750, and [v] 850.

research groups like Malik et al. (2007), Sadighi-Bonabi and Rahmatollahpur (2010), and Li et al. (2015), are looking into the general bubble shape and the connection between the bubble strength and laser pulse characteristics.

Perhaps the very first experimental evidence of plasma wake generation due to intense laser-plasma interaction was given by Hamster et al. (1993), when they detected sub-picosecond terahertz pulses on focussing optical pulses of sub-picosecond width and terawatt power onto both solid and gas targets. Though their purpose was not explicitly to demonstrate particle acceleration, they did manage to confirm the existence of the wakefield, as predicted by theory.

A more detailed report dealing directly with LWFA was presented by Nakajima et al. (1992, 1994) at the Linear Accelerator Conference in Canada. They used an Nd:glass laser system capable of producing ultrashort pulses of 1 ps width and 30 TW power to generate a plasma in He and to generate external electrons from a solid Al target. The electron trajectory was curved using strong dipoles, and the beam was injected into the focussed plasma with a time delay produced by adjusting the optical path of the two lasers. An electron energy spectrometer and scintillation counter were placed at the end to detect the accelerated electrons. For a detailed description of experimental setup, please see Nakajima et al. (1992). This group was able to generate a plasma wave with large amplitude and a peak acceleration gradient of 2.5 GeV/m, as shown in Figure 6.15.

Other groups also provided experimental evidence of LWFA, such as Marques et al. (1996) at the Ecole Polytechnique in France and Siders et al. (1996) at the University of Texas in Austin, Texas. Leemans et al. (2006) also have demonstrated the generation of monoenergetic GeV electrons experimentally by using LWFA. Their experiments produced results in good agreement with theoretical predictions. Another application of the LWFA is in producing light sources, which has been discussed by Albert and Thomas (2016).

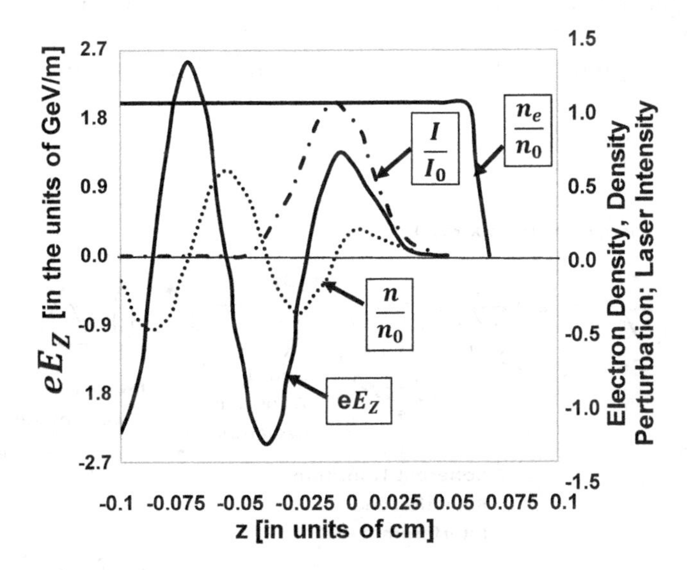

FIGURE 6.15

Behaviour of axial electric field, normalized electron density $\left(\frac{n_e}{n_0}\right)$, normalized density perturbation $\left(\frac{n}{n_0}\right)$, and normalized laser intensity $\left(\frac{I}{I_0}\right)$ as a function of distance propagated in plasma.

6.4.4 Plasma Wakefield Accelerator (PWFA)

As discussed in the LWFA mechanism, the plasma wave is driven by a single laser pulse where the ponderomotive force induces a charge separation for the acceleration of charged particles. However, the plasma waves can also be driven by a relativistic electron / positron bunch where the Coulomb field induces the charge separation, and the acceleration process takes place in that wakefield. As mentioned earlier, this acceleration mechanism is known as PWFA. Here, a relativistic beam of charged particles, such as electrons or positrons, having bunch length σ_z (approximate to half of the plasma wavelength, i.e. $\lambda_p \sim 2\sigma_z$) excites the large amplitude plasma wave. The electric field of the plasma wave is given by $eE_p = 240(\text{MeV / m})\left(\frac{N}{4\times10^{10}}\right)\left(\frac{0.6\text{ mm}}{\sigma_z}\right)^2$ where N is the number of particles per bunch (Joshi et al. 2002). In the case of the excitation of plasma wakefield by electron bunch, the Coulomb force of the space charged particles expels the plasma electrons and sets up the plasma oscillations as the electrons rush back in after the beam gets passed. In another case, if the driving beam is a bunch of positrons, the Coulomb force would pull out the background plasma electrons that will overshoot and set up the plasma oscillations trailing behind the beam.

The electron / positron beam–driven PWFA scheme was recognized in 1988, while performing numerous experiments at Argonne National Laboratory by Rosenzweig et al. (1988) that introduced the world to the concept of the 'blow-out' or bubble regime. Further testing of the PWFA mechanism was done in the 1990s at the 3 km long Stanford Linear Accelerator Center (SLAC) linac housed in the final 150 m of the beamline, as shown in Figure 6.16. This section of the beamline, named the Final Focus Test Beam (FFTB) facility, was used to test the magnetic-focussing system for a linear collider. They arranged two bends in the beamline, one to achieve different total path lengths for different electron energies and another in the transport line from the damping ring to the linac. Both of these bends have negative 'group delay' or negative 'R56', attained by the addition of the four-dipole chicane magnet at the ~1 km mark of the 3 km long linac with a well-chosen positive R56. It could help to remove the chirp in the first one-third of the linac and compress the

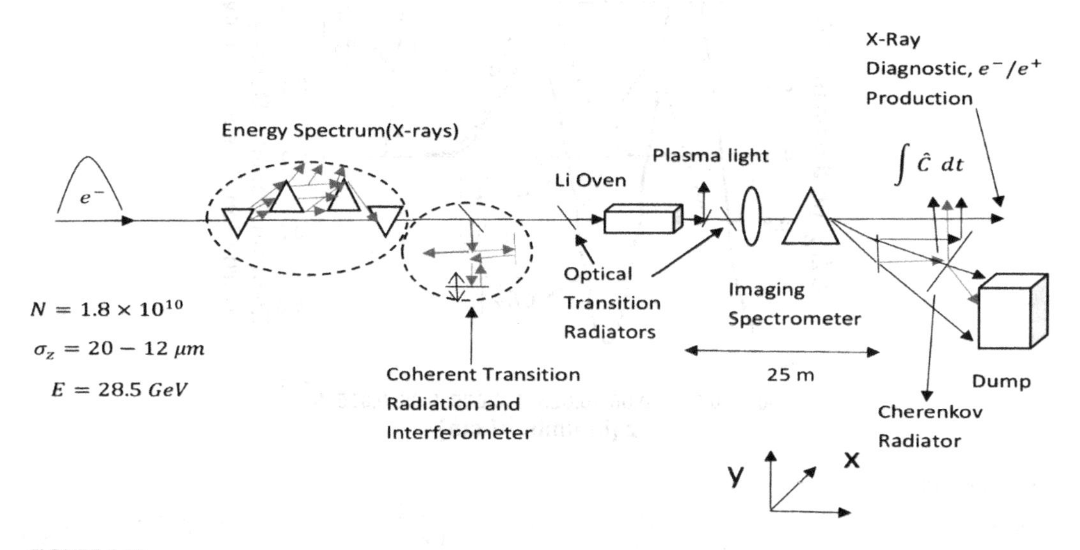

FIGURE 6.16
Schematic of PWFA experiment in FFTB beamline.

pulse to 50 μm RMS (root-mean square). Finally, the new, additional chirp due to the wakefields in the final two-thirds of the linac could be nearly entirely removed in the FFTB, allowing for even more pulse compression. With the addition of this chicane, the pulses went from an RMS bunch length of about 600 μm down to as low as 12 μm.

As shown in Figure 6.16, the experiment consists of 1-μm-thick titanium foil used to generate coherent transition radiation for bunch-length measurements. Then this radiation gets focussed down into a column of lithium vapor contained in a heat-pipe oven at about 1000 °C. Furthermore, the utilization of an imaging spectrometer analyzed the final energy of the beam before heading to the beam dump. At a distance of 1 m from each end of the lithium oven, there is another 1-μm-thick foil, which was located for optical transition radiation (OTR) measurements of the beam size and position. To analyze the spectral measurements of plasma and Cherenkov light, another titanium foil was used downstream of the lithium oven. Finally, the residual energy was determined by an imaging spectrometer followed by a beam dumper. The efficient conversion of bunch energy into a plasma wave would take place when the bunch length (RMS) approximates to half of the plasma wavelength, i.e. $\lambda_p \sim 2\sigma_z$, followed by the enhancement of resonant electron density. This type of increment in operating density maximizes the accelerating field by the relation $E_z \approx 96\varepsilon_{epw}\sqrt{n_e}$, where ε_{epw} denotes the field amplitude of electron plasma waves. Now, the focussed electron beam not only attains the required electric field to tunnel-ionize the lithium, but it also reduces the need for photo-ionization of the plasma. The notation $\int \hat{C}dt$ designated in Figure 6.16 corresponds to the imaging of the Cherenkov light onto a CCD camera with no time resolution. In the energy-dispersion plane of the spectrometer, the Cherenkov emitter (a piece of aerogel or a length of ambient air) acts as an electron detector. In the spectrometer, the vertical dispersion of electrons takes place based on their energies with the installation of a long dipole magnet. The additional presence of quadrupole magnets images the exit of the plasma onto the dispersion plane, distinguishing between the exit energy and exit angle.

In the beam-driven wakefield, a large number of particles present in the bunch core transfer their energy to particles present in the back of the same bunch. Thus, the wakefield can double the energy of the accelerator, as proposed by Lee et al. (2001). It was suggested that if we use ten times shorter bunches for the PWFA, then the accelerating gradient of a linear collider can be doubled, for example, a 50-GeV SLAC linear collider would convert into a 100-GeV machine. In an experiment, the energy doubling of electrons at ~ 40 GeV has been achieved by using the PWFA scheme (Blumenfeld et al. 2007), indicating the impact of such plasma accelerators on other fields of technology concerning high-energy particles. Moreover, this beam-driven wakefield can also be used to accelerate positrons and muons.

At the Facility for Advanced Accelerator Experimental Tests (FACET) in Menlo, California, several experiments based on PWFA are going on. These involve the acceleration of positrons using wakes in uniform plasma and the achievement of electron acceleration with ultrahigh gradients. For the advancement in PWFA, a facility for the 10-GeV electron beam is under development (Joshi et al. 2018).

6.5 Photon Acceleration

In plasma-based acceleration processes, an electrical disturbance (for example, an intense laser pulse in PBWA or relativistic electron beam in PWFA) propagating through a plasma sets up the plasma wave, which once generated is able to perform 'beam loading'. Beam loading means that when we inject a trailing bunch of electrons into the accelerating phase of the plasma wave, as in PWFA, then the energy transfers from the wave to the trailing electrons. If we replace the trailing bunch of electrons with a short electromagnetic pulse with a wavelength 1.5 times of the plasma wavelength, then the second injected pulse creates an identical wake, which is out of phase with the first one by 180° (Wilks et al. 1989). Now, behind the second pulse, the superposition of two wakes lowers the amplitude of the plasma wave, indicating the absorption of energy by the second pulse from the wake created by the first laser pulse. Hence, the frequency of the second pulse will be continuously upshifted, as suggested by the conservation of the photon number density.

The rate of upshifted frequency was estimated in the presence of relativistic Langmuir waves, propagating in homogenous plasma with velocity v in the x-direction. Considering δn to be the amplitude (in terms of density) of Langmuir waves and n_0 as the initial plasma density, the initial plasma frequency (ω_{p0}) was modified in density perturbations by the relation

$$\omega_p^2(x,t) = \omega_{p0}^2\left\{1-\varepsilon\sin\left[k_p(x-vt)\right]\right\} \tag{6.4}$$

where $k_p = 2\pi/\lambda_p$ is the plasma wavenumber and $\varepsilon = \dfrac{\delta n}{n_0}$ is the normalized amplitude. Now, the second laser pulse is injected into the accelerating phase of the wave with frequency $\omega > \omega_p$ and width $\Delta x \le \lambda_p/2$. The variation in the frequency of an electromagnetic wave can be estimated from the dispersion relation $\omega^2 = \omega_p^2(x,t) + c^2k^2$ as

$$2\omega\delta\omega - 2c^2k\delta k = \delta\omega_p^2 \tag{6.5}$$

Eq. (6.5) can be written in the following form in the plasma-wave frame, setting $\omega' = \gamma_p(\omega - vk)$

$$\delta\omega_p^2 = 2\omega\delta\omega(1 - c^2k/\omega v) \tag{6.6}$$

The plasma frequency, in the laboratory frame, is assumed to be a function of $(x - vt)$, which varies as

$$\delta\omega_p^2 = \frac{\partial\omega_p^2}{\partial x}(\delta x - v\delta t) \tag{6.7}$$

The combination of Eqs. (6.6) and (6.7) gives rise to

$$\frac{\delta\omega}{\delta x} = \omega_p^2\varepsilon k_p/2\omega \tag{6.8}$$

This amount of increment in the frequency of accelerated laser packet reveals the energy loss of the accelerating plasma wave. The wave velocity of an accelerating wave is $v < c$; hence, the accelerated photons face a phase slip with respect to the wave and prevent the energy gain measured by Eq. (6.8). To estimate the limit on the frequency upshift, we make Lorentz transformation in a moving frame, where the dispersion relation remains constant under transformation and can be written in terms of moving (primed) quantities as

$$c^2k'^2 = \omega'^2 - \omega_{p0}^2\left\{1-\varepsilon\sin\left[k_p'x'\right]\right\} \tag{6.9}$$

It was found that the injection of the laser pulse into the trough of the plasma wave will increase k' and maximize the frequency upshift. This happens when the pulse just turned around at the peak of density (where $k' = 0$ or $\omega' = \omega_{p0}(1+\varepsilon)^{1/2}$) and then travels back until it reaches the trough one more time. With all these assumptions and Eq. (6.9), the initial wave number turns out to be $k_i' = -(\omega_{p0}/c)(2\varepsilon)^{1/2}$ in a moving frame, i.e. the photons are initially travelling backward. Then the gradually accelerating photons slip back in the wave until they are reflected at the peak and reach the bottom of the trough and attain the final momentum with an opposite sign to $k_i' = +(\omega_{p0}/c)(2\varepsilon)^{1/2}$. Again, using the Lorentz transformation and assuming the phase velocity of the plasma wave to be approximately equal to c, the frequency in the laboratory frame comes out to be

$$\omega_f \approx \omega_i\left[1+2(2\varepsilon)^{1/2}+4\varepsilon+O(\varepsilon^{3/2})\right] \tag{6.10}$$

The obtained frequency attains its maximum value on satisfying the following condition imposed on the wave phase velocity

$$\gamma_{\text{wave}} \le \left(\omega_i/\omega_p\right)\left[(1+\varepsilon)^{1/2}-(2\varepsilon)^{1/2}\right]^{-1} \tag{6.11}$$

The peak density and energy for a relativistic electron beam are assigned as ε and γ_{wave}, respectively. The dephasing of accelerated photons can be controlled by using a ramped plasma density so that the phase velocity of the plasma wave varies continuously and keeps the photons in the same phase, provided the density gradient scales as

$$L_n = \left(\frac{1}{n}\frac{dn}{dx}\right)^{-1} \approx m\lambda_p\gamma_{\text{wave}}^2\frac{1+\sqrt{\varepsilon}}{\varepsilon} \tag{6.12}$$

where m denotes the number of wavelengths behind the back of the driver. These density ramps enhanced the frequency upshifts by a factor of 10 or more.

The frequency upshift for the injection of an Nd:glass laser into ramped plasma density $n_0 = 10^{18}\,\text{cm}^3$, containing 30% plasma wave ($\varepsilon = 0.3$), has been plotted in Figure 6.17. It was found that the wavelength decreases as the plasma density varies according to Eq. (6.12). The pulse length of the injected laser is $\lambda_p = 1.4\ \mu\text{m}$ and characteristic lengths for dispersion and diffraction are $L_{\text{disp}} = 0.9\text{m}$ and $L_{\text{diff}} = 0.74$ m, respectively.

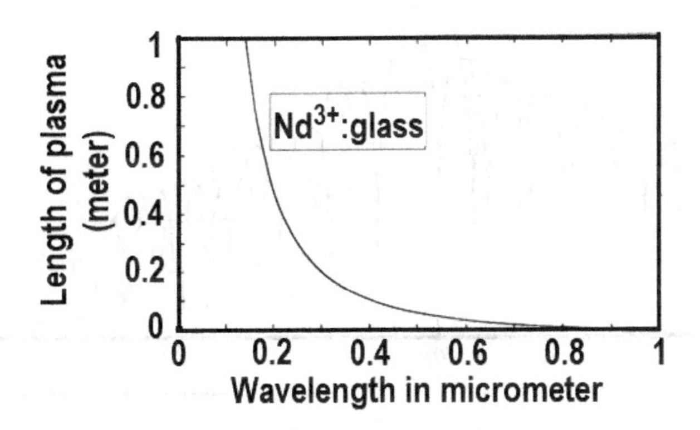

FIGURE 6.17
Amount of upshift possible for an Nd:glass laser injected into ramped density plasma.

Frequency upshift for a fully relativistic electromagnetic pulse was analyzed by PIC simulations, where the plasma wave was set up by a wakefield driver with an initial vale of γ as 22. The driver set up density perturbation with value $\varepsilon \sim 0.25$ and produced a wake into which a laser pulse was injected. The initial frequency of the laser pulse was $\omega_i = 18\omega_p$ whereas the pulse width was $L_{\text{pulse}} \approx 2.5c / \omega_p$. An upshift of up to 10% (i.e. $\omega_f = 19.8\omega_p$) was found in the laser pulse after a distance of $237c / \omega_p$, which was closer to the energy gain predicted in a linear model by Eq. (6.8), which was $\omega_f = 19.5\omega_p$. This small difference occurred due to the nonlinear wave steepening arising in the wake, even at the density perturbations of 25% (Figure 6.18; Wilks et al. 1989).

The frequency shift of the second laser pulse can also be analyzed by using ray theory with the photon equation of motion (Mendonça and Oliveira E Silva 1994), as

$$\frac{d\vec{r}}{dt} = \frac{\partial \omega}{\partial \vec{k}}, \quad \frac{d\vec{k}}{dt} = -\frac{\partial \omega}{\partial \vec{r}} \tag{6.13}$$

where the photon position $\vec{r}$ and wave vector $\vec{k}$ determine the photon frequency as $\omega = \left(k^2c^2 + \omega_p^2\right)^{1/2}$.

Photon acceleration can occur both in non-stationary as well as in other time-varying media, which also has been confirmed by experiment (Dias et al. 1997). Assuming the interaction of a laser beam with relativistic ionization fronts with velocity v_f, using Eq. (6.13), the frequency shift of the photons comes out be

$$\Delta\omega = \frac{\omega_p^2}{2\omega} \frac{v_f}{c \pm v_f} \tag{6.14}$$

The plus / minus sign gives the propagation direction of photons parallel or anti-parallel to the ionization front, respectively. The frequency shifts up to 144 nm have been already observed for ionization fronts propagating with the velocity 0.99c (Lopes et al. 2004).

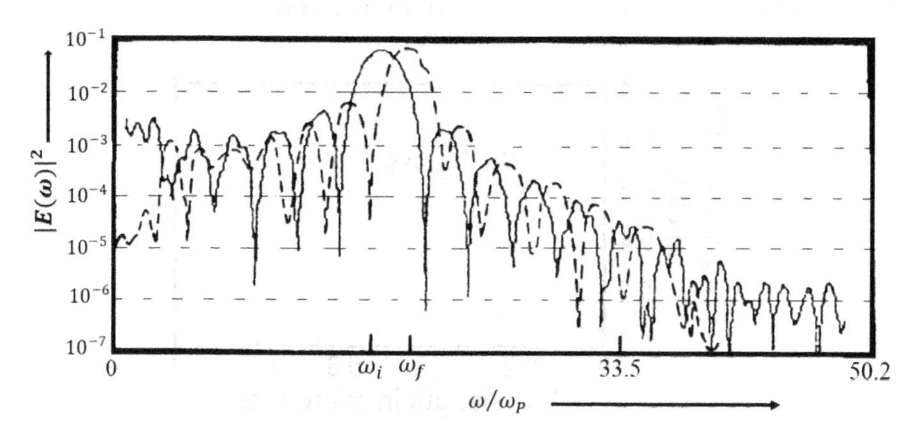

FIGURE 6.18
The initial (solid line), and final (dotted line) pulse frequency shows frequency upshift of laser packet.

6.6 Plasma-Based Ion Acceleration

In this section, we talk about the interaction of an intense laser beam with underdense plasma to accelerate ions. The ponderomotive force of the laser beam expels the plasma electrons and induces charge separation, which sets up the space-charge field. During such interaction, the focussed laser beam was found to be in a 'cigar'-like shape, which implies that the ions will experience primarily a radial electric field (Krushelnick et al. 1999; Wei et al. 2004). Hence, the energetic ions were measured along the perpendicular direction to the laser propagation direction. The above technique used to accelerate the ions is termed as 'Coulomb explosion' or 'ponderomotive shock acceleration'. The ions accelerated in the laser-plasma interaction region have been reported to gain energy roughly equal to the quiver energy. The analysis of ion acceleration in laser-plasma interaction is quite essential as it unveils a great deal of important information related to self-focussing and channeling occurred due to relativistic and charge displacement effects. Also, many researchers have successfully related the acceleration of ions with the production of neutrons within hot channel formation (Tabak et al. 1994; Roth et al. 2001).

Experiment concerning ion acceleration in underdense plasma has been performed at Rutherford Appleton Laboratory, where the laser pulse with a wavelength of 1.054 μm, energy 180 J, and pulse duration of $0.5-0.7$ ps was focussed on the supersonic gas jet (with a 2 mm nozzle diameter) by the $f/3$ off-axis parabolic mirror to a focal spot size of 10 μm in a vacuum. Using He as the working gas, the backing pressure in the gas nozzle was able to produce plasma density in the range $(0.04-1.4)\times 10^{20}\,\mathrm{cm}^{-3}$. The energy spectrum of ions was measured with a Thomson parabola ion spectrometer and 1-mm-thick CR39 nuclear track detector. The spectrometer was placed 80 cm away from the interaction region and at 100° from the laser propagation direction. After placing a stack of several radiochromic film (RCF) strips 6 cm away from the interaction region, the angular distribution was measured. Typical He ion spectra at a high and a low plasma density are shown in Figure 6.19. There were two characteristics of ions that came to light after obtaining their energy spectra: the energy of ions gets enhanced at high density, and at high energy a plateau was observed. At higher densities, the additional shock acceleration enhanced the ion acceleration.

The observation of plateau structure in Figure 6.19 indicates that laminar shock waves are driven by the laser induced acceleration process (Silva et al. 2004). Additionally, the radial momentum of ions as a function of radial distance points out the acceleration of

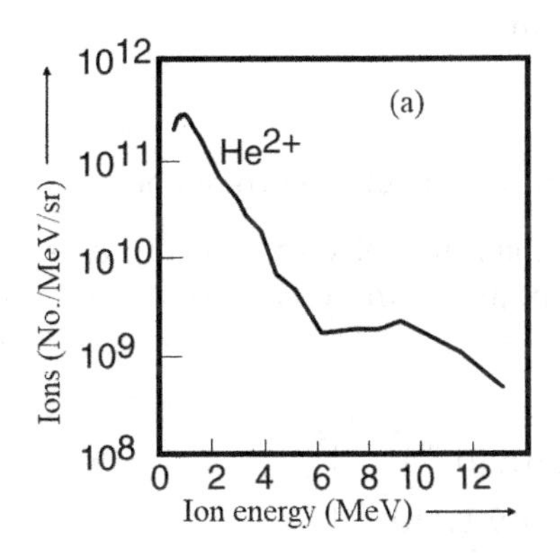

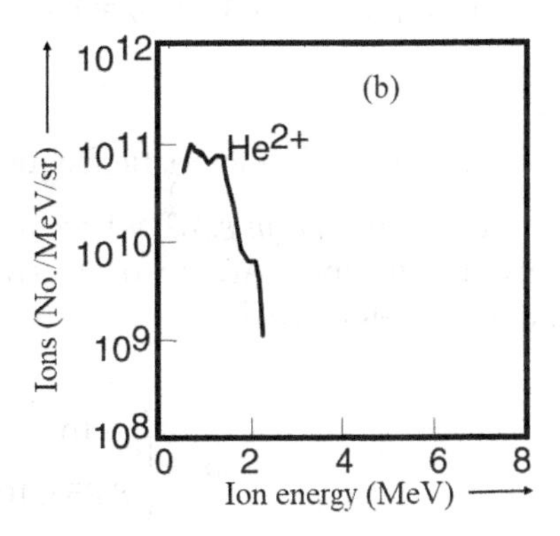

FIGURE 6.19
Spectra of He^{2+} ion at (a) high and (b) low density of plasma.

ions due to the interaction of collisionless shocks. Moreover, the ion acceleration due to the ponderomotive expulsion of plasma was found to be enhanced by collisionless shock acceleration.

6.7 Conclusions

The main advantage of a plasma-based accelerator is its ability to support ultrahigh acceleration gradients without a risk of material breakdown, since the plasma is already an ionized medium. In PBWA, all the electrons with energies more than 16 MeV were found to trap in the plasma waves. Also, it was found that the plasma wave amplitude is maximum for a slight mismatch between the beat frequency and plasma frequency. In LWFA, between the energy interval of 300–360 MeV, there are approximately 3.5×10^{10} electrons after 750 laser cycles. The peak accelerating field of around 130 GV/m and peak particle energy of around 440 MeV have been achieved in the method of SM-LWFA. If we use ten times shorter bunches for the PWFA, then the accelerating gradient of a linear collider can be doubled. The energy doubling of electrons at ~40 GeV has been achieved in this scheme. In plasma-based ion acceleration, the energy of ions (He^{2+}) gets enhanced at high plasma density, and at high energy, a plateau is observed. At higher densities, the additional shock acceleration is found to enhance the ion acceleration.

Since the initial experiments were quite interesting in the sense of their development stages and those provide a description that inspires researchers and students, we had focussed on such experiments only. Otherwise there are several advancements in the said experiments and also some new concepts were taken up in the coming time. It was not possible to discuss all the experiments and the techniques in view of the scope of the book.

6.8 Selected Problems and Solutions

PROBLEM 6.1

Calculate the plasma frequency and the 'cold wave-breaking limit' in the non-relativistic case for the given plasma density $n_e = 1 \times 10^{18}\,\text{cm}^{-3}$.

SOLUTION

Plasma oscillates at its characteristic frequency, i.e. the plasma frequency, given (in angular units) by $\omega_p = ((n_e e^2)/(m_e \varepsilon_0))^{\frac{1}{2}}$, where n_e is the initial electron density, ε_0 is the vacuum permittivity, and m_e and e are, respectively, the mass and charge of an electron. For the given set of parameters, we have

$$\omega_p = \left(\frac{10^{24} \times \left(1.6 \times 10^{-19}\right)^2}{8.85 \times 10^{-12} \times 9.1 \times 10^{-31}} \right)^{\frac{1}{2}}$$

$$\omega_p = 5.6 \times 10^{13}\,\text{rad s}^{-1}$$

The separation of electrons and ions in the plasma wave sets up huge electric fields, which can be used to accelerate charged particles. The magnitude of the electric field in a plasma wave can reach the order of the cold wave-breaking limit, given by

$$E_{wb} = (m_e c\omega_p)/e$$

Hence, this field is found as follows

$$E_{wb} = \frac{9.1\times 10^{-31}\,\text{kg}\times 3\times 10^{8}\,\text{ms}^{-1}\times 5.6\times 10^{13}\,\text{rad s}^{-1}}{1.6\times 10^{-19}\,\text{C}}$$

$$E_{wb} = 95.55\ \text{GV m}^{-1}$$

PROBLEM 6.2

For the density of electrons in a plasma as $2\times 10^{18}\,\text{cm}^{-3}$, find out the value of the acceleration gradient obtained using the field of plasma wave excited in the linear regime.

SOLUTION

The acceleration gradient is given by

$$E \cong c\sqrt{\frac{m_e n_e}{\varepsilon_0}}$$

$$E \cong 0.096\sqrt{n_e}\ (\text{m}^{-3})$$

$$E = 0.096\sqrt{2\times 10^{24}} \sim 136\ \text{GVm}^{-1}$$

PROBLEM 6.3

Calculate the magnitude of the electric field, intensity, and power of a Gaussian laser beam having spot size $r_0 = 10\ \mu\text{m}$ and wavelength $\lambda = 1\ \mu\text{m}$ with laser strength parameter $a_0 = 1$.

SOLUTION

Laser strength parameter is given in terms of the electric field E_L of the laser as $a_0 \simeq eE_L/\omega m_e c$ with frequency $ck = 2\pi c/\lambda$.

The electric field of the laser beam is calculated as

$$E_L = \frac{2\pi m_e c^2}{e\lambda} = 3.21\times 10^{12}\ \text{V/m}$$

The intensity I and power P can be obtained by using the following relations

$$I = \frac{1}{2}\varepsilon_0 c E_L{}^2,\ P = I\left(\frac{\pi r_0^2}{2}\right)$$

Alternatively, the laser strength parameter can also be written as

$$a_0 \simeq 0.85\times 10^{-9}\,\lambda[\mu\text{m}](I[\text{W/cm}^2])^{\frac{1}{2}}$$

which gives

$$I = 1.38 \times 10^{18}\,\mathrm{W/cm^2}$$

And,

$$P(\mathrm{GW}) \approx 21.5(a_0 r_0/\lambda)^2$$

$$P = 21\ \mathrm{GW}$$

$$E_L[\mathrm{TV/m}] \approx 3.2 a_0/\lambda[\mu\mathrm{m}] = 3.2\ \mathrm{TV/m}$$

PROBLEM 6.4

Consider LWFA driven by a circularly polarized laser pulse with normalized intensity $a^2 = a_0^2 \exp\left(-\dfrac{2r^2}{r_0^2}\right)\sin^2(\pi\varsigma/L)$ for $0 < \varsigma < L$ with maximum pulse length $L \approx \lambda_p$. Find the axial electric field behind the pulse, $\varsigma < 0$ for $E_0 = 100$ V/m, $r = 3$ cm, $r_0 = 6$ cm, $a_0 = 40.2$, and phase region of wake $\varsigma = \pi/4k_p$. What would be the value if the laser is linearly polarized?

SOLUTION

The axial electric field is

$$\frac{E_z}{E_0} = -\frac{\pi}{4} a_0^2 \exp\left(-\frac{2r^2}{r_0^2}\right)\cos(\varsigma k_p)$$

$$= -\frac{\pi}{4}(40.2)^2 \exp\left(-\frac{1}{2}\right) cos(\pi/4)$$

Putting the value of E_0

$$E_z = -54407.5\ \mathrm{V/m}$$

For linear polarization a_0^2 is replaced by $a_0^2/2$. Hence, one finds

$$E_z = -27203\ \mathrm{V/m}$$

PROBLEM 6.5

If the wave velocity of plasma wave in LWFA is $v_p = 0.8c$ and maximum amplitude of nonlinear wave is E_0 = 100 V/m, then find the amplitude of wake generated in plasma. Consider the relativistic case.

SOLUTION

Relativistic wave breaking field is given by

$$E_{\max} = \sqrt{2}(\gamma - 1)^{1/2} E_0$$

where $\gamma = \left(1 - \frac{v_p^2}{c^2}\right)^{-1/2} = 1.67$. Hence, the field

$$E_{max} = \sqrt{2}(1.67 - 1)^{1/2} \times 100$$
$$= 115.75 \text{ V/m}$$

PROBLEM 6.6

Calculate plasma wave potential required for trapping an electron in warm plasma when the phase velocity of the plasma wave is 0.8*c* in the case of wakefield excitation.

SOLUTION

The minimum potential at the wave breaking amplitude or at the singularity of density is given by

$$\varphi_{min} = 1/\gamma_p - 1$$

The relativistic Lorentz factor associated with the phase velocity of the plasma wave is

$$\gamma_p = \left(1 - \frac{v_p^2}{c^2}\right)^{-1/2} = 1.67$$

Hence, $\varphi_{min} = 0.40$ V

PROBLEM 6.7

Consider a laser with frequency 0.2×10^{15} Hz and velocity 0.6*c* propagating in a plasma having density $n_e = 10^{18}$ cm^{-3}, and excites the wakefield. Calculate

a. The normalized wave breaking electric field for the wakefield acceleration in terms of maximum amplitude of nonlinear wave, E_0.
b. Maximum length over which electrons are accelerated, i.e. the dephasing length.

SOLUTION

a. Wave breaking electric field is given by

$$E_{WB} = \sqrt{2}(\gamma_p - 1)^{1/2} E_0$$

where $\gamma_p = \left(1 - \frac{v_p^2}{c^2}\right)^{-1/2} = 1.25$

$$E_{WB} = \sqrt{2}(1.25 - 1)^{1/2} = 0.707E_0$$

b. Dephasing length is given by $L_d = \gamma_p^2 \lambda_p$, where λ_p is the plasma wavelength given by

$$\lambda_p(\mu\text{m}) = 3.3 \times \frac{10^{10}}{\sqrt{n_e(\text{cm}^{-3})}} = 33\ \mu\text{m}$$

Hence, the dephasing length is

$$L_d = 51.56\ \mu\text{m}$$

Suggested Reading Material

Albert, F., & Thomas, A. G. (2016). Applications of laser wakefield accelerator-based light sources. *Plasma Physics and Controlled Fusion, 58*(10), 103001.

Andreev, N. E., Gorbunov, L. M., Kirsanov, V. I., Pogosova, A. A., & Ramazashvili, R. R. (1992). Resonant excitation of wakefields by a laser pulse in plasma. *JETP Letters, 55*(10), 551–555.

Antonsen, T. M. Jr, & Mora, P. (1992). Self-focusing and Raman scattering of laser pulses in tenuous plasmas. *Physical Review Letters, 69*(15), 2204.

Blumenfeld, I., Clayton, C. E., Decker, F. J., ... Zhou, M. (2007). Energy doubling of 42 GeV electrons in a metre-scale plasma wakefield accelerator. *Nature, 445*(7129), 741.

Bulanov, S. V., Pegoraro, F., & Pukhov, A. M. (1995). Two-dimensional regimes of self-focusing, wake field generation, and induced focusing of a short intense laser pulse in an underdense plasma. *Physical Review Letters, 74*(5), 710.

Clayton, C. E., Everett, M. J., Lal, A., Gordon, D., Marsh, K. A., & Joshi, C. (1994). Acceleration and scattering of injected electrons in plasma beat wave accelerator experiments. *Physics of Plasmas, 1*(5), 1753–1760.

Clayton, C. E., Marsh, K. A., Dyson, A., ... Joshi, C. (1993). Ultrahigh-gradient acceleration of injected electrons by laser-excited relativistic electron plasma waves. *Physical Review Letters, 70*(1), 37.

Dias, J. M., Stenz, C., Lopes, N., ... Mendonça, J. T. (1997). Experimental evidence of photon acceleration of ultrashort laser pulses in relativistic ionization fronts. *Physical Review Letters, 78*(25), 4773.

Esarey, E., Sprangle, P., Krall, J., & Ting, A. (1996). Overview of plasma-based accelerator concepts. *IEEE Transactions on Plasma Science, 24*(2), 252–288.

Esarey, E., Ting, A., & Sprangle, P. (1989). Optical guiding and beat wave phase velocity control in the plasma beat wave accelerator. In *AIP Conference Proceedings*, Vol. 193, No. 1, American Institute of Physics, pp. 71–86.

Gorbunov, L. M., & Kirsanov, V. I. (1987). Excitation of plasma waves by an electromagnetic wave packet. *Sov. Phys. JETP, 66*(2), 290–294.

Hamster, H., Sullivan, A., Gordon, S., White, W., & Falcone, R. W. (1993). Subpicosecond, electromagnetic pulses from intense laser-plasma interaction. *Physical Review Letters, 71*(17), 2725.

Joshi, C., Blue, B., Clayton, C. E., ... Lee, S. (2002). High energy density plasma science with an ultrarelativistic electron beam. *Physics of Plasmas, 9*(5), 1845.

Joshi, C., & Katsouleas, T. (2003). Plasma accelerators at the energy frontier and on tabletops. *Physics Today, 56*(6), 47–53.

Joshi, C., Adli, E., An, W., ... & Mori, W. B. (2018). Plasma wakefield acceleration experiments at FACET II. *Plasma Physics and Controlled Fusion, 60*(3), 034001.

Katsouleas, T., & Dawson, J. M. (1983). Unlimited electron acceleration in laser-driven plasma waves. *Physical Review Letters, 51*(5), 392.

Kaur, M., & Gupta, D. N. (2018). Excitation of plasma wave by lasers beating in a collisional and mild-relativistic plasma. *Journal of Physics: Conference Series, 1067*(4), 042014.

King, P. M., Lemos, N., Shaw, J. L., ... & Albert, F. (2019). X-ray analysis methods for sources from self-modulated laser wakefield acceleration driven by picosecond lasers. *Review of Scientific Instruments, 90*(3), 033503.

Kostyukov, I., Pukhov, A., & Kiselev, S. (2004). Phenomenological theory of laser-plasma interaction in "bubble" regime. *Physics of Plasmas, 11*(11), 5256–5264.

Krall, J., Ting, A., Esarey, E., & Sprangle, P. (1993). Enhanced acceleration in a self-modulated-laser wake-field accelerator. *Physical Review E, 48*(3), 2157.

Krushelnick, K., Clark, E. L., Najmudin, Z., ... Danson, C. (1999). Multi-MeV ion production from high-intensity laser interactions with underdense plasmas. *Physical Review Letters, 83*(4), 737.

Lee, S., Katsouleas, T., Hemker, R. G., Dodd, E. S., & Mori, W. B. (2001). Plasma-wakefield acceleration of a positron beam. Physical review E - statistical physics. *Plasmas, Fluids, and Related Interdisciplinary Topics, 64*(4), 045501.

Leemans, W. P., Nagler, B., Gonsalves, A. J., ... Hooker, S. M. (2006). GeV electron beams from a centimetre-scale accelerator. *Nature Physics*, 2(10), 696.

Li, X. F., Yu, Q., Gu, Y. J., Huang, S., Kong, Q., & Kawata, S. (2015). Bubble shape and electromagnetic field in the nonlinear regime for laser wakefield acceleration. *Physics of Plasmas, 22*(8), 83112.

Lopes, N. C., Figueira, G., Dias, J. M., ... Stenz, C. (2004). Laser pulse frequency up-shifts by relativistic ionization fronts. *Europhysics Letters, 66*(3), 371.

Mahmoud, S. T., Gauniyal, R., Ahmad, N., Rawat, P., & Purohit, G. (2018). Electron acceleration by beating of two intense cross-focused hollow gaussian laser beams in plasma. *Communications in Theoretical Physics, 69*(1), 86.

Malik, H. K., Kumar, S., & Nishida, Y. (2007). Electron acceleration by laser produced wake field: Pulse shape effect. *Optics Communications, 280*(2), 417–423.

Marques, J. R., Geindre, J. P., Amiranoff, F., & ... Grillon, G. (1996). Temporal and spatial measurements of the electron density perturbation produced in the wake of an ultrashort laser pulse. *Physical Review Letters, 76*(19), 3566.

Marsh, K. A. (1994). Trapped electron acceleration by a laser-driven relativistic plasma wave. *Nature, 368*, 7.

Mendonça, J. T., & Oliveira E Silva, L. (1994). Regular and stochastic acceleration of photons. *Physical Review E, 49*(4), 3520.

Nakajima, K., Enomoto, A., Nakanishi, H., ... Tajima, T. (1992). A proof of principle experiment of laser wakefield accelerator. In *Proceedings of the 1992 Linear Accelerator Conference*, 1992, Vol. 1, pp. 332–334.

Nakajima, K., Kawakubo, T., Nakanishi, H., ... & Zhang, T. (1994). A proof-of-principle experiment of laser wakefield acceleration. *Physica Scripta, 1994*(T52), 61

Nakajima, K., Fisher, D., Kawakubo, T., ... Shiraga, H. (1995). Observation of ultrahigh gradient electron acceleration by a self-modulated intense short laser pulse. *Physical Review Letters, 74*(22), 4428.

Pukhov, A., & Meyer-ter-Vehn, J. (2002). Laser wake field acceleration: The highly nonlinear broken-wave regime. *Applied Physics B, 74*(4–5), 355–361.

Rosenzweig, J. B., Cline, D. B., Cole, B., ... Simpson, J. (1988). Experimental observation of plasma wake-field acceleration. *Physical Review Letters, 61*(1), 98.

Roth, M., Cowan, T. E., Key, M. H., ... Powell, H. (2001). Fast ignition by intense laser-accelerated proton beams. *Physical Review Letters, 86*(3), 436.

Sadighi-Bonabi, R., & Rahmatollahpur, S. H. (2010). A complete accounting of the monoenergetic electron parameters in an ellipsoidal bubble model. *Physics of Plasmas, 17*(3), 33105.

Schmid, K. (2011). Electron Acceleration by Few-Cycle Laser Pulses: Theory and Simulation. In *Laser Wakefield Electron Acceleration* (pp. 83–107). Springer, Berlin, Heidelberg.

Schroeder, C. B., Benedetti, C., Esarey, E., Chen, M., & Leemans, W. P. (2018). Two-color ionization injection using a plasma beatwave accelerator. *Nuclear Instruments and Methods in Physics Research Section A: Accelerators, Spectrometers, Detectors and Associated Equipment, 909*, 149–152.

Seryi, A. (2015). *Unifying physics of accelerators, lasers and plasma*, Boca Raton, FL: CRC Press.

Siders, C. W., Le Blanc, S. P., Babine, A., ... Downer, M. C. (1996). Plasma-based accelerator diagnostics based upon longitudinal interferometry with ultrashort optical pulses. *IEEE Transactions on Plasma Science, 24*(2), 301–315.

Silva, L. O., Marti, M., Davies, J. R., & ... Mori, W. B. (2004). Proton shock acceleration in laser-plasma interactions. *Physical Review Letters, 92*(1), 015002.

Sprangle, P., Joyce, G., Esarey, E., & Ting, A. (1988, November). Laser wakefield acceleration and relativistic optical guiding. In *AIP Conference Proceedings*, Vol. 175, No. 1, American Institute of Physics, pp. 231–239.

Strickland, D., & Mourou, G. (1985). Compression of amplified chirped optical pulses. *Optics Communications, 56*(3), 219–221.

Tabak, M., Hammer, J., Glinsky, M. E., … Mason, R. J. (1994). Ignition and high gain with ultrapowerful lasers. *Physics of Plasmas, 1*(5), 1626.

Tajima, T., & Dawson, J. M. (1979). Laser electron accelerator. *Physical Review Letters, 43*(4), 267.

Tang, C. M., Sprangle, P., & Sudan, R. N. (1984). Excitation of the plasma waves in the laser beat wave accelerator. *Applied Physics Letters, 45*(4), 375–377.

Wei, M. S., Mangles, S. P. D., Najmudin, Z., … Krushelnick, K. (2004). Ion acceleration by collisionless shocks in high-intensity-laser- underdense-plasma interaction. *Physical Review Letters, 93*(15), 155003.

Wilks, S. C., Dawson, J. M., Mori, W. B., Katsouleas, T., & Jones, M. E. (1989). Photon accelerator. *Physical Review Letters, 62*(22), 2600.

7

X-Ray Lasers

7.1 Introduction

Ever since the lasers were first discovered in 1960, research has been ongoing to enable lasing at the shortest possible wavelength and to make the size of the laser drivers smaller with a higher repetition rate. X-ray lasers with wavelengths of a few nanometers are therefore the ultimate short-wave lasers. In the last two decades there has been considerable development in the field of X-ray lasing owing to its promising prospects as a practical X-ray source for important applications in the fields of medicine, biology, chemistry, and physics. Significant progress has been made in the field of X-ray lasing by enabling higher intensity, shorter pulse duration, and higher coherence, and by making the laser compact in size with a high repetition rate.

X-rays, which are electromagnetic waves with a characteristic wavelength in the range of 0.1–100 Å, were discovered in the 1890s. They are generated by a sudden collision of highly accelerated electrons with a solid target. The subsequent radiation due to atomic inner-shell transitions produced short wavelength X-rays. X-ray lasers, which are a coherent source of X-rays, were, however, conceptualized in the 1970s. The conventional lasers generated until then were based on emitting a large number of coherent and monochromatic photons on the decay of excited electrons in a gaseous medium to ground state. The concept of X-ray lasers was envisioned on the realization that the use of ions instead of gaseous medium could enable the generation of laser beams with much higher energies.

Initially there were two problems with the short wavelength laser based on the conventional lasing techniques. The first problem was the non-availability of a concentrated pumping pulse with sufficiently high energy to enable population inversion. The second problem was the lack of appropriate material capable of reflecting and transmitting light in the deep or extreme ultraviolet (EUV or XUV) region. The need for higher energy arises because, in the case of a laser, an increase in transition energy lowers the lifetime of the excited state (Hecht 2008). In view of these obstacles, two different approaches were adopted for X-ray laser generation: the use of an undulator, which is a gain medium of a free-electron laser (FEL), and the use of plasma as a gain medium.

7.2 Need of Plasma for X-Ray Lasers

Ronald Andrew was the first person to predict the X-ray laser. The X-ray laser envisioned by him was based on a concept of highly ionized plasma in which the free electrons drop back to the lower level of the plasma ions, thus causing the generation of a stimulated emission of radiation having a wavelength of 0.12–1.2 nm in a travelling wave amplifier.

The generation of short wavelength X-ray lasers requires a highly dense plasma with highly ionized ions. The basis of X-ray laser generation using plasma as a gain medium is a three-step process, namely ionization, excitation, and decay. The first step, ionization, involves the creation of ions due to stripping off of electrons of the atoms of the target material, when struck by a light pulse. The second step, excitation, takes place by energizing the ions with the incident light pulse ('exciting' or 'amplifying' the ions). The decay of the excited ions into the ground state results in the emission of photons. An X-ray laser is generated by emitting millions of photons at the same wavelength. Amongst the other methods of generating X-rays, a high harmonic generation, which is based on nonlinear frequency conversion and can be employed for the amplification of certain harmonics in plasma, is also popular. This method, however, has a major limitation of being able to achieve a fairly low optical power.

So far, X-ray laser has been demonstrated in a hot plasma column that is a few centimetres long with a diameter of about 100 pm, which is created by irradiation of a target material such as a thin foil, fiber, or solid material. The generation of the laser could happen due to the transition that takes place between the excited states of highly charged ions (Bleiner et al. 2014).

7.2.1 Collision Excitation and Recombination Methods

Two principle methods used for excitation, when plasma is used as an X-ray laser active medium, are the collision excitation method and the recombination method.

In the collision excitation method, Ne-like ions or Ni-like ions are excited from the ground states. The Ne-like ions are created when the stripping off of electrons from lasing material atoms results in a closed-shell stable ion having ten electrons. Ni-like ions are also closed shell and stable having 28 electrons. These ions are highly stable in plasma. A plasma having a large fraction of these stable ions is said to be a stable lasing plasma.

In the recombination scheme, a hot plasma consisting of highly ionized bare nuclei or He-like ions is used. In a plasma, when a free electron combines with ion on collision, energy comprised of the kinetic energy of the electron and the binding energy of the ion is emitted as photons. As a result of recombination, excited states of H-like or Li-like ions are created and excess energy comprised of the kinetic energy of the electron and the binding energy of the ion is emitted as photons. X-ray lasers are generated when millions of such monochromatic photons with energy corresponding to the X-rays are emitted. The energy requirement for achieving population inversion in the case of recombination scheme (tens of electronvolts) is much less than the energy requirement for the collision excitation scheme (several hundreds of electronvolts). X-ray laser generation therefore needs a large laser facility capable of generating a sub-nano second laser pulse that has energy ranging from several hundred of Joules to several kilojoules.

Currently, X-ray lasers are widely used as a source for studying high-density plasma. These lasers are designed to produce soft X-rays (X-rays with a long wavelength). The applications of the X-rays are limited due to the non-achievability of a pulse length below 50 ps and the inability to focus a large number of photons on a small sample. These limitations are overcome by the use of X-ray FELs. The use of X-ray FELs has enabled new experiments to be carried out that require focusing a large number of photons in X-ray pulses on a sample with dimensions comparable to that of a molecule for a very short duration (a few femtoseconds). These experiments were aimed at understanding the dynamics of atomic and molecular processes. The X-ray FELs have therefore enabled the study of the ultrasmall and ultrafast processes.

7.3 Some Aspects of X-Ray Lasers

X-ray lasers are the ideal tools for probing dense plasmas because of salient features that make them especially conducive for the study. The short wavelength lasers are characterized by a decreased refraction as well as a larger penetration compared with the other longer wavelength optical probes.

X-ray lasers have high brightness, which makes them exceptionally suitable for plasma imaging (Haessler et al. 2011). Yttrium X-ray lasers are unparalleled in this regard. The coherence property of the X-ray probe can be exploited by using the X-ray laser as a density diagnostic. The use of collisionally pumped X-ray lasers for density diagnostics would yield blurred images as they have longer pulse lengths (approximately hundreds of picoseconds). During this duration, significant motion of electrons can take place, resulting in blurred images. Generation of short pulses (with durations of less than 50 ps) became crucial for extending diagnostic techniques.

The generation of X-rays in the collisionally pumped X-ray lasers is caused by the transition of electrons within the ions in the plasma. The transition is between the ground state and the various higher energy levels. The ionization of the atoms is caused by the shear strength of the electric field of the laser. When the field strength is high, the electron is ejected from the atom. This happens because the electron orbit gets distorted by an amplitude exceeding the atomic diameter because of the high field strength, as shown in Figure 7.1. High field limit, which is the threshold for ionization, corresponds to a power density of around 10^{16} W/cm^2.

On focusing, the Linac Coherent Light Source (LCLS) X-rays, which have an oscillation period in the range of ~10^{18} s, create a high-power density (>10^{16} W/cm^2) owing to their shorter wavelengths (~0.3 nm). The rapidly oscillating electric field of the high-frequency X-rays causes the electrons to shake, but they still are not able to break free from the atom (Figure 7.1). This happens because of the smaller distortion amplitude of the electron orbit (described by $A \sim E/v^2$) in the case of X-rays. The distortion amplitude is caused by the oscillating electric field and is about 10^6 times smaller in the case of X-rays compared with the case of visible laser light. Therefore, the high frequency of X-ray fields results in atoms remaining stable against ionization. This concept, however, remains to be verified experimentally.

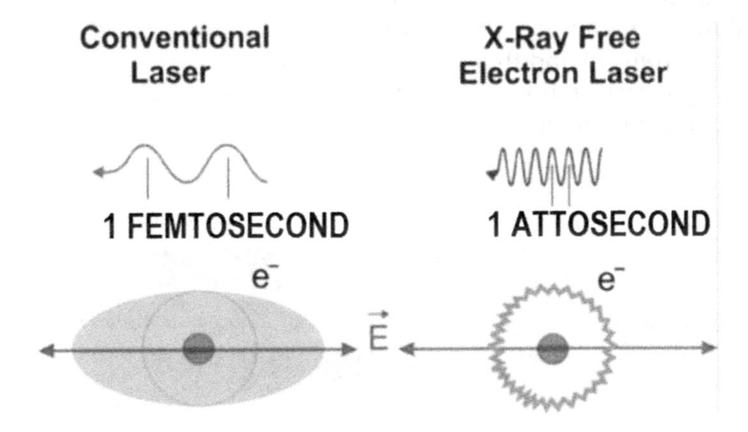

FIGURE 7.1
The distortion in valence electron orbit caused by a large amplitude oscillating electric field with a different frequency.

7.4 Generation of Coherent Soft X-Rays

The two fundamental processes used for the generation of coherent soft X-rays are free-electron lasers (FELs) and plasma-based soft X-ray lasers (plasma SXRLs).

7.4.1 Free-electron Lasers (FELs)

FELs are light sources based on spontaneous and / or induced emissions from a beam of relativistic electrons on interaction with a static and periodic magnetic field generated by an undulator, as discussed by Geloni et al. (2017). The undulators work on the principle that a charged particle, when accelerated, emits electromagnetic radiation. The FEL emits X-rays because, on acceleration of electrons to relativistic velocities, there is Doppler frequency upshifting of the radiation emitted. The FELs cause the generation of lasers by accelerating electrons to relativistic velocities and guiding them through an undulator, which is a periodic lattice of alternating magnetic dipolar fields. An electron beam that initially had random position distribution of electrons undergoes a change, and the electrons appear regularly spaced within the beam approximately at the wavelength of the X-rays. The electrons are generated by knocking off metal particles by a conventional laser. The magnetic field of the undulator is sinusoidal and induces the accelerated electrons to move in this field in a zig-zag trajectory; oscillating in a plane perpendicular to the direction of the magnetic field (Figure 7.2).

While racing through the undulator, each electron emits X-ray radiation that gets amplified during its passage through the undulator. The amplification is caused by the interaction of the radiated X-rays with the electrons. Because X-ray radiation is faster, it surpasses the electrons that generated them and interacts with the electrons travelling ahead in the undulator; this is called slippage. This interaction causes some electrons to speed up while some slow down, gradually resulting in the organization of electrons into a multitude of thin discs, called bunching (Figure 7.3). The electron bunches emit flashes of laser-like radiation while flying through the undulator. The total X-ray field generated by an ensemble of electrons is the sum of fields generated by all the electrons. Initially, the fields generated by the electrons lack coherence because there is partial cancellation on superimposition of the fields. Intensity of the radiation in this case is proportional to N_e, the number of electrons. The radiation from the electrons that are bunched is synchronized and they each emit coherent X-ray radiation, which is intense and of extremely short duration. The number of periods of the emitted wave train are the same as that of the undulator.

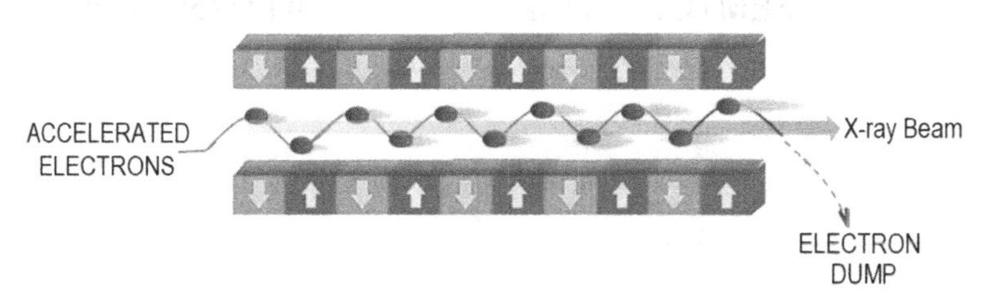

FIGURE 7.2
Schematic representation of a free-electron laser.

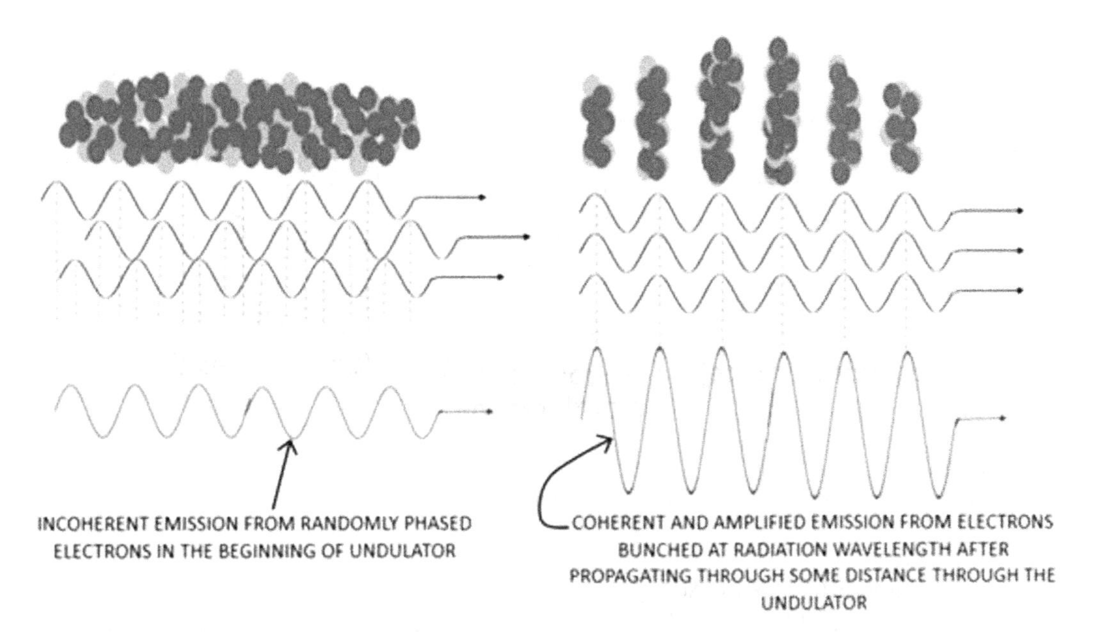

FIGURE 7.3
Schematic representation of the bunching of electrons while propagating through the undulator.

7.4.1.1. Repetition Rate, Wavelength, and Coherence of X-Rays

The number of periods and the wavelength are shorter than the undulator field wavelength by a relativistic contraction factor, which is inversely proportional to the square of its energy. Thus radiation with desired wavelength can be emitted by increasing the energy of the electron beam. Coherence is induced in the fields generated by the electrons due to 'free-electron laser collective instability'. There is co-propagation of the electron-beam and the radiation along the undulator. The co-propagating radiation affects the position of the electrons, thus increasing the coherence in the emitted radiation. The redistribution of the electrons is caused by the FEL collective instability. The collective instability causes an orderly arrangement of the electrons in the electron beam due to the interaction of the electrons moving through the undulator with the electromagnetic field generated by other electrons. This interaction modifies the electron energy, which is dependent on the characteristics of the undulator, the electron density, and the distance between the intra-beam electrons. The trajectories of higher energy electrons are bent less compared with the trajectories of the lower energy electrons. The modification of the electron energy results in the regularly spaced electrons (bunching) within the beam at a distance that is approximately equal to the radiation wavelength of the undulator, λ_W, as shown in Figure 7.4.

The bunching of electrons causes in-phase superimposition of the electromagnetic fields of the electrons, which results in amplification of the field. The amplification of the field causes it to grow exponentially at a rate called the 'gain length', which is the most important instability parameter. The effect is collective as the phase of the propagating electrons is also affected by the fields from radiation generated by other electrons, increasing coherence (collective synchronization) in radiation from the electron beam. The intensity of the coherent radiation is proportional to the square of the number of electrons (N_e^2).

Examples of X-ray laser generation based on the FEL process are the Free-Electron Laser in Hamburg (FLASH) and the Linac Coherent Light Source (LCLS) facilities at the Stanford

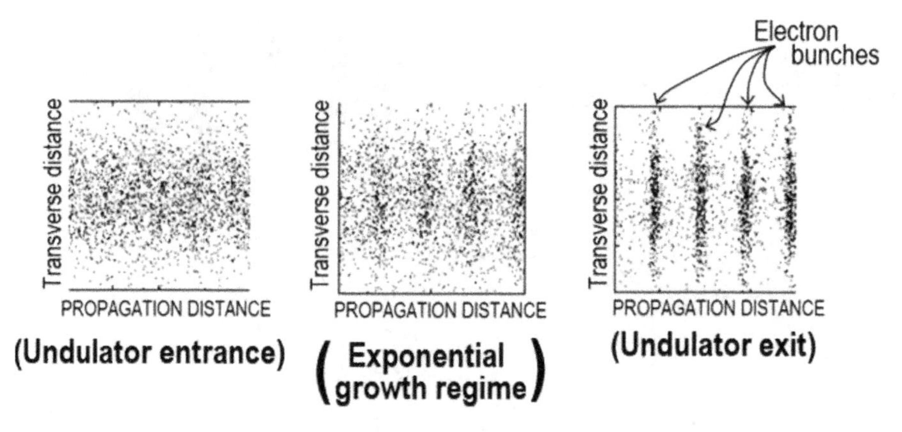

FIGURE 7.4
Distribution of electrons at various positions of the undulator.

Linear Accelerator Center (SLAC) National Accelerator Laboratory. Generally, the pulses of a soft X-ray from FLASH with a tunable wavelength of 7–50 nm attain an energy of 10–100 μJ in a duration of 10–100 fs compared with the X-ray pulses from LCLS, whose wavelength comes down to 0.1 nm at the same pulse features.

7.4.1.1.1 Limitations and Their Solution

The prime limitations experienced during the generation of the X-ray laser are the short life span of inner quantum energy levels of excited atom core and the large energy requirement (1–10 kV) for the excitation of the inner atomic quantum energy levels. The energy requirement in the case of visible lasers is ~1 eV. The intense pumping power that is required to cause population inversion is also impractically high. Difficulty in the development of low loss optical cavities for X-ray laser oscillators is yet another limitation.

In the case of the FELs, the problem relating to the short life span of inner quantum energy levels of the excited atom was addressed by the use of the SASE-FEL (Dell'Angela et al. 2015). The SASE-FEL enables X-ray generation from an intense relativistic electron beam. The radiation is nearly diffraction limited and is spatially coherent.

The 'collective instability', which causes coherence in radiation emitted from an electron beam, can be initiated with the use of a high-gain FEL amplifier or with the use of a SASE-FEL. In the case of the high-gain FEL amplifier, an external electromagnetic field starts the instability, whereas in the case of the SASE-FEL, the instability is started by the random synchrotron radiation noise that is generated by the electron beam at the entrance of the undulator.

The propagation of the X-ray radiation is at a faster rate compared with that of the electrons in the SASE-FEL. This causes the electrons to lag behind the X-ray radiation by one wavelength per undulator period, which is called slippage at resonance. The electrons of the beam interact with the radiation, which are originated from the electrons that preceded them in the beam. This interaction plays a crucial role in causing a smoothing out of temporal distribution of the intensity of the radiation, which was initially varying on the scale of λ (spontaneous radiation) to an ultimate temporal distribution variation on the scale of the cooperation length at saturation. Creation of instability in the SASE-FEL takes place if two conditions are fulfilled. First, the electrons should have an energy, radius, and angular spread similar to that of the X-ray beam. Second, the gain should be large enough to compensate for the losses.

One unique feature of X-ray free-electron lasers (XFELs) is their peak flux or brightness, i.e. flux per source size and per emission angle. The XFEL emits the X-rays in ultrafast bursts of 100 fs. Because of this extreme shortness of time, this kind of laser can be used in investigating matters even at nanoscale. This is the time with which the atoms vibrate. Therefore, XFELs have the ability to 'see' the invisible, down to the size of an atom, as well as to take snapshots of the motion of atoms and ultrasmall objects like molecules and clusters. In the near future, one would be able to record the movie of the dissociation of the larger molecule into a smaller molecule with the help of XFELs.

High-order Harmonic (HOH) is the ideal seed laser source because of the superior beam quality of the laser generated, which has very good spatial coherence (Johnson et al. 2019). The HOHs, which are the odd multiples of the laser intensity (frequency), are produced due to the nonlinear effects that occur during the interaction of the linearly polarized short laser pulses with gases or solids. The injection of these HOHs into high gain plasma-based soft X-ray laser amplifiers causes generation of higher order harmonics (Haessler et al. 2011). The pulse duration is in the range of 100 ns and the wavelengths get shortened down to 1 nm. There is lowering of photon yield during this process (Shiner et al. 2009).

7.4.2 Plasma-Based Soft X-Ray Lasers (SXRL)

Plasma SXRLs are the most important of different types of plasma-based lasers. The role of the plasma is as a gain medium for causing population inversion, which is crucial for generation of the laser. The mechanisms that are important for the plasma-based lasers are the recombination mechanism and the collisional excitation mechanism, which involve the highly charged ions present in a dense plasma filament. A plasma, densely populated with high-energy electrons and a large fraction of lasing ions, acts as a medium with a large optical gain and enables generation of SXRLs.

7.4.2.1 Electrical Discharges and Pump Laser Pulses

Several options such as pump laser pulses or electrical discharges are used for achieving the highly ionized plasma. Variation in features of the generated SXRLs such as the output energy, optical coherence, and duration can be achieved. When electrical discharge is used for obtaining the optical gain in plasma, the laser generated is the fast capillary discharge SXRL whereas the use of long laser pulses results in quasi-steady state (QSS) SXRLs. The generation of an SXRL is also seen by the interaction of circularly polarized pulses of ultrashort duration with plasma. The methods that are commonly used for plasma excitation, which is crucial for the generation of the X-ray laser, are the capillary discharge SXRL and the transient collisional excitation (TCE) SXRL.

The capillary discharge SXRL, also known as the electric discharge method, is a compact and efficient method for plasma excitation leading to the creation of SXRLs. On passage of high-voltage electrical discharge through a capillary filled with a gas results in the formation of a cylindrical plasma. The plasma gets compressed towards its axis due to the Lorentz force that comes into play due to the passage of high current through the plasma. The plasma compression causes an increase in the temperature and density essential for X-ray laser generation. This cylindrical plasma is characteristically a high-density plasma having a high axial uniformity, which enables a high pumping efficiency. Several tens of centimetres long Ar-filled capillary discharge lasers are operated by currents around ~100 kA. A Ne-like Ar laser with a 0.8 ns pulse with 4 Hz frequency has been demonstrated for a wavelength of 46.9 nm. This scheme has not been investigated for the wavelengths shorter than this.

For tabletop operations in small laboratories for generation of SXRL using pump laser pulses, the pump laser energy needs to be low. The pump energy in the case of QSS SXRLs has been brought down as low as 100 J but is still high for tabletop operations. The first low pump energy (15 J) SXRL was produced as a result of transient excitation. In this method, called the TCE method, two different laser pulses are used. Both pulses differ from each other in terms of their parameters. The pulse, which interacts first with the plasma, is a nanosecond pulse and is called the pre-pulse. Its function is to generate a plasma with the required ionization stage of Ne-like or Ni-like ions, which is similar to the function of a long laser pulse in the QSS method. The second pulse, called the main pulse, is a picosecond pulse. It causes a transient population inversion in the plasma due to the electron collisional excitation. The ionization balance in this case is not affected by the quick heating process because the hydrodynamic expansion and the kinetic relaxation are both negligible. By use of this technique, a high gain is attained in a very small duration of several picoseconds. The large gain of up to 100 enables a saturated output from plasma less than 10 mm long and with a pump energy of less than 10 J, which is within the allowable limit for tabletop SXRLs. The first such tabletop SXRL, working on the TCE mechanism, is the Ni-like Pd at 14.7 nm; 7.3 nm is the shortest wavelength achieved in this regime from Ni-Sm pumped with 70 J pump laser energy.

The initial TCE SXRLs were based on the $2p^5 3p - 2p^5 3s$ lasing transition of Ne-like ions on electron impact collisional excitation from the ground state to the upper lasing level. Later, more efficient $3d^9 4d - 3d^9 4p$ lasing transitions of Ni-like ions were adopted (Robinson et al. 2010; Piracha et al. 2011). Lasing transitions in Ne-like and Ni-like X-ray lasers are shown in Figure 7.5.

The grazing incidence pumping (GRIP) was implemented to reduce the pumping energy required for TCE SXRLs. The pre-plasma is created by a normal incident pre-pulse, but the main pulse, which is a short pulse beam, is incident at grazing angle and refracted in the plasma. This enables enhanced pumping efficiency as there is an increase in the path length of the main pulse, which results in better deposition of energy. This method also enables achieving the optimum pumping density, because by adjusting the GRIP angle

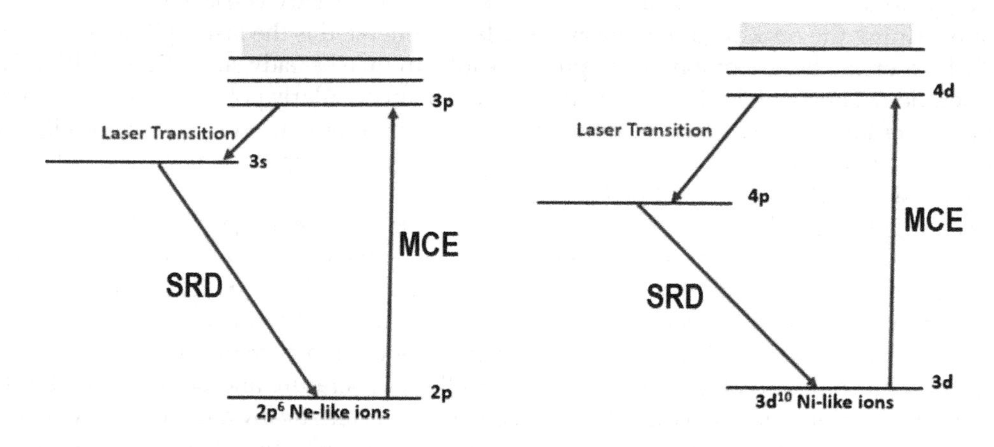

FIGURE 7.5
Lasing transitions in Ne-like and Ni-like lasers. SRD, strong radiation decay; MCE, monopole collisional excitation.

electron density at the energy absorbed can be controlled. In the case of Ni-like Mo, lasing was enabled at 150 mJ compared with the ~1 J pumping energy requirement in the saturation regime. SXRLs at 10 Hz with an energy up to 2 μJ have been generated using the GRIP technique (Tümmler et al. 2005).

7.5 Chirped Pulse Amplification and Mach-Zehnder Techniques

X-ray lasers can be generated by other techniques, namely, chirped pulse amplification (CPA) and Mach-Zehnder techniques.

7.5.1 Chirped Pulse Amplification (CPA) Technique

This technique is used to amplify an ultrashort laser pulse to a higher level, i.e. up to petawatt level, by stretching out the pulse temporally and spectrally before amplification. The CPA technique was adopted to get around the problem of overheating of optical components such as amplifiers, lenses, and mirrors because the short duration of the extremely short pulses results in high-power density, even at lower energies. The overheating of the optical components limited the repetition rate of the lasers. The use of the CPA technology enabled generation of intense very short pulses by stretching out a low-energy pulse to more than 10,000 times its initial duration prior to amplification, which resulted in laser optics being exposed to lower power density pulses. The amplified pulse is recompressing it back to its initial duration, thus generating very short laser pulses with extremely high peak powers. In CPA, a short pulse is first stretched by a combination of grating pulse stretcher and mirrors and then amplified by an amplifier. This amplified stretched pulse is further compressed by pulse compressor grating, which results in an amplified short pulse. These detailed components and steps in CPA techniques are shown in Figure 7.6, where the whole process is illustrated in a flowchart.

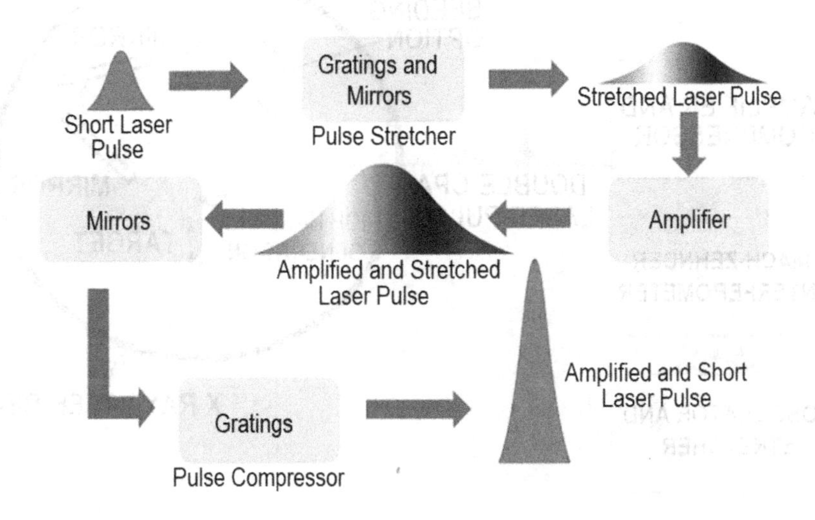

FIGURE 7.6
Chirped pulse amplification process.

7.5.2 Mach-Zehnder Technique

Mach-Zehnder is an interferometer that contains the beam splitter and mirrors to split and transmit the beam and recombine the superimposed beams. Using this Mach-Zehnder interferometer scheme, the phase change is converted into intensity change.

Initially a stretch oscillator pulse is fed into the Mach-Zehnder interferometer unit to generate an optimized double pulse that is further amplified and compressed to get a final temporal shape. Then this double CPA laser pulse is injected to the target chamber and focussed at a line by a spherical mirror to produce X-ray laser emission. Apart from this, HOHs are generated by a short pulse laser focussed in a gas tube for seeding the X-ray laser. Similarly, GRIP of the main pulse in the pre-plasma is carried out to optimize the energy deposition. The schematic of X-ray laser generation using Mach-Zehnder interferometer and CPA technique is shown in Figure 7.7.

7.5.2.1 Working

Conventionally, a laser works by excitation (pumped) of the lasing material's atoms by causing excitation of the electrons of the outer shell to a metastable upper lasing level. The atoms of the lasing material are then stimulated into emitting coherent photons of laser wavelength. The emitted photons can induce emission of photons from other pumped atoms, thus causing an exponential increase in the number of coherent photons and forming the laser. To generate lasers in the X-ray regime, transitions of inner shells of atoms would be required, which is made possible by pumping of ions instead of atoms, thus causing transitions at the X-ray wavelengths by enabling the excitation of the ions in the remaining outer shell. The laser active media play a crucial role in the generation of the X-ray lasers. The highly ionized plasma is the most used active medium for the X-ray laser, which is created by the capillary discharge or by irradiating a solid target

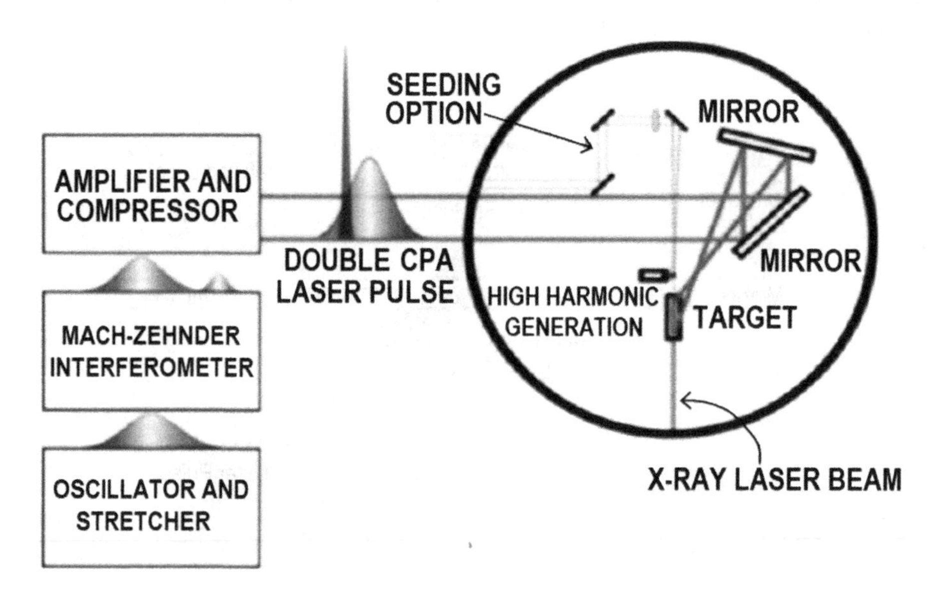

FIGURE 7.7
X-ray laser generation using Mach-Zehnder interferometer and CPA.

with a linearly focussed optical pulse. As mentioned earlier, the Ne-like (with 10 electrons remaining) and Ni-like (with 28 electrons remaining) ions are found to be the most stable configuration. Hundreds of electronvolts are required for the electron transition energy in highly ionized plasma. There are three common methods for the generation of the laser active media, i.e. capillary plasma discharge media, solid-slab target media, and plasma excited by optical field.

An example of X-ray lasing material is the Ne-like yttrium ion. Irradiation of yttrium with an intense optical laser light leads to the formation of a very hot (~10^7 K) cylindrical plasma consisting of ionized yttrium from which 29 electrons are stripped off, leaving 10 electrons similar to neon. This is therefore called the Ne-like yttrium laser. The Ne-like yttrium ions are metastable owing to their closed-shell structure.

Ionization of the lasing material is therefore the first step in X-ray laser generation followed by pumping, which is enabled using various techniques such as with energetic unbound electrons colliding with the ions (collisionally pumped X-rays), with plasma recombination, optical field ionization, and inner shell excitation. The lasing action occurs along the direction of the plasma, which is long and 3D. The real challenge in X-ray laser generation lies in optimization of the correct combination of the lasing material, ionization state, lasing levels, temperature, and electron density.

The extent of the plasma created on irradiation of the solid target with high-intensity laser light is determined by the laser spot size. The diameter of the created plasma, which is cylindrical, can vary from hundreds of micrometres to a few millimetres, so an irradiated foil can create plasma that can span orders of magnitude in density and temperature at a given time. (Figure 7.8).

7.5.2.2 Coherence Enhancement

Coherent X-ray application requires the pulse to be of shorter duration and there is a need to have coherence in photons per pulse. Normally, the coherence of the optical laser is achieved by multiple amplification in a resonant cavity; however, in the X-ray region the construction of the resonant cavity is very difficult. Due to the large gain in both the plasma-based X-ray lasers as well as XFELs, single pass amplification is enough to achieve

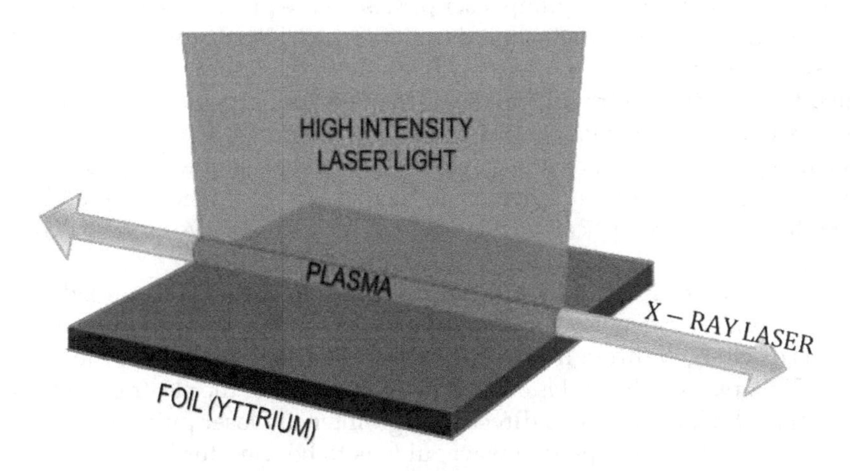

FIGURE 7.8
X-ray laser generation using high-intensity laser.

the intense and directional X-ray beam. This kind of laser amplification is called spontaneous emission amplification or self-amplified spontaneous emission (SASE). This has enabled significant improvement in the radiation coherence without the use of a resonant cavity. One possible solution to sort out this trouble is the injection of high-quality X-ray seed pulse into the amplification medium, causing generation of highly brilliant and coherent soft X-rays. In the case of the laser plasma-based SXRL, higher order harmonic light is used as the seed, while the self-seeding light from the first part of the undulator is used to generate keV X-rays in XFELs. To generate ultra intense, fully phase coherent soft X-ray output, a stretched and a chirped high-quality X-ray seed pulse is injected into a large-scale plasma amplifier.

In a capillary plasma discharge media setup, a several centimetres long capillary, which is made of resistant material, such as Alumina, confines a high-current submicrosecond electrical pulse in a low pressure gas. Further compression of the plasma discharge is caused by Lorentz force. Capillary Ne-like Ar^+ laser is a good example of this. In solid-slab target media, highly excited plasma is emitted by the target on hitting by an optical pulse. Longer pre-pulse as well as shorter and high energetic pulses are generally used for plasma creation and further excitation in plasma volume, respectively. For short lifetime, a sheared excitation pulse (i.e. GRIP) is required. Since the refractive index decreases above the resonance frequency, the amplified pulse is bent from the target surface by gradient in the refractive index of the plasma. Last, in optical field excited plasma, at high optical densities, the effective electron tunnelling is caused by or the potential barrier ($>10^{16}$ W/cm^2) is suppressed by highly ionized gas without contact with target or capillary. For synchronization of pump and signal pulse, a collinear setup is used.

An X-ray laser with a high average power in the extreme UV (EUV or XUV) region can be used to produce a new technique for high-quality nanoscale laser machining and detecting phase defects in extreme UV lithography masks.

7.5.3 Lasing in Plasma Waveguide

Generation of SXRLs requires a large pump power density, which is achieved by the use of picosecond or femtosecond pump laser pulses. These pulses enable X-ray lasing even at moderate energies owing to their short duration, which also results in the reduction in size of the lasers and an increase in the repetition rate of the lasers. However, the gain length is limited to a few millimetres due to the diffraction and refraction of the pump beam that makes it essential for the gain coefficient to be high enough to enable lasing. The limitation of small gain medium is taken care of by use of a plasma waveguide. This results in an increase in gain length, and there is an increase in the efficiency in high-power laser pumping because of the optical waveguiding of the pump pulse as well as the generated soft X-rays.

In the case where optical waveguiding is the sole use of a gas-filled capillary discharge waveguide, a current pulse of duration ~1 ms and a magnitude of several hundred amperes is discharged through a hydrogen gas-filled capillary, as depicted schematically in Figure 7.9. A plasma channel is created by the passage of the electric discharge through the capillary, which is capable of directing high-intensity laser pulse over a length of several centimetres. When the plasma waveguide is to be used for lasing, to enable driving the short wavelength laser within the plasma channel, the hydrogen is doped with the lasing gas.

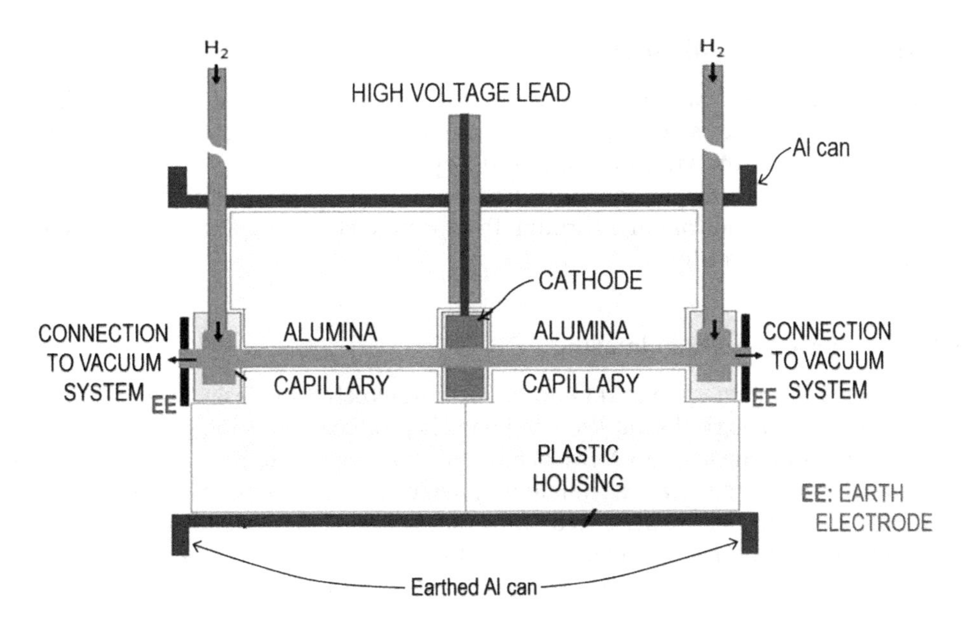

FIGURE 7.9
Schematic diagram of a gas-filled capillary discharge wave-guide.

7.6 Applications of the X-Ray Laser

X-ray lasers produce pulses that are ultrashort, ultrafast, and possess ultrahigh brightness because of which they have carved a niche for themselves in the area of nanoscopy, such as for observing tiny structures at atomic scale and for filming extremely fast processes like making or breaking of chemical bonds. X-ray lasers therefore are of great importance currently in the area of atomic and molecular physics, surface physics and chemistry, materials science, and current technology (Lee et al. 2012).

7.6.1 Molecular Movie

The use of intense, ultrashort (approximately femtosecond) pulses has made it possible to construct a molecular movie of a chemical reaction by taking images of the atoms during a reaction and arranging them in a sequence. Ultrafast laser spectroscopy making use of the ultrashort laser pulses makes it possible by enabling the study of dynamics at very small timescales. This has opened up vast opportunities to understand and hence control reactions, catalysis, and so forth.

7.6.2 Coherent Diffraction Imaging (CDI)

Coherent diffraction imaging (CDI) is another important imaging in which X-ray lasers are often used. It enables generation of a 3D image of a nanostructure. An X-ray laser beam when incident on a material gets scattered and produces a diffraction pattern, which helps generation of a 3D image by the use of an iterative feedback algorithm. The images produced using CDI are aberration free and have very high resolution.

7.6.3 High-Resolution Microscopy

The SRXLs and XUV lasers have become indispensable to high-resolution microscopy owing to their compact size, superior beam quality, and very reliable operation. X-ray microscopy works on the same principle as the optical and electron microscopes to provide magnified images. The X-rays penetrate most of the objects, and the image is produced by detecting the X-rays after passing through the specimen by using a charge-coupled device (CCD) or by exposing a film.

7.6.4 Phase-Resolved X-Ray Imaging

Phase-resolved X-ray imaging is yet another area in which X-ray lasers are preferred, especially for imaging biological samples. The image is produced by using the information concerning changes of phase of an X-ray beam after passing through a specimen. The phase shift is transformed into variations in intensity, which is subsequently recorded by detectors. The use of X-rays enables viewing structures that are of micrometre range. By scanning an object section by section, even 3D images can be obtained using this process. The extreme shortness of the X-ray pulses can be used to advantage to capture very fast processes.

7.6.5 Diagnostic Tool for Highly Dense Plasmas

One of the latest uses of the X-ray laser is as a probe tool for the imaging and exploration of high-density plasmas. X-ray lasers have made it possible to develop diagnostic techniques that are very useful for evaluating these plasmas.

The success of X-ray lasers as probes for studying dense plasma is due to their short wavelength, because of which there is reduced refraction and they are able to penetrate deeper in plasma. The better penetration by the X-ray lasers in the plasma provides images with better resolution. There is a decrease in refraction because of reduced electron density gradient due to shorter wavelength. The spatial resolution is better because of the reduced refraction resulting in enhanced data interpretation. Along with this, its high brightness makes it a very useful tool in imaging applications.

7.6.5.1 Mechanism of Imaging

X-ray lasers have become extremely important diagnostic tools for probing plasmas because of the advancement in the field of X-ray lasers and in mirror technology. There are three principle techniques that are used for imaging high-density plasmas: high-resolution imaging of the fine structure in plasma; Moire deflectometry, which is used to measure the density gradient in plasma; and interferometry, which is used to measure the direct electron density.

The critical issues in the high resolution are the wavelength of the probe and refraction in medium. Shorter wavelength light generally penetrates much farther in to plasma and is less affected by the density gradient, which also results in reduced refraction (Figure 7.10).

Optical elements such as mirrors are incorporated in the setup using X-ray laser as a plasma diagnostic for improved output. This setup, which consists of a sequence of two multilayer mirrors, is shown in Figure 7.11. The first multilayer mirror collects the X-ray beam and converges or diverges it significantly. This collimated beam, after passing through mirror, backlights the plasma, which was generated by laser irradiation on

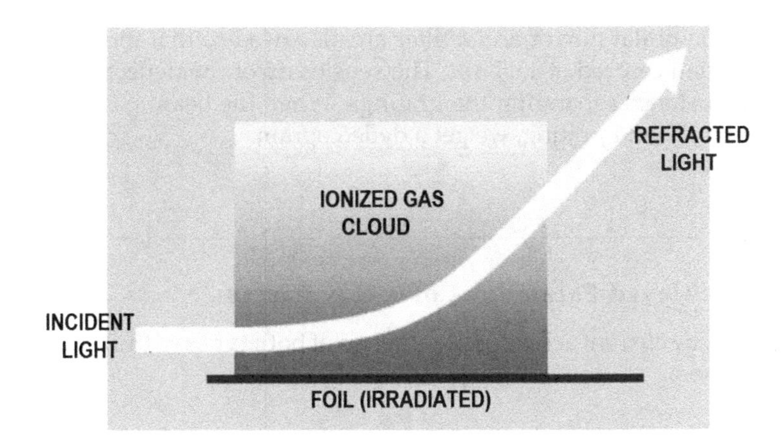

FIGURE 7.10
Internal process of imaging of plasma through an X-ray laser.

a target. The second spherical multilayer mirror focusses the image of the plasma onto a high dynamic range CCD detector, which is highly sensitive to X-rays.

To minimize damage to the multilayer mirrors used, the highly reflecting mirrors are located around 50 cm away from the end by using a small part of the mirrors, while keeping the remainder of the unused part of the mirrors shielded. The multilayer mirrors used in this arrangement consist of 15 layer pairs of molybdenum and silicon having a measured reflectivity of approximately 60% at normal incidence at a wavelength of 15.5 nm. High reflectivity is essential for higher efficiency of the mirrors in a complex optical system. A pair of gratings is used in this setup that is placed just before CCD detector. Along with

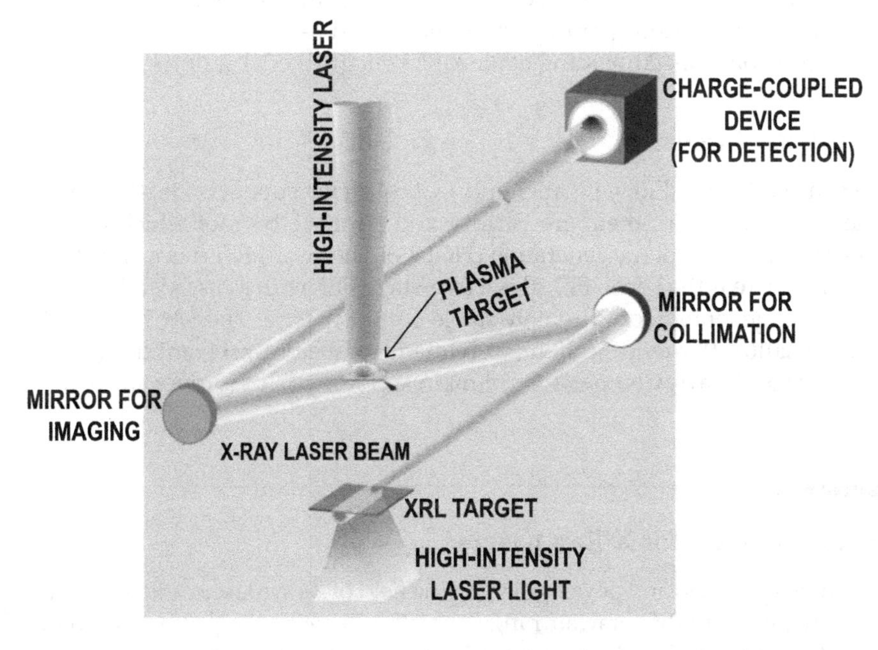

FIGURE 7.11
Experimental setup for direct imaging plasma XRL stands for X-ray laser.

this a combination of flat mirror and a filter are also used so that the detector sees a narrow range of radian centred at 15.5 nm. The sensitivity of the deflectometer is controlled by varying the distance separating the gratings. When the beam passes through all this arrangement of CCD and grating, we get a deflectogram.

7.7 Current Achieved Parameters of X-Ray Lasers

We will mention the current achieved parameters of both types of the X-ray lasers, i.e. hard and soft X-ray lasers.

7.7.1 Hard X-Ray Lasers

The key features of the X-ray free-electron laser are defined by pulse duration, intensity, linewidth, and wavelength range of the X-ray laser. So far, the most advanced next-generation light sources for hard X-rays are the European XFEL, PAL-XFEL, SwissFEL, LCLS, and SPRing-8 Angstrom compact FEL (SACLA), all of which are SASE-FELs. Even though the underlying operating principles for all these systems is the same (by generation of high-intensity X-ray laser by passage of highly accelerated high-energy electrons through undulators), the European XFEL and the LCLS HRX undulator are the most advanced of all, as they are driven by a superconducting linear accelerator. The superconducting linear accelerator utilizes the superconductive behavior of Niobium (Nb) at extremely low temperatures. Use of supercold Nb eliminates energy loss in the form of heat, when energy is put into it. In the European XFEL, the accelerator utilizes Nb cavities that are chilled to –271°C, which makes them lose all the electrical resistance. The acceleration process is therefore made very efficient, as electrons receive all the electromagnetic signals that enter the cavity. The various parameters of these hard X-ray lasers have been compared in Table 7.1 (Patterson et al. 2010; Kang et al. 2017; Pellegrini et al. 2016).

7.7.2 Soft X-Ray Lasers

Amongst the soft X-ray FELs, FLASH and LCLS-II have super-conducting linear accelerators. The most efficient of them are Fermi-I and Fermi-II, both of which are seeded FELs based on high-gain harmonic generation (HGHG). The Fermi FEL-I is a one-stage harmonic upshift FEL, whereas the Fermi FEL-II is a two-stage FEL with a cascaded HGHG. The use of HGHG enables generation of extremely stable output spectra. The SACLA SXFEL was commissioned in 2016, after modifications were made to one of the beamlines (beamline-1), for generation of soft X-rays. Comparisons of main soft X-ray FELs is given in Table 7.2.

7.8 Future Aspects for X-Ray Lasers

The continuous progress and development in the field of X-rays lasers will result in an increase in their peak power to the terawatt range, which will cause the pulse duration to decrease below femtosecond level and an increase in the brightness. These improvements will lead to improved X-ray optics, data acquisition, and analysis. These upgrades in the X-ray lasers

TABLE 7.1

Comparison of Hard X-ray FELs.

	European XFEL	PAL-XFEL	SwissFEL	LCLS	SACLA	LCLS-II CuRF	LCLS-II SCRF(HXR Und)
Abbreviation	European X-Ray Free-Electron Laser	Pohang Accelerator Laboratory X-Ray Free-Electron Laser	Switzerland's Free-Electron Laser	Linac Coherent Light Source	SPRing-8 Angstrom Compact Free-Electron Laser	Linac Coherent Light Source-II	Linac Coherent light Source-II
Location	Germany	Korea	Switzerland	USA	Japan	USA	USA
Year of commissioning	2016	2016	2016	2009	2011	2019	2020
Accelerator technology	Super-conducting	Normal-conducting	Normal-conducting	Normal-conducting	Normal-conducting	Normal-conducting	Super-conducting
Minimum laser wavelength (nm)	0.05	0.1	0.1	0.15	0.08	0.05	0.25
Maximum electron energy (GeV)	17.5	10	5.8	13.6	8.5	15	4.15
Number of flashes/second	27,000	60	400	120	60	120	1,000,000
Length of the facility (km)	3.4	5	0.7	3	0.75	3	3
Peak brilliance (photons s^{-1} $mrad^{-2}$ mm^{-2} per 0.1%bw)	5×10^{33}	5×10^{31}	5×10^{32}	2×10^{33}	1×10^{33}	2×10^{33}	2×10^{32}

TABLE 7.2

Comparison of Main Soft X-ray FELs.

	FLASH	Fermi FEL-I	Fermi FEL-II	LCLS-II SXR Und	SACLA SX-FEL
Abbreviation	Free-Electron Laser in Hamburg	Free-Electron Laser radiation for multidisciplinary Investigations	Free-Electron Laser radiation for multidisciplinary Investigations	Linac Coherent Light Source	SPRing-8 Angstrom Compact Free-Electron Laser
Location	Germany	Italy	Italy	USA	Japan
Year of commissioning	2005	2012	2012	2009	2016
Accelerator technology	Super-conducting	Normal-conducting	Normal-conducting	Super-conducting	Normal conducting
Laser wavelength (nm)	4.2–52	20–100	4–20	1–6	12.4
Maximum electron energy (GeV)	1.25	1.5	1.5	4	0.78
Minimum pulse duration (FWHM) (fs)	15	30	30	70	300
Number of flashes/second	10	10–50	10–50	120	60

may even provide new opportunities in the field of atomic and molecular science, imaging of nonperiodic systems, and the evolution of systems in non-equilibrium states.

There is a great deal of emphasis currently to reduce the cost of X-ray lasers and to make them more compact, without compromising the efficiency. Pulse shaping is one of the methods being explored to increase the efficiency and the brightness of the X-ray lasers. Use of an injector, an amplifier double-target configuration, or a travelling wave excitation are some of the techniques that are being worked on for developing X-ray lasers with enhanced brightness and coherence. Recently, there has been advancement in the sub-picosecond laser technology that has resulted in an increase in the power densities of small-scale X-ray lasers to exceed 10^{18} to 10^{20} W/cm^2.

Improvement in X-ray lasers will result in improved plasma imaging as the spatial resolution of plasma image is dependent on the period of the X-ray laser pulse. Unlike the transmission electron microscopes where the samples are required to be less than 0.4 μm in thickness, the samples for imaging using X-ray lasers can be 2 to 10 μm thick with no radiation damage to the samples.

With the constant research activity going on in the field of X-ray lasers, highly efficient, low-priced, compact tabletop lasers will be a reality in the near future.

7.9 Conclusions

It is clear that highly ionized and dense plasma is required for the generation of short wavelength X-ray lasers, where the three-step process, i.e. ionization, excitation, and decay, is involved. The collision excitation method and the recombination method are the two principle methods used for excitation. The transition of electrons between the ground state and various higher energy levels within the ions in the plasma leads to the generation of X-ray lasers.

The FEL emits X-rays because of the Doppler frequency upshifting of radiations emitted on acceleration of the electrons to relativistic velocities. These accelerated electrons are guided through a periodic lattice of alternating magnetic dipolar fields, called an undulator. Here the radiation with desired wavelength can be emitted by increasing the energy of the electron beam. Coherence is induced in the fields generated by the electrons due to 'free-electron laser collective instability'. On the other hand, in plasma SXRLs the role of plasma is to work as a gain medium for causing population inversion. The capillary discharge SXRL, also known as the electric discharge method, is a compact and efficient method for plasma excitation leading to the creation of SXRLs. In addition, CPA and Mach-Zehnder techniques are also used to generate X-ray lasers. CPA technique is used to amplify an ultrashort laser pulse to a higher level, i.e. up to the petawatt level, by stretching out the pulse temporally and spectrally before amplification. On the other hand, the phase change is converted into intensity change in the Mach-Zehnder interferometer technique, where initially a stretch oscillator pulse is fed into the interferometer unit to generate an optimized double pulse, which is further amplified and compressed to get a final temporal shape. Then the double CPA laser pulse is injected into the target chamber and focussed at a line by a spherical mirror to produce X-ray laser emission.

The interesting application of X-ray lasers is in the area of nanoscopy, i.e. for observing tiny structures at atomic scale and for filming extremely fast processes, such as making or breaking of chemical bonds. These lasers are important for surface physics and chemistry, materials science, and current technology.

7.10 Selected Problems and Solutions

PROBLEM 7.1

In plasma dominated by free-free absorption, calculate the absorption coefficient α for plasma with 1 keV temperature, and wavelength $\lambda = 155$ Å, and average ionization 30 (mid Z plasma).

SOLUTION

We know that the expression for the absorption coefficient is given by

$$\alpha = 2.44 \times 10^{-37} < Z^2 > \frac{n_e n_i}{\sqrt{KT}\,(h\nu)^3}\left[1 - \exp\left(-\frac{h\nu}{KT}\right)\right] \text{cm}^{-1}$$

Since 1 eV = 11,600 K, $1keV = 10^3 \times 11600$ K. Also putting Z = 30 (for mid Z plasma), $\lambda = 155$ Å, we obtain the following value of α

$$\alpha \approx 2.6 \times 10^{-43} n_e^2 (\text{for } n_e = n_i)$$

PROBLEM 7.2

Calculate the line brightness of the Planckian line with radiation temperature T_{rad}.

SOLUTION

The expression for the line brightness of the Planckian line with radiation temperature T_{rad} is given by

$$I = 2h\nu^3 n_p / c^2 \text{Wcm}^{-2}\text{Hz}^{-1}\text{sr}^{-1}$$

where n_p is the number of photons per mode. This number is given as

$$n_p = \left[e^{\frac{h\nu}{kT_{rad}}} - 1 \right]$$

Hence

$$I = 2 \times 6.63 \times 10^{-34} \times \left(\frac{3 \times 10^8}{155 \times 10^{-10}} \right)^3 \times \frac{0.8196}{\left(3 \times 10^8\right)^2}$$

$$= 2.1 \times 10^{15} \text{Wcm}^{-2}\ \text{Hz}^{-1}\text{sr}^{-1}$$

Suggested Reading Material

Bleiner, D., Arbelo-Pena, Y., Masoudnia, L., & Ruiz-Lopez, M. (2014). Table-top X-ray lasers using a plasma gain-medium: Limits and potentials. *Physica Scripta, 2014*(T162), 014050.

Dell'Angela, M., Malvestuto, M., & Parmigiani, F. (2015). Time resolved X-ray absorption spectroscopy in condensed matter: A road map to the future. *Journal of Electron Spectroscopy and Related Phenomena, 200*, 22–30.

Geloni, G., Huang, Z., & Pellegrini, C. (2017). The physics and status of x-ray free-electron lasers. In: Uwe Bergmann, Vittal Yachandra, Junko Yano, eds. *X-ray free electron lasers: applications in materials, chemistry and biology*. London: The Royal Society of Chemistry, (pp. 1–44).

Haessler, S., Caillat, J., & Salières, P. (2011). Self-probing of molecules with high harmonic generation. *Journal of Physics B: Atomic, Molecular and Optical Physics, 44*(20), 203001.

Hecht, J. (2008). The history of the X-ray laser. *Optics and Photonics News, 19*(5), 26–33.

Johnson, A. S., Austin, D. R., Avni, T., Larsen, E. W., & Marangos, J. P. (2019). Attosecond soft X-ray high harmonic generation. *Philosophical Transactions of the Royal Society A, 377*(2145), 20170468.

Kang, H. S., Min, C. K., Heo, H., Kim, C., Yang, H., Kim, G., ... & Park, B. R. (2017). Hard X-ray free-electron laser with femtosecond-scale timing jitter. *Nature Photonics, 11*(11), 708–713.

Lee, J., Nam, C. H., & Janulewicz, K. A. Eds. (2012). X-ray lasers 2010. *Proceedings of the 12th International Conference on X-Ray Lasers, 30 May-4 June 2010*, Vol. 136, New York: Springer Science & Business Media.

Patterson, B. D., Abela, R., Braun, H. H., Flechsig, U., Ganter, R., Kim, Y., ... & Rivkin, L. (2010). Coherent science at the SwissFEL x-ray laser. *New Journal of Physics, 12*(3), 035012.

Pellegrini, C., Marinelli, A., & Reiche, S. (2016). The physics of x-ray free-electron lasers. *Reviews of Modern Physics, 88*(1), 015006.

Piracha, N. K., Ali, R., Baig, M. A., Duncan-Chamberlin, K. V., Kalyar, M. A., Mehmood, M., & Ahmed, R. (2011). Laser excited population redistribution in the 2p53p multiplet in neon. *Optics Communications*, *284*(12), 2872–2875.

Robinson, I., Gruebel, G., & Mochrie, S. (2010). Focus on X-ray beams with high coherence. *New Journal of Physics*, *12*(3), 035002.

Shiner, A. D., Trallero-Herrero, C., Kajumba, N., Bandulet, H. C., Comtois, D., Légaré, F., & Villeneuve, D. M. (2009). Wavelength scaling of high harmonic generation efficiency. *Physical Review Letters*, *103*(7), 073902.

Tümmler, J., Janulewicz, K. A., Priebe, G., & Nickles, P. V. (2005). 10–Hz grazing–incidence pumped Ni-like Mo x-ray laser. *Physical Review E*, *72*(3), 037401.

8

High Harmonic Generation

8.1 Introduction

The invention of the ruby laser by Dr. T. Maiman indeed has brought a revolution to the field of optics. Operation of the first laser in 1960 enabled us to study the behavior of light in optical materials at higher intensities that otherwise would not have been possible using conventional sources of light. In the pre-laser era, the response of the medium to light under the ordinary conditions was linear. Consequently, the optical phenomena were described in terms of linear absorption and linear refractive index. However, after the invention of lasers, nonlinear optics (NLO) appeared experimentally when Maiman constructed the first operating optical ruby laser. Lasers provide highly coherent optical waves, with strong electromagnetic field amplitude reaching a level where deviation in the response of the medium from linear behavior is started. It was observed that the optical response of the material no longer remains linear at sufficiently high light intensities. In contrast, it shows nonlinear dependence on the electric field strength that forms the basis of NLO and further led to the discovery of various fascinating phenomena, such as two-photon absorption (2PA), nonlinear refraction (NLR), saturable absorption (SA), reverse SA (RSA), second- and third-harmonic generations, sum and difference frequency generations, degenerate four-wave mixing (DFWM), and so forth. Nonlinear optical effects have been examined in the applications of optical detectors, optical sensors, optical switches, and optical limiters.

8.2 Nonlinear Absorption

The phenomenon of a nonlinear change (increment or reduction) in the absorption with enhancing intensities of the radiation is called nonlinear absorption. At significantly immense radiation intensities, an enhancement in the probability of the material to absorb more than one photon can be made before relaxing to ground state. Therefore, after the invention of lasers, the investigation of multiphoton absorption (MPA) has been extensively made. Also, using intense laser fields, a redistribution in the population is induced, which results in stimulated absorption and emission, the generation of free carriers, and complex energy transitions in the complicated molecular systems. Such phenomena are depicted optically in increased (reverse saturable) or reduced (saturable) absorptions (Sutherland 2003). It is a nonlinear process because this type of absorption is proportional to the square of intensity and higher orders.

The effects generated from the study of nonlinear absorption can be used in multiple fields like optical limiting and nonlinear spectroscopy. In fact, the phenomena originating from NLO are interesting for various applications either in technology or in science.

After the invention of the high-intensity laser source, it becomes mandatory to have complete information of the absorption characteristics of the substance before using it for some applications. Some of these applications are discussed below.

8.2.1 Saturable Absorption (SA)

SA is the process of reduction in the absorption coefficient of the material with increasing intensity of the incident radiations / lasers. Mostly, materials exhibit the process of SA but generally at considerable high optical intensities, i.e. intensities near the optical damage. The process of SA takes place when the absorption cross section of the ground state is greater than that of the excited state. The materials that show SA generally have a negative nonlinear absorption coefficient.

This process can be understood as follows. When a considerable highly intense light is incident on a saturable absorber material, the atoms present in its ground state / level get excited to some upper energy states / levels with a rate such that there is an insufficient decay time to come back to the ground level before the ground level becomes depleted. Afterwards, the absorption is saturated. For such materials, the excited state cross section is lower than that of the ground state. Also, when the system is pumped with laser beams of high intensities an enhancement in the transmission of such systems is recorded.

A general expression to show the relationship between absorption coefficient and saturation intensity can be obtained by considering the steady-state approximation. In such approximations, the population of energy states remains constant for the period of the pulse. Hence

$$\alpha_{SA} = \frac{\alpha_{0SA}}{1 + \frac{I}{I_{sat}}} \tag{8.1}$$

where I and α_{0SA}, respectively, are the intensity and linear absorption coefficients within the medium. The saturation intensity is represented by I_{sat}, where a reduction in the absorption coefficient to half of its linear value is achieved. Mathematically, saturation intensity is given by $I_{sat} = \frac{h\upsilon}{\sigma\tau}$ where σ and τ, respectively, refer to the absorption cross section of the ground state and lifetime of the excitations.

Saturable absorbers are beneficial in laser cavities and are generally adopted for passive Q-switching in lasers. Sometimes, they are also called good optical limiters.

8.2.2 Reverse Saturable Absorption (RSA)

The process of RSA occurs in those materials whose absorption cross section for the ground state is smaller than that of excited states. Such materials show a positive nonlinear absorption coefficient. Under high fluence illuminations or high intensities, a reduction in the transmission is recorded for those materials that exhibit RSA, in accordance with

$$\alpha_{RSA}(I) = \alpha_{ORSA} + \sigma N(I) I \tag{8.2}$$

where σ, $N(I)$ and α_{ORSA} are, respectively, the absorption cross section, excited states' population density, and linear absorption coefficient.

Reverse saturable absorbers have a variety of applications in optical pulse processing and computing. Also, a reverse saturable absorber can be used for construction of molecular spatial light modulators, which act as an input-output device in optical computers.

8.2.3 Two-Photon Absorption (2PA)

2PA is the process of transition from the ground level / state to a higher excited level / state of a system through the absorption of the two photons either of different or the same frequencies. Generally, two types of 2PA mechanisms exist, namely resonant (sequential) and non-resonant (simultaneous) excitation. In sequential excitation, transition to the excited state occurs via an intermediate state. Then transition to a further higher excited state from the intermediate state is made through the absorption of a second photon. In simultaneous or non-resonant excitation, the transition to a higher excited state is made through the simultaneous absorption of two photons without any intermediate state (Figure 8.1). The second process, i.e. the simultaneous absorption of two photons, is generally referred to as 2PA. Since the intermediate states are not real, the simultaneous absorption of two photons by the system is required. The benefit of adopting the materials, which show a pure 2PA process, is their quick response to any modification in the input optical signal's intensity.

2PA is a third-order process whose magnitude at the low light intensities is a few orders lower than the linear absorption. The rate of atomic transition of 2PA depends on the square of the light intensities; therefore, 2PA is a nonlinear absorption process. Also, at higher intensities, this process can predominate on the linear absorption (Tkachenko 2006). The cross section for single-photon absorption is greater than 2PA in orders of magnitude because, to generate the 2PA process, the requirement of a significant high intensity laser is mandatory.

In 2PA process, the nonlinear absorption is proportional to the square of input intensity, stated as follows

$$\frac{dI}{dZ} = -(\alpha + \beta I)I \tag{8.3}$$

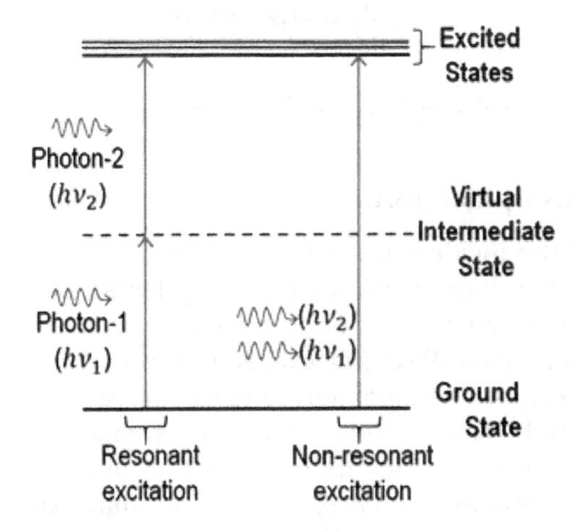

FIGURE 8.1
Sequential or resonant excitation and simultaneous or non-resonant excitation.

where β and α are, respectively, 2PA coefficient and linear absorption coefficient, and Z and I are the position coordinate along the axis of propagation and input intensity in the medium, respectively. The relationship between the 2PA cross section and 2PA coefficient is given by

$$\sigma = \frac{\hbar \upsilon \beta}{N} \tag{8.4}$$

where υ and N are, respectively, the frequency of incident radiation and number density of the molecules in the system.

Some of the essential features of 2PA are stated as follows:

i. Single-photon absorption takes place between those levels / states that have opposite parity. However, in the nonlinear medium, 2PA occurs between those levels / states that have identical parity. In other words, the transitions, which were forbidden for single-photon absorption, are allowed for 2PA.
ii. The probability of 2PA is proportional to the square of the intensity, i.e. to the fourth power of the electric field.

8.2.4 Multiphoton Absorption (MPA)

MPA is the process of the interaction of electromagnetic radiations with matter accompanied by the absorption of several photons in a single elementary event. The main difficulty in observing multiphoton processes is their extraordinary low probability compared with that of single-photon processes. Only after the invention of lasers did it become possible to observe the process of multiphoton excitation (MPE) in which several photons of exciting radiation are absorbed simultaneously via virtual states in a medium. The process of MPA is referred to as the simultaneous absorption of n-photons either from multiple beams or from a single beam. Intensity propagation equation inside the medium for the absorption of $n + 1$ photons from a single optical beam is stated as

$$\frac{dI}{dZ} = -\left(\alpha + \gamma I^{n}\right) I \tag{8.5}$$

where γ corresponds to the absorption coefficient of $(n + 1)$ photons.

8.2.5 Excited State Absorption (ESA)

When the intensity of the incident radiation is enough to deplete the ground level considerably, the excited levels become populated. In systems like semiconductors and polyatomic molecules, there is an existence of highly dense levels close to the levels that have participated in excitation (Shen 1984). The excited electrons, before undergoing transitions back to the ground level, can immediately undergo a transition to one of these highly dense levels. There may be several higher-lying states that might be coupled with these intermediate levels and for which the differences in the energy with the energy of the incident photon are in near-resonance. Hence, the electrons may experience absorption before they relax to the ground level completely. It will promote the electrons to a higher-lying state. This is called the excited state absorption (ESA) process.

2PA is an irradiance (flux received by a surface per unit area)-dependent process whereas ESA is fluence (total number of particles per unit area with which a material is irradiated) dependent. This means that if the mechanism is ESA then the same nonlinear absorption will result from the two different pulses with the same fluence. It is feasible to validate whether 2PA or ESA dominates in the induced absorption through the measurement of nonlinear absorption for several pulse durations. As a rule, the transmittance changes at a given pulse energy will be independent of the pulse width if the mechanism is ESA, but it will depend on pulse width if it is 2PA.

8.2.6 Free Carrier Absorption (FCA)

In semiconductors, an electron will promote to conduction band if it absorbs a photon whose energy is more than the band gap. At the conduction band, since electrons are the free carriers, when a field is applied, they may contribute to the current flow. The excited electrons will relax to the bottom of the conduction band by thermalizing them. After a characteristic recombination time, in the valance band, they will recombine with the excited holes. However, if the electron is still in the conduction band and the incident intensity is considerably high, then they have a high probability or high tendency to absorb another photon. Such processes are called free carrier absorptions (FCAs).

The intensity of the propagating beam if only free carrier and linear absorption exist is written as

$$\frac{dI}{dZ} = -(\alpha + \sigma N)I \tag{8.6}$$

where σ and α are, respectively, the cross section of FCA and linear absorption. The intensity-dependent carrier density is denoted by N.

8.3 Harmonic Generation

When a nonlinear material interacts with the photons of incident laser radiation, radiations that correspond to harmonic frequencies are generated. Such nonlinear optical processes are designated as harmonic generation (HG) processes. These processes usually occur at the optical intensities with the magnitude of 10^{14} W/cm^2 or greater. The common HG processes are second-harmonic generation (SHG) and third-harmonic generation (THG). However, today, the analysis of higher-order harmonic generations (HHGs) are also conducted to reveal considerable applications of HGs.

In SHG, when two similar frequency photons interact with a nonlinear material, after their effective combination there is a generation of a new photon whose energy is twice that of the incident photons. In other words, new photon is generated with half the wavelength and twice the frequency of the incident photons (Figure 8.2). Therefore, SHG is also designated as the frequency doubler process. The tendency of occurrence of SHG in a medium is characterized by its second-order nonlinear susceptibility ($\chi^{(2)}$). There are some cases in which the approximate complete convergence of light energy into second-harmonic frequency is observed. In such cases, intense pulsed laser beams are passed through large crystals and phase matching is attained by attentive alignment. Whereas

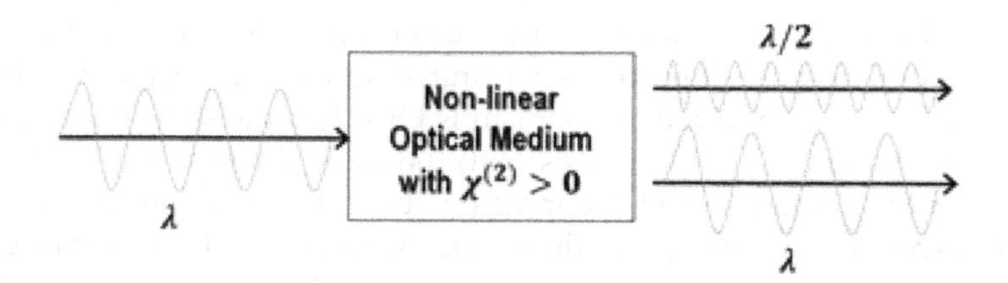

FIGURE 8.2
Conversion of an exciting wave into SHG (Boyd 1992) when it interacts with a nonlinear optical medium with finite $\chi^{(2)}$.

in some cases such as second-harmonic imaging microscopy, a small amount of light energy is transformed to second-harmonic frequency. Good SHG-generating materials are peripheral nerves and collagen fibers.

The basic phenomenon of THG is similar to that of SHG. In THG, three similar frequency photons are interacted with a nonlinear material, and there is an involvement of third-order nonlinear susceptibility ($\chi^{(3)}$). The mathematical model required to theoretically describe the THG is explained in detail in Section 8.7.1, and their detailed applications are discussed in Section 8.7.2. If the frequency of incident photons is different, then the preferred term is four-wave mixing (FWM; see Section 8.8). Materials used for THG are nonlinear crystals like β-BaB_2O_4 (Barium Borate: BBO). Also, the generation of third harmonics is possible based on the membranes in microscopy. The higher-order HGs also are theoretically feasible, but due to the requirement of interaction of a greater number of the photons, it has a minuscule probability of occurrence. A considerable different mechanism is adopted for the generation of higher-order harmonics. The *N*th HG process is depicted in Figure 8.3.

The generation of harmonics can be possible in those media that depict an intense nonlinear response corresponding to the incident laser light. As long as the shape of atomic potential is parabolic, the electrons bound in this atomic potential will behave like a simple harmonic oscillator. These electrons will oscillate and emit light at the incident light wave frequency only. At low intensities, the polarization has a linear

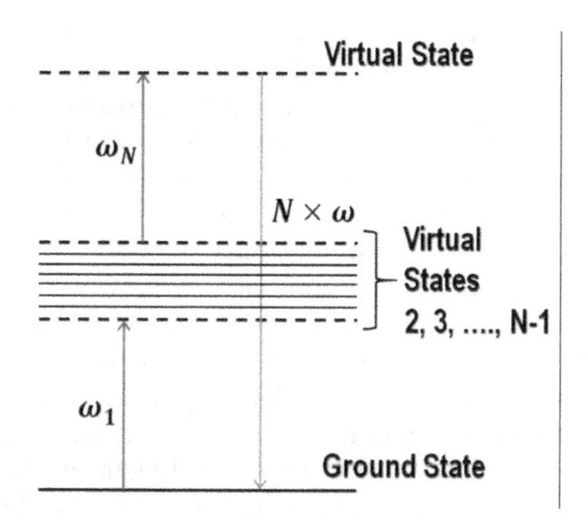

FIGURE 8.3
Schematic description of *N*th harmonic generation process.

behaviour with an applied electric field, i.e. $P^{(1)} = \varepsilon_0 \chi^{(1)} E$. This is because for adequately lower amplitudes (cf. Taylor expansion) all of the atomic potential wells are parabolic. However, the shape of atomic potential usually differs from an ideal parabolic at the higher intensities. Therefore, higher-order terms must be included in the polarization of the medium, i.e. $|P^{(n)}| \sim \varepsilon_0 \chi^{(n)} E$. This clearly indicates that the displacement of electrons should be explained as a series composed of higher-order frequencies like $2\omega, 3\omega, \ldots, N\omega$ to illustrate the emitted field exactly and thus the emission of higher integer multiples of the laser frequency (harmonics).

Ganeev et al. (2011) and Ganeev (2018) have investigated the harmonic generations in carbon nanotubes (CNTs) containing plasma pumps. With the help of ultrashort pulses, which are produced through a Cr:forsterite laser at a wavelength of 1.25 μm, the experimental studies of THG from the solid samples of CNTs have been performed (Stanciu et al. 2002). The results of these experiments showed an exceptional non-perturbative behaviour of third harmonic at the low input laser intensity of around 10^{10} W/cm^2. With the help of fundamental radiations of 1064 nm, investigations of SHG and THG in single-walled carbon nanotube films have been conducted (De Dominicis et al. 2004). The recorded analyses were achieved on both the samples of CNTs developed with a catalyst-free method and commercially available CNTs. The occurrence of THG was detected in both the samples, whereas SHG was detected only on the samples of CNTs developed with a catalyst-free method. HHGs in CNTs containing plasma plumes have been shown and HGs up to the 29th order were produced by Ganeev et al. (2011).

The harmonic spectrum displays several new characteristics, if we add a field to the fundamental field. Through the addition of a fundamental field and its second-order harmonic, we can generate both odd and even harmonics by difference and sum frequency mixing (Aubanel and Bandrauk 1994; Zuo et al. 1995). Several harmonics can be generated in a bi-chromatic field, which cannot be generated in the case of a monochromatic field. Today, in the field of laser physics, the process of HG is widely adopted to obtain a shorter wavelength through basic laser radiations.

8.4 Frequency Mixer/Generation: Theoretical Understanding

One of the most essential uses of nonlinear processes is to generate new frequencies from a fixed wavelength laser. To analyze this, let us consider an optical electric field incident on a non-centrosymmetric optical medium, which is composed of two different frequency components. This can be represented as

$$E(t) = E_1 e^{-i\omega_1 t} + E_2 e^{-i\omega_2 t} + c.c. \tag{8.7}$$

Now, the polarization of the second order is of the following form

$$P^{(2)}(t) = \varepsilon_0 \chi^{(2)} E^2(t) \tag{8.8}$$

$$P^{(2)}(t) = \varepsilon_0 \chi^{(2)} \left[E_1 e^{-i\omega_1 t} + E_2 e^{-i\omega_2 t} + c.c. \right]^2$$

$$P^{(2)}(t) = \varepsilon_0 \chi^{(2)} \left[E_1 e^{-i\omega_1 t} + E_2 e^{-i\omega_2 t} + E_1^* e^{i\omega_1 t} + E_2^* e^{i\omega_2 t} \right]^2$$

$$P^{(2)}(t)=\varepsilon_0\chi^{(2)}\Big[E_1^2e^{-2i\omega_1 t}+E_1E_2e^{-i(\omega_1+\omega_2)t}+E_1E_1^*+E_1E_2^*e^{-i(\omega_1-\omega_2)t}+E_2^2e^{-2i\omega_2 t}+E_1E_2e^{-i(\omega_1+\omega_2)t}$$
$$+E_1^*E_2e^{i(\omega_1-\omega_2)t}+E_2E_2^*+E_1E_1^*+E_1^*E_2e^{i(\omega_1-\omega_2)t}+E_1^{*2}e^{2i\omega_1 t}+E_1^*E_2^*e^{i(\omega_1+\omega_2)t}$$
$$+E_1E_2^*e^{-i(\omega_1-\omega_2)t}+E_2E_2^*+E_1^*E_2^*e^{i(\omega_1+\omega_2)t}+E_2^{*2}e^{2i\omega_2 t}\Big]$$

$$P^{(2)}(t)=\varepsilon_0\chi^{(2)}\Big[\left(E_1^2e^{-2i\omega_1 t}+E_2^2e^{-2i\omega_2 t}+2E_1E_2e^{-i(\omega_1+\omega_2)t}+2E_1E_2^*e^{-i(\omega_1-\omega_2)t}\right)+\left(2E_1E_1^*+2E_2E_2^*\right)\Big] \tag{8.9}$$

Here the terms carrying $e^{+i\omega t}$ have been neglected as the field cannot grow. It is clear that the nonlinear polarization contains terms at double frequencies, sum frequencies, difference frequencies, and constant terms, which govern the phenomena of optical rectification. These terms are represented as follows:

$$P(2\omega_1)=\varepsilon_0\chi^{(2)}E_1^2 \text{..........................SHG}$$

$$P(2\omega_2)=\varepsilon_0\chi^{(2)}E_2^2 \text{..........................SHG}$$

$$P(\omega_1+\omega_2)=2\varepsilon_0\chi^{(2)}E_1E_2 \text{...............Sum HG}$$

$$P(\omega_1-\omega_2)=2\varepsilon_0\chi^{(2)}E_1E_2^* \text{...............Difference HG}$$

$$P(0\omega)=2\varepsilon_0\chi^{(2)}(E_1E_1^*+E_2E_2^*) \text{..........Optical rectification}$$

As seen, different non-zero components are present in the nonlinear polarization. However, in actual practice, the output generated by the nonlinear optical interaction is comprised of only one of the above defined frequency components with the appropriate intensity. This is because only two frequency components, for which the certain phase matching condition is satisfied, can be efficiently produced by the nonlinear polarization. The condition of phase matching cannot be fulfilled by all the frequency components. Experimentally one can produce the desired frequency component by properly selecting the orientation of the nonlinear crystal and the polarization of the input incident beam.

8.5 Second-Harmonic Generation (SHG): Theoretical Understanding

In SHG when two same frequency photons are interacted with a nonlinear material, after their effective combination there is generation of a new photon whose energy is twice that of the incident photons. There are usually three types of second harmonics: types 0, I, and II. In types 0, I, II, two photons are with extraordinary polarization, ordinary polarization, and orthogonal polarization, respectively, with respect to the crystal. Only one of these second harmonics will occur for a given crystal orientation. SHG is an even order nonlinear optical effect that will be allowed only in the media without inversion symmetries. The process of SHG is produced in biological tissues or birefringent crystals with noncentrosymmetric structures like microtubules or collagen structures.

The first demonstration of SHG was conducted by Franken et al. (1961) after the invention of lasers. Since the required highly intense coherent beam was created only after the discovery of lasers, this demonstration was made achievable only through the discovery of lasers. In their experiment, a ruby laser of wavelength 694 nm was focussed onto the quartz sample. Then, the output light was sent through a spectrometer to record the spectrum on the photographic paper. This depicted the generation of light at a wavelength of 347 nm. Bloembergen and Pershan (1962) have initially described the formulation of SHG.

Let us consider a laser beam whose electric filed strength is expressed as

$$E(t) = Ee^{-i\omega t} + c.c. \tag{8.10}$$

Let it be incident on a crystal for which $\chi^{(2)} \neq 0$. The nonlinear polarization in such crystal is given by

$$P^{(2)}(t) = \varepsilon_0 \chi^{(2)} E^2(t) \tag{8.11}$$

Now, $E^2(t) = (Ee^{-i\omega t} + E^* e^{i\omega t} + c.c.)^2 = E^2 e^{-2i\omega t} + E^{*2} e^{2i\omega t} + 2EE^*$

Hence

$$P^{(2)}(t) = \varepsilon_0 \chi^{(2)} \left(E^2 e^{-2i\omega t} + E^{*2} e^{2i\omega t} + 2EE^*\right)$$

$$P^{(2)}(t) = 2\varepsilon_0 EE^* \chi^{(2)} + \varepsilon_0 \chi^{(2)} \left(E^2 e^{-2i\omega t}\right) \tag{8.12}$$

The second-order nonlinear polarization term consists of two terms. The first term depicts the contribution at zero frequency, which results in the process of optical rectification in which a static electric field is developed across the nonlinear crystal. The second term depicts the contribution at frequency 2ω, which leads to the generation of radiation at second-harmonic frequency, which is known as SHG.

8.6 Applications of Second-Harmonic Generation

In the laser industry, a green laser of 532 nm is produced from a source of 1064 nm via SHG. Because of constraints on intrinsic symmetry, SHG contributes the analytical advantages over the linear spectroscopic methods like UV-visible and fluorescence absorbance spectroscopies. Also, because of the high sensitivity of SHG, it is able to detect the analytes with low concentrations like peptides, small molecules, and proteins. The process of SHG does not photodamage the cells because it is a coherent process and does not excite the dipoles in terms of the energy states. In medical and biological science, the process of SHG is adopted in high-resolution optical microscopy. A combination of SHG and MPE imaging has been studied and published for breast and ovarian cancer. The tissue structure can be directly visualized with the help of SHG microscopy. Also, the high imaging resolution up to the depth of several 100 microns can be achieved by SHG microscopy.

8.7 Third-Harmonic Generation (THG)

At low intense light sources, the polarization (P) depends linearly on the electric field (E). Today, high-intensity laser sources are available in which the polarization (P) depends not only on the first power of the electric field (E) but also on the higher powers of E. The polarization in terms of the electric field can be expressed as

$$P = \varepsilon_0(\chi^{(1)}E + \chi^{(2)}E^2 + \chi^{(3)}E^3 + \ldots) \tag{8.13}$$

The demonstration of SHG is not possible in centrosymmetric materials because of their inversion symmetries in the electric field and polarization. Since in such symmetries only odd terms are present, the process of SHG is forbidden. THG in principle can occur in all matters because it is not restricted to the condition of non-centrosymmetric materials. Depending on the symmetry, incident light, and material properties, all materials have shown some nonlinearity of the third order. Third-order behavior of the process of the THG presents an essential benefit over SHG in the sense that it permits full access to the phase information and amplitude of a pulse without any direction-of-time ambiguity (Blount and Klauder 1969). For the non-invasive microscopy, third-harmonic signals are adopted, especially in the biological materials, without the requirement of labeling of fluorescence. The process of THG has also been employed in the imaging of the subcellular structure in neurons, lipids of tissue and cells, and cell nuclei (Yelin and Silberberg 1999; Débarre et al. 2006; Yu et al. 2008).

8.7.1 Theory of Third-Harmonic Generation

In THG, the generated frequency is three times that of the fundamental frequency and wavelength is one-third of the fundamental wavelength. It is not possible to produce third harmonic directly from the fundamental harmonic because the probability of reaction of three photons is much lower compared with that of two photons. THG is generally a two-step process. To generate a third harmonic from the fundamental harmonic, first we must generate a second harmonic by two-photon reaction ($\omega + \omega = 2\omega$). After that, the second harmonic at 2ω and fundamental harmonic at ω are combined together to give a third harmonic at 3ω ($2\omega + \omega = 3\omega$).

In the cases of dispersive media, the process of HG becomes a non-resonant process because of the mismatching of wave number of Nth harmonics with the N-times of the wave number of the fundamental harmonic. Therefore, one must employ density ripple to make it a resonant process.

The theory of THG in plasma (dispersive medium) with density ripple is explored in this section. A plane polarized laser beam is assumed to propagate along the density ripples in plasma. Due to the laser field, electrons start to vibrate at relativistic velocities and electron mass gets modified when the laser has a much higher intensity. Due to the laser, electrons also experience second-harmonic ponderomotive and quasi-static forces. Electrons are pushed out in the radial direction and a space charge field is created due to the ponderomotive force. In steady-state condition, the space charge force balances the ponderomotive forces and a channel of depressed electron density is created. Electron density perturbations are induced by the second-harmonic ponderomotive force. This, in turn, will beat with the oscillatory electron velocity, which is due to pump to generate the third-harmonic current and, thus, the third-harmonic radiation. The process turns into resonant when the difference between three times the wave number of the pump laser (pulse) and wave number of the third harmonic matches through the wave number of the density ripples. The mathematical elaboration of the THG is given below.

Let us consider a collisionless plasma having density ripples $n_q = n_{q0}e^{iqz}$ and electron density, $n = n_0 + n_q$. The laser beam propagates through the rippled density plasma. The laser electric field for super-Gaussian profile can be expressed as

$$\vec{E} = \hat{x}E_0(r)e^{-i\omega t} \tag{8.14}$$

where $E_0 = E_{00}\exp\left(-r^4/2r_0^4\right)$ and the width of the laser beam is represented by r_0. In the field of the laser, electrons start to oscillate at the velocity given by

$$\vec{v} = \frac{e\vec{E}}{mi\omega\gamma} \tag{8.15}$$

The ponderomotive force on the electrons due to the laser is given by

$$\vec{F}_P = -\frac{m}{2}\vec{v}\cdot\vec{\nabla}(\gamma\vec{v}) - \frac{e}{2}\vec{v}\times\vec{B} \tag{8.16}$$

Solving this expression, it comes out to be

$$\vec{F}_P = e\vec{\nabla}\Phi_P \tag{8.17}$$

where Φ_P is given by $\Phi_P = -\frac{mc^2}{e}(\gamma - 1)$ and γ as $\gamma = \left(1 + \frac{a^2}{2}\right)^{\frac{1}{2}}$. Electrons are pushed out in the radial direction due to the ponderomotive force and a space charge field is created. The modified electron density using the Poisson's equation is given by

$$\frac{n_e}{n_0} = 1 - \frac{\vec{\nabla}\cdot\vec{F}_P}{4\pi n_0 e^2} \tag{8.18}$$

Using Eq. (8.16), the above expression can be rewritten in the form of normalized laser field amplitude, $a = \frac{e|E|}{m\omega c}$ as

$$\frac{n_e}{n_0} = 1 + \frac{c^2}{\omega_p{}^2}\nabla^2\left(1 + \frac{a^2}{2}\right)^{\frac{1}{2}} \tag{8.19}$$

On solving Eq. (8.19), we have

$$\frac{n_e}{n_0} = 1 - \frac{c^2}{\omega_p{}^2 r_0^2}\left[a^4\left(1 + \frac{a^2}{2}\right)^{-\frac{3}{2}} x^{\frac{3}{2}} + 4a^2 x^{\frac{1}{2}}\left(1 + \frac{a^2}{2}\right)^{-\frac{1}{2}}(1 - x)\right] \tag{8.20}$$

where $a^2 = a_0^2 e^{-x}$ together with $x = \frac{r^4}{r_0^4}$ and $a_0 = \frac{eE_{00}}{m\omega_0 c}$. The normalized laser field amplitude at $r = 0$ is given by a_0.

We can write the following for the wave number

$$k = \frac{\omega}{c}\eta = \frac{\omega}{c}\sqrt{\varepsilon} \tag{8.21}$$

where $\varepsilon = 1 - \frac{\omega_{PO}{}^2}{\omega^2}\frac{n_e}{n_0\gamma}$ and η is the refractive index.

The electrons also experience second-harmonic ponderomotive force due to the laser at 2ω, $2k$, stated as

$$\vec{F}_{2\omega,\,2k} = -\frac{e}{2c}\left(\vec{v}\times\vec{B}\right) = -\frac{e^2E_0^2\eta}{2cmi\omega\gamma}e^{-2i(\omega t - kz)}\hat{z} \tag{8.22}$$

Due to this force, the velocity perturbation at $\vec{v}_{2\omega,\,2k}$ is given by the equation of motion, stated as follows

$$m\frac{\partial}{\partial t}\left(\gamma\vec{v}_{2\omega,\,2k}\right) = \vec{F}_{2\omega,\,2k} - e\vec{E}_{2\omega} \tag{8.23}$$

Here $\vec{E}_{2\omega}$ is the self-consistent field at 2ω frequency. Through equation of motion and Poisson's equation, we can write

$$\vec{v}_{2\omega,\,2k} = -\frac{e^2E_0^2\eta}{4c\varepsilon_{2\omega}m^2\omega^2\gamma^2}e^{-2i(\omega t - kz)}\hat{z} \tag{8.24}$$

where $\varepsilon_{2\omega} = 1 - \frac{\omega_{PO}{}^2}{4\omega^2}\frac{n_e}{n_0\gamma}$.

Using the continuity equation, the oscillatory velocity ($\vec{v}_{2\omega,\,2k}$) is coupled with n_q to produce density perturbation at 2ω, $2k+q$, as follows

$$\frac{\partial}{\partial t}\left(n_{2\omega,\,2k+q}\right) + \vec{\nabla}\cdot\left(\frac{1}{2}n_q\vec{v}_{2\omega,\,2k}\right) = 0, \text{which gives}$$

$$n_{2\omega,\,2k+q} = -n_{q0}\frac{e^2E_0^2\eta(2k+q)}{16c\varepsilon_{2\omega}m^2\omega^2\gamma^2}e^{-i(2\omega t - (2k+q)z)} \tag{8.25}$$

To generate third-harmonic nonlinear current density, this perturbation beats with $\vec{v}_{\omega,\,k}$.

$$\vec{j}_{3\omega}^{NL} = -n_{q0}\frac{e^2E_0^2\eta(2k+q)}{16c\varepsilon_{2\omega}m^2\omega^2\gamma^3}e^{-i(3\omega t - (3k+q)z)}\hat{x} \tag{8.26}$$

Now, we can write the linear current density as

$$\vec{j}_{3\omega}^{L} = \frac{n_0e^2\vec{E}_{3\omega}}{3mi\omega\gamma} \tag{8.27}$$

The wave equation corresponding to the third-harmonic field is given below

$$\nabla_\perp^2\vec{E}_{3\omega} + \frac{\partial^2\vec{E}_{3\omega}}{\partial z^2} - \frac{1}{c^2}\frac{\partial^2\vec{E}_{3\omega}}{\partial t^2} - \frac{4\pi}{c^2}\frac{\partial\vec{j}_{3\omega}^{L}}{\partial t} = \frac{4\pi}{c^2}\frac{\partial\vec{j}_{3\omega}^{NL}}{\partial t} \tag{8.28}$$

Taking $\vec{E}_{3\omega} = \hat{x}A_{3\omega}(z,r)e^{-i(3\omega t - k_3 z)}$ where $k_3 = \frac{3\omega}{c}\left(1 - \frac{\omega_{P0}^2}{9\omega^2}\frac{n_e}{n_0\gamma}\right)^{1/2}$ in Eq. (8.28), it can be written as

$$\frac{\partial^2 A_{3\omega}}{\partial r^2} + \frac{1}{r}\frac{\partial A_{3\omega}}{\partial r} + 2ik_3\frac{\partial A_{3\omega}}{\partial z} - k_3^2 A_{3\omega} + \frac{9\omega^2}{c^2}A_{3\omega} - \frac{\omega_{p0}^2}{c^2\gamma_0}A_{3\omega}$$
$$= \frac{12\pi n_{q0}ec^2a^3\eta(2k+q)}{16c^2\gamma_0^3\varepsilon_{2\omega}}e^{-i(k_3-(3k+q))z} \tag{8.29}$$

On the right-hand side, the source term consists of an exponential factor that is the function of $[k_3 - (3k+q)]z$. Corresponding to the phase matching condition $k_3 = (3k+q)$, the response of the third-harmonic field $E_{3\omega}$ is maximum. Phase matching condition $k_3 = (3k+q)$ gives rise to $q = \frac{4\omega}{3c}\frac{\omega_{P0}^2}{\omega^2}\frac{n_e}{n_0\gamma}$.

Taking $A_{3\omega} = e^{-\frac{3r^4}{2r_0^4}}F_3(z)$ and integrating the resulting equation from 0 to ∞ followed by the multiplication with the factor of $e^{-\frac{3r^4}{2r_0^4}}rdr$, we get the following equation

$$\frac{dF_3}{d\varsigma} = -i\frac{k_0\sqrt{3}}{k_3\sqrt{\pi}}\left(F_3 + \frac{\omega_{p0}^2}{c^2\gamma_0}F_3\beta_3 + \frac{3\omega_{p0}^2\eta E_{00}(2k+q)}{16c\omega}\frac{n_{q0}}{n_0}a_0^2\beta_4\right) \tag{8.30}$$

where

$$\beta_3 = \int_0^\infty rdr\left[\alpha - \left(1 + \frac{a_0^2}{2}e^{-\frac{r^4}{r_0^4}}\right)^{-\frac{1}{2}}\right]e^{-\frac{3r^4}{r_0^4}} \text{ and } \beta_4 = \int_0^\infty rdr\left(1 + \frac{a_0^2}{2}e^{-\frac{r^4}{r_0^4}}\right)^{-\frac{3}{2}}e^{-\frac{3r^4}{r_0^4}}$$

The third-harmonic field amplitude has been obtained through the numerical solution of the differential Eq. (8.30) with the initial conditions $F_3 = 0$ at $\varsigma = 0$.

The behaviour of normalized amplitude of the third-harmonic field as a function of normalized distance of propagation (ς) is depicted in Figure 8.4. From this figure, an enhancement in the amplitude of the third-harmonic power is clearly seen with increasing distance. This can be understood as follows. For self-guided beams, the laser field amplitude remains constant with the propagation distance. Therefore, the condition of phase matching is fulfilled for all z. Hence, the amplitude of the phase-matched third-harmonic field varies linearly with z. Further details can be found in Devi and Malik (2018).

The behaviour of $\frac{qc}{\omega_{p0}}$ with $\frac{\omega}{\omega_{p0}}$ is shown in Figure 8.5. With an increase in the laser frequency, the phase mismatching is seen to decrease. This behaviour can be understood based on the refractive index, which is a function of laser frequency in the plasma. The refractive index approaches to unity at higher laser frequency. Therefore, the phase mismatch and dispersion effects are decreased. The phase mismatch is decreased with an increment in the laser beam amplitude. More details of this mechanism can be found in Devi and Malik (2018).

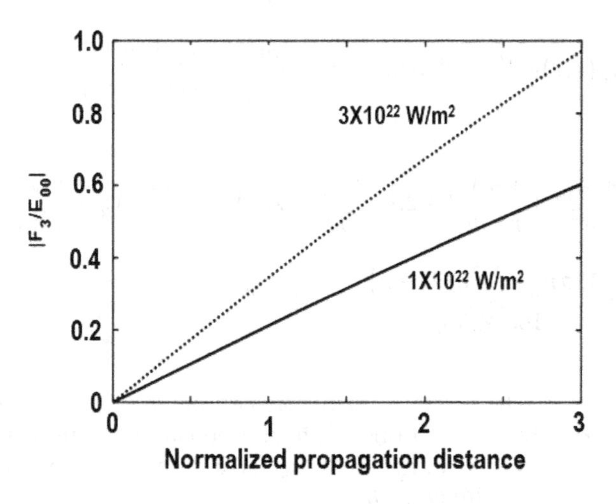

FIGURE 8.4

Behaviour of normalized amplitude of the third-harmonic field $\left(\left|\frac{F_3}{E_{00}}\right|\right)$ as a function of normalized distance of propagation (ς) when $r_0 = 20\ \mu\text{m}$ and $\lambda = 1.06\ \mu\text{m}$, showing the effect of laser intensity.

8.7.2 Application of Third-Harmonic Generation

THG finds application in the production of high-power laser sources and in nonlinear microscopy. In high-power laser interaction such as inertial confinement fusion, THG is used to convert the laser light into ultraviolet light. The most common way to generate ultraviolet light is by using an Nd:YAG laser with a wavelength of 1064 nm. The generated light has a wavelength of 355 nm by the third-harmonic process. The schematic diagram for this is shown in Figure 8.6. We can use two LBO (Lithium Triborate: LiB_3O_5) crystal or LBO and BBO crystals. The first case is the frequency doubler, which gives second-harmonic light. The second case is the sum frequency generation, which gives the third-harmonic light.

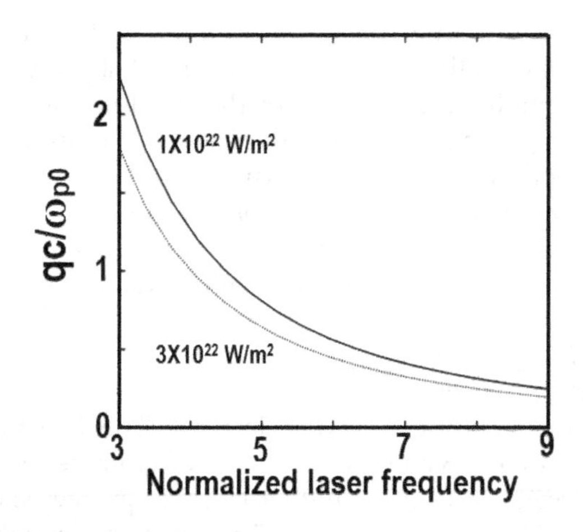

FIGURE 8.5
Variation of qc/ω_{p0} with normalized laser frequency ω/ω_{p0} when $r_0 = 20\ \mu\text{m}$ and $\lambda = 1.06\ \mu\text{m}$, showing the effect of laser intensity.

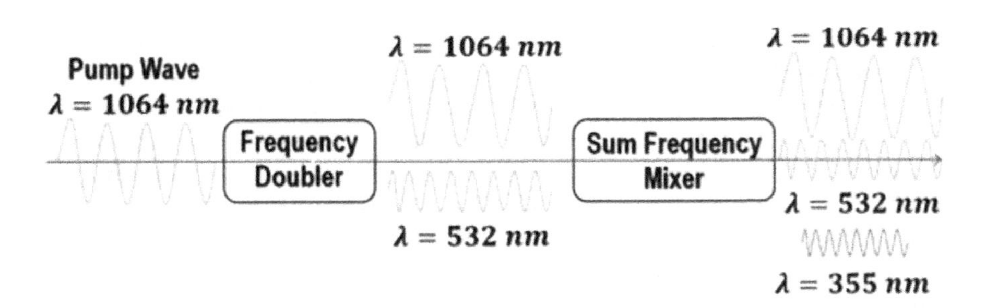

FIGURE 8.6
Schematic diagram for the generation of ultraviolet light by the THG process.

The techniques based on THG processes offer crucial benefits over other degenerate NLO processes like the Kerr effect, such that they are not susceptible to scattering at fundamental wavelengths. This is because signals of THG can be comfortably filtered both spectrally and spatially. When the signal beams have wavelength in the eye-safe near-infrared range, then these applications are also adaptable with silicon-based electronics through the use of the frequency conversion process.

Today, a significant concern is placed toward enhancing the values of third-order nonlinear susceptibility ($\chi^{(3)}$) and to develop the materials that generate considerable high conversion efficiencies of the THGs with large angular and spectral bandwidths. For this, a considerable decrease in the absorption is required at the third-harmonic wavelength. In the branched molecules, hyperbranched and dendrimers polymer-based in octupolar units (Beljonne et al. 2002), the immense nonlinearities are predicted. Therefore, THG can be used to produce centrosymmetric materials.

8.8 Four-Wave Mixing (FWM)

FWM is also one of the nonlinear effects that has arisen from the third-order optical nonlinearity, as is illustrated by the $\chi^{(3)}$ coefficient. This can appear if at least two different components of frequency are propagating simultaneously in a nonlinear medium like optical fiber. In FWM interactions, three lights of different wavelengths produce a fourth wavelength light. The fundamental point to describe all the FWM processes is the concept of the interaction of three electromagnetic fields to generate the fourth field. Physically, this process may be understood by assuming a dielectric medium in which fields are interacted individually. In the dielectric, an oscillating polarization is generated by the first input field, which in turn, re-radiates with some phase shift. This phase shift is calculated through the damping of individual dipoles. The polarization of the dielectric will also be driven by the second field. Now, the harmonics in the polarization at the differences and sum frequencies will be generated by the interference of two waves. The third field will also generate the polarization. This will beat with both the difference and sum frequencies as well as the other input fields. This beating with difference and sum frequencies is the 'fourth field' in the FWM process. The number of fields and interactions produced from this fundamental process are not fixed because each of the produced beating frequencies could also perform as the new source field.

In FWM processes, the interactions strongly depend on the relative phases of all the beams. Therefore, these processes are also described as phase-sensitive processes. If the condition of phase matching is satisfied, then the effect of FWM can be significantly accumulated over a greater distance, e.g. in a fiber. A considerable suppressing of the FWM processing is obtained in the cases of strong phase mismatching. In bulk media, the condition of phase matching can also be attained with the use of suitable angles between the beams.

The concept of FWM is appropriate in various circumstances / situations, and some of the examples are stated as follows:

1. FWM can have crucial destructive effects in the optical fiber communications, especially in the framework of wavelength division multiplexing. Here, an imbalance of channel power and / or crosswalk between the different wavelength channels can be caused. One method to restrain this is preventing the equidistant spacing of the channels.
2. The concept of FWM is employed in laser spectroscopy, usually in the form of coherent anti-stokes Raman spectroscopy (CARS). In CARS, two input waves produce a detected signal with a little greater optical frequency. One can detect the dephasing rates and excited-state lifetimes by varying the time delay between the input beams.
3. In fiber amplifiers, FWM can be implicated in strong spectral broadening; for example, for nanoscale pulses. FWM is specifically imperative in the circumstances with long pump pulses.
4. FWM can also be implemented for optical image processing, phase conjugation, and holographic imaging.

8.9 Harmonic Generation Control

Both theoretical and experimental analyses of various important features of HGs have been studied. The methods to control the coherent radiation output for practical purposes are a matter of interest. We can modify the spectrum according to our usefulness by controlling HGs. This can be achieved either by the expansion of the harmonic spectrum to the orders higher than that of predicted by the well-familiar cutoff law $U_i + 3U_P$ (Krause et al. 1992) or by the enhancement of a particularly small range of harmonics in the total spectrum near the desired energy. Here, U_P is the ponderomotive energy, given by $\frac{E^2}{4\omega^2}$ in atomic units; U_i and ω, respectively, correspond to the ionization potential of the atomic states and the driving laser field's frequency. In HG processes, the benefit of classical returning trajectories of the electrons has been studied by the semi-classical model of HG (Krause et al. 1992; Corkum 1993). This model advises that a change in the spectrum can be made by the modifications in these trajectories. Watanabe et al. (1994) in their early work have considered the simultaneous interaction of two different commensurate frequency lasers with an atom that reinforces the above idea. The plateau structure can be expended to $(U_i + kU_P)$ with $k > 3$ for the combinations of the fields $E_1\sin(\omega t)+$ $E_2\sin(3\omega t+\varphi)$. The magnitude of k is decided by the fields' intensities and their relative phase. The returning trajectories are modified by a two-colour field, and a correct prediction of the position of cutoff is obtained from the classical investigation of these trajectories. The benefit of superimposition of a second colour is not only limited to the extension of the spectrum to higher orders (Protopapas et al. 1995) but also in the increment of two to three orders of magnitude in the low-energy regime of the spectrum (with energies lower than $U_i + U_P$).

8.10 Conclusions

The interesting nonlinear phenomena are realized when the intensity of the laser becomes sufficiently high and it interacts with matter / plasma. With enhancing intensities of radiation, the phenomenon of a nonlinear change (increment or reduction) in the absorption is called nonlinear absorption. At significantly immense intensities of radiation, an enhancement in the probability of the material to absorb more than one photon can be made before relaxing to the ground state. In this situation, 2PA, MPA, ESA, SA, RSA, and FCA become important. Other important phenomena are related to the generation of high harmonics, which means SHG, THG, and so forth. Since the second- and third-harmonics are important, the theoretical formulation and applications of these harmonic generations were discussed in greater detail.

8.11 Selected Problems and Solutions

PROBLEM 8.1

Derive the expression of axial electron density inside the plasma channel formed by a Gaussian laser beam for a non-relativistic case.

SOLUTION

Let us consider the intensity of the laser given by

$$\left|E_0E_0^*\right| = A_0^2\, e^{-\frac{x^2}{x_0^2}}$$

and the density of electron given by

$$n_e = n_0 + n_1 \tag{8.31}$$

where n_0 and n_1 are the unperturbed and perturbed parts of the electron density, respectively.

Under the steady state, the electrostatic force due to charge separation balances the ponderomotive force; stated as

$$-e|E_s| = |F_p| \tag{8.32}$$

where $\vec{F}_p = \frac{-e^2}{4m_e\omega_0^2}\vec{\nabla}|E_0E_0^*|$ is the ponderomotive force for the non-relativistic case and $\vec{E}_s$ is governed by Poisson's equation, as given below

$$\vec{\nabla}\cdot\vec{E}_s = -\frac{e(n_e - n_0)}{\varepsilon_0}$$

$$= -\frac{en_1}{\varepsilon_0} \text{ (using Eq. 8.31)}$$

Taking the divergence of Eq. (8.32), we have

$$-e\vec{\nabla}\cdot\vec{E}_s = \vec{\nabla}\cdot\vec{F}_p$$

$$\frac{e^2n_1}{\varepsilon_0} = \frac{-e^2}{4m_e\omega_0^2}(\vec{\nabla}\cdot\vec{\nabla})|E_0E_0^*|$$

$$n_1 = -\frac{\varepsilon_0 A_0^2}{4m_e\omega_0^2}\nabla^2 e^{-\frac{x^2}{x_0^2}}$$

$$n_1 = \frac{-\varepsilon_0 A_0^2}{2m_e\omega_0^2}\left(\frac{2x^2}{x_0^4} - \frac{1}{x_0^2}\right)e^{-\frac{x^2}{x_0^2}}$$

Density on the axis, i.e. $x = 0$, is given as

$$n_1 = \frac{\varepsilon_0 A_0^2}{2m_e\omega_0^2}\left(\frac{1}{x_0^2}\right)$$

PROBLEM 8.2

Derive the expression of axial electron density inside the plasma channel formed by a Gaussian laser beam for the relativistic case.

SOLUTION

In the relativistic case, ponderomotive force is modified as $\vec{F}_p = m_ec^2\vec{\nabla}(1-\gamma)$ with $\gamma = \left(1 + \frac{e^2|E_0E_0^*|}{m_e^2\omega_0^2c^2}\right)^{\frac{1}{2}}$.

Therefore,

$$-e\vec{\nabla}\cdot\vec{E}_s = \vec{\nabla}\cdot\vec{F}_p$$

$$\frac{e^2 n_1}{\varepsilon_0} = m_e c^2 \vec{\nabla}\cdot\vec{\nabla}(1-\gamma)$$

$$n_1 = \varepsilon_0 m_e c^2 \nabla^2 \left[1 - \left(1 + \frac{e^2 |E_0 E_0^*|}{m_e^2 \omega_0^2 c^2}\right)^{\frac{1}{2}}\right]$$

$$n_1 = \frac{\varepsilon_0 A_0^2}{\gamma m_e \omega_0^2}\left(\frac{1}{x_0^2} - \frac{2x^2}{x_0^4} + \frac{2x^2}{x_0^4}\left(\frac{\gamma^2 - 1}{\gamma}\right)e^{-\frac{x^2}{x_0^2}}\right)e^{-\frac{x^2}{x_0^2}}$$

Density on the axis, i.e. $x = 0$, is given as

$$n_1 = \frac{\varepsilon_0 A_0^2}{\gamma_0 m_e \omega_0^2 x_0^2}$$

Here γ_0 is the value of γ at $x = 0$.

PROBLEM 8.3

For a BBO crystal, find the sum intensity I_3 for $L_{NL}^2 = 0.008\ \text{m}^2$, $\frac{\Delta kz}{2} = 1.35$, $z = 0.215$ m, and $I_1(0) = 10^{17}\ \text{Wm}^{-2}$.

SOLUTION

Sum intensity is given by

$$I_3(z) = I_1(0)\frac{z^2}{L_{NL}^2}\sin\left(\frac{\Delta kz}{2}\right)$$

Putting in the respective values, one gets

$$I_3(z) = 10^{17} \times \frac{(0.215)^2}{0.008}\sin(1.35)$$

$$I_3(z) = 1.36 \times 10^{16}\ \text{Wm}^{-2}$$

PROBLEM 8.4

Calculate the wave number of density ripples in units of $\frac{\omega_{p0}}{c}$ (ω_{p0} is equilibrium plasma frequency and c is speed of light) required for THG from laser-plasma interaction for the following data: $\frac{\omega_{p0}}{\omega_0} = 0.01$, $\frac{n_e}{n_0} = 0.45$, and $\gamma = 1.4$.

SOLUTION

The wave number of density ripples required for THG in terms of $\frac{\omega_{p0}}{c}$ is

$$q = \frac{4}{3}\left(\frac{n_e \omega_{p0}}{\gamma n_0 \omega_0}\right)\frac{\omega_{p0}}{c}$$

$$= \left(\frac{4 \times 0.45 \times 0.01}{3 \times 1.4}\right)\frac{\omega_{p0}}{c}$$

$$= 0.00428\, \frac{\omega_{p0}}{c}$$

PROBLEM 8.5

Find the nonlinear susceptibility of phenol at the air-water interface, when the adsorbate density of phenol is 0.8 mmol/g, nonlinear polarizability of phenol molecule is 5.15, and the angle which the surface makes with the normal *z*-axis is 45°.

SOLUTION

The nonlinear susceptibility is given by

$$\chi^{(2)} = N_s \left\langle \cos^3\theta \right\rangle \alpha^{(2)}$$

where N_s is the adsorbate density and $\alpha^{(2)}$ is nonlinear polarizability. Putting in the values, one gets

$$\chi^{(2)} = 0.8 \times \cos^3 45° \times 5.15$$
$$= 1.45$$

PROBLEM 8.6

What will be the intensity of SHG ($I_{2\omega}$) for a focussed beam for which gain coefficient (C) is estimated as 0.29, beam width $\omega_0 = 0.01$ cm, length of crystal $L_c = 0.3$ mm, and intensity of laser (I_ω) is 10^{17} W/cm^2.

SOLUTION

$I_{2\omega}$ for a focussed beam is given by

$$I_{2\omega} = \frac{C^2 I_\omega^2 L_c^2}{\pi \omega_0^2}$$

Putting in the values, one gets

$$I_{2\omega} = \frac{(0.29)^2 \times \left(10^{17}\right)^2 \times \left(0.3 \times 10^{-1}\right)^2}{3.14 \times (0.01)^2}$$

$$= 2.41 \times 10^{33}\ \text{W}/\text{cm}^2$$

Suggested Reading Material

Aubanel, E. E., & Bandrauk, A. D. (1994). Orbital alignment and electron control in photodissociation products by two-color laser interference. *Chemical Physics Letters, 229*(1–2), 169–174.

Beljonne, D., Wenseleers, W., Zojer, E., ...Brédas, J. L. (2002). Role of dimensionality on the two-photon absorption response of conjugated molecules: The case of octupolar compounds. *Advanced Functional Materials, 12*(9), 631–641.

Bloembergen, N., & Pershan, P. S. (1962). Light waves at the boundary of nonlinear media. *Physical Review, 128*(2), 606.

Boyd, R. W. (1992). Chapter 1 - The Nonlinear Optical Susceptibility, *Nonlinear Optics*, Academic Press, Pages 1–55.

Blount, E. I., & Klauder, J. R. (1969). Recovery of laser intensity from correlation data. *Journal of Applied Physics, 40*(7), 2874–2875.

Corkum, P. B. (1993). Plasma perspective on strong field multiphoton ionization. *Physical Review Letters, 71*(13), 1994.

Débarre, D., Supatto, W., Pena, A. M., ... Beaurepaire, E. (2006). Imaging lipid bodies in cells and tissues using third-harmonic generation microscopy. *Nature Methods, 3*(1), 47–53.

De Dominicis, L., Botti, S., Asilyan, L. S., ... Appolloni, R. (2004). Second-and third-harmonic generation in single-walled carbon nanotubes at nanosecond time scale. *Applied Physics Letters, 85*(8), 1418–1420.

Devi, L., & Malik, H. K. (2018). Resonant third harmonic generation of super-Gaussian laser beam in a rippled density plasma. *Journal of Theoretical and Applied Physics, 12*(4), 265–270.

Franken, E. P., Hill, A. E., Peters, C. W., & Weinreich, G. (1961). Generation of optical harmonics. *Physical Review Letters, 7*(4), 118.

Ganeev, R. A. (2018). *Nanostructured nonlinear optical materials: formation and characterization*, San Diego, CA: Elsevier. (Ganeev, R. A. (2018). Chapter 7 - High-order harmonic generation in carbon-containing nanoparticles, In Advanced Nanomaterials, Nanostructured nonlinear optical materials: formation and characterization. Elsevier, Pages 267–308).

Ganeev, R. A., Naik, P. A., Singhal, H., ... Gupta, P. D. (2011). High-order harmonic generation in carbon-nanotube-containing plasma plumes. *Physical Review A, 83*(1), 13820.

Krause, J. L., Schafer, K. J., & Kulander, K. C. (1992). High-order harmonic generation from atoms and ions in the high intensity regime. *Physical Review Letters, 68*(24), 3535.

Protopapas, M., Sanpera, A., Knight, P. L., & Burnett, K. (1995). High-intensity two-color interactions in the tunneling and stabilization regimes. *Physical Review A, 52*(4), R2527.

Shen, Y. R. (1984). *The principles of nonlinear optics*. New York: Wiley-Interscience, p. 575.

Stanciu, C., Ehlich, R., Petrov, V., ... Rotermund, F. (2002). Experimental and theoretical study of third-order harmonic generation in carbon nanotubes. *Applied Physics Letters, 81*(21), 4064–4066.

Sutherland, R. L. (2003). Handbook of nonlinear optics second edition, revised and expanded. *Optical engineering*. New York: Marcel Dekker, p. 82. (Sutherland, R. L. (2003). *Handbook of nonlinear optics*. CRC press; this is a book not a journal article).

Tkachenko, N. V. (2006). *Optical spectroscopy: methods and instrumentations*. Boston: Elsevier, Appendix C: Two photon absorption, pp. 293–294. (Tkachenko, N. V. (2006). Appendix C: Two photon absorption, *Optical spectroscopy: methods and instrumentations*. Elsevier, Pages 293-294).

Watanabe, S., Kondo, K., Nabekawa, Y., Sagisaka, A., & Kobayashi, Y. (1994). Two-color phase control in tunneling ionization and harmonic generation by a strong laser field and its third harmonic. *Physical review letters*, 73(20), 2692.

Yelin, D., & Silberberg, Y. (1999). Laser scanning third-harmonic-generation microscopy in biology. *Optics Express, 5*(8), 169–175.

Yu, C. H., Tai, S. P., Kung, C. T., & ... Sun, C. K. (2008). Molecular third-harmonic-generation microscopy through resonance enhancement with absorbing dye. *Optics Letters, 33*(4), 387–389.

Zuo, T., Bandrauk, A. D., Ivanov, M., & Corkum, P. B. (1995). Control of high-order harmonic generation in strong laser fields. *Physical Review A, 51*(5), 3991.

9

Attosecond Laser Generation

9.1 Introduction and Update on Laser Intensity

The exceptional advancement in ultrafast laser technology and particularly the very high instantaneous field intensity that it provides have led diverse areas in the field of science such as light source technology, accelerator technology, inertial confinement fusion, and laboratory astrophysics to the study of warm dense matter (WDM). Figure 9.1 shows the development of laser intensity with ascending years. Today, Extreme Light Infrastructure (ELI), pan-European, and Institute of Laser Engineering (ILE), Japan produce the highest peak laser power. They are generally the facilities of 100 and 10 PW (1 PW = 10^{15} Watts), which contribute to the ultrarelativistic interaction region. The production of exawatt and probably zettawatt pulses are possible through the combination of cascaded conversion compression (C^3) with a large-scale pump laser. This will expand the laser matter interactions to high-energy particle physics and vacuum nonlinear physics.

In the ultrarelativistic region, i.e. when I_L, the laser intensity, is more than 10^{23} W/cm^2, the first infrastructure to conduct the investigation of laser material interactions is ELI. The infrastructure will also contribute to the analysis of a new generation of the compact accelerators, which are providing radiation beams and energetic particles from the duration of 10^{-15} s (femtosecond) to 10^{-18} s (attosecond). The laser intensities greater than 10^{25} W/cm^2 will lead to a new avenue from the ultrafast attosecond to zepto-second (10^{-21} s) investigations of laser matter interactions.

9.2 Need for Attosecond Lasers

Due to diverse applications, the field of ultrafast laser physics has attracted considerable interest not only in chemical and atomic physics, but also in various areas of fundamental and applied science (Krausz and Ivanov 2009; Li et al. 2016). Because of the very small size of the molecule and its fast movement, we require a special variety of light to detect the molecular level's activities. The solution to capture the motion at the molecular level is an ultrafast laser light, called an attosecond laser. The dynamics of electrons in atoms and molecules can be recorded on the timescale of an attosecond (10^{-18} s) by pump-probe experiments, because of the very precise measurement of time delay between attosecond pulse and its driving laser pulse synchronized with the attosecond laser pulse, which serves to an abundance of pump (Uiberacker et al. 2007; Goulielmakis et al. 2010). In a visible wavelength regime, the optical period of light fields is of the order of a few femtoseconds; therefore, the attosecond science is essentially appropriate for the fundamental interpretation of light matter interaction. The generation of attosecond laser pulse during high-order harmonics (HOHs) has enabled its progress (Sansone et al. 2006; Goulielmakis

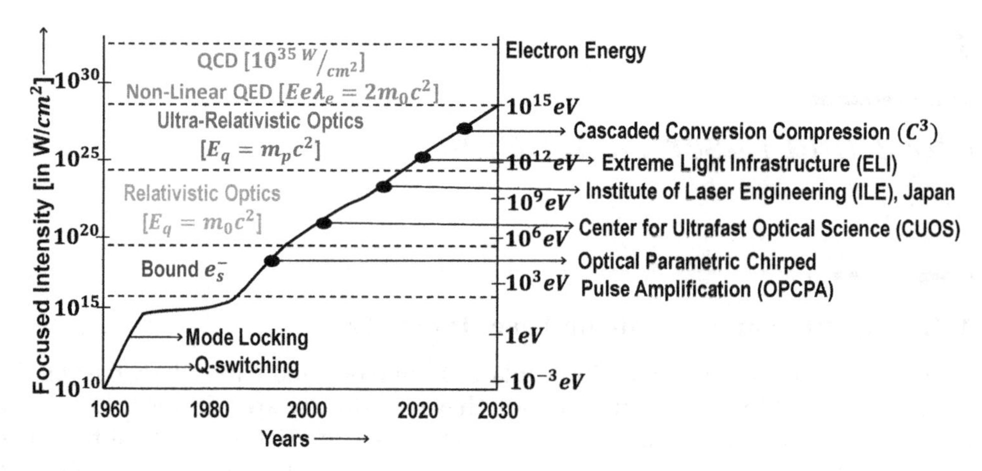

FIGURE 9.1
Diagram of maximum laser intensity plotted for ascending years.

et al. 2008; Ferrari et al. 2010). The major limitation of conventional laser-based HOHs is low conversion efficiency, which gives rise to low photon flux for the photon energy and attosecond pulse in the extreme ultraviolet regime.

9.3 X-Ray Free-Electron Laser (XFEL)

A free-electron laser (FEL) has mostly the same characteristic as that of a conventional laser beam like narrow bandwidth, low diffraction, and high power. The major difference between FEL and a conventional laser is in their amplification medium. In a conventional laser, light amplification occurs as a result of stimulated emission of the electrons, which are bound to atoms. In the case of FEL, the light amplification occurs as a result of free (unbound) electrons. This amplification is designated as a self-amplified spontaneous emission (SASE) mechanism. Based on the accelerator energy and undulator parameters, the wavelength of the FEL is tunable over a broad range because the electrons are free and can make any transition.

9.3.1 Self-Amplified Spontaneous Emission (SASE)

SASE is a process with FEL in which a high-energy electron beam is used to create a laser beam. In the SASE process, an electron bunch with uniform density distribution and having velocity comparable to speed of light is injected into an undulator. The undulator is a periodic arrangement of the magnetic dipoles. These can be superconducting magnets or permanent magnets. Along the length of the undulator, the static magnetic field is alternating with a wavelength λ_μ. When electrons traverse in the undulator, they are forced to oscillate and radiate energy within a certain energy bandwidth. The emitted photons having velocities slightly greater than the electrons interact with the electrons in each undulator period. The electron bunch oscillating through the undulator is interacting with its owned created electromagnetic field through the spontaneous emission. Based on the phase differences among radiation and the electrons' oscillation, the electrons will accelerate or decelerate. The electrons in phase with the radiation will lose their energy

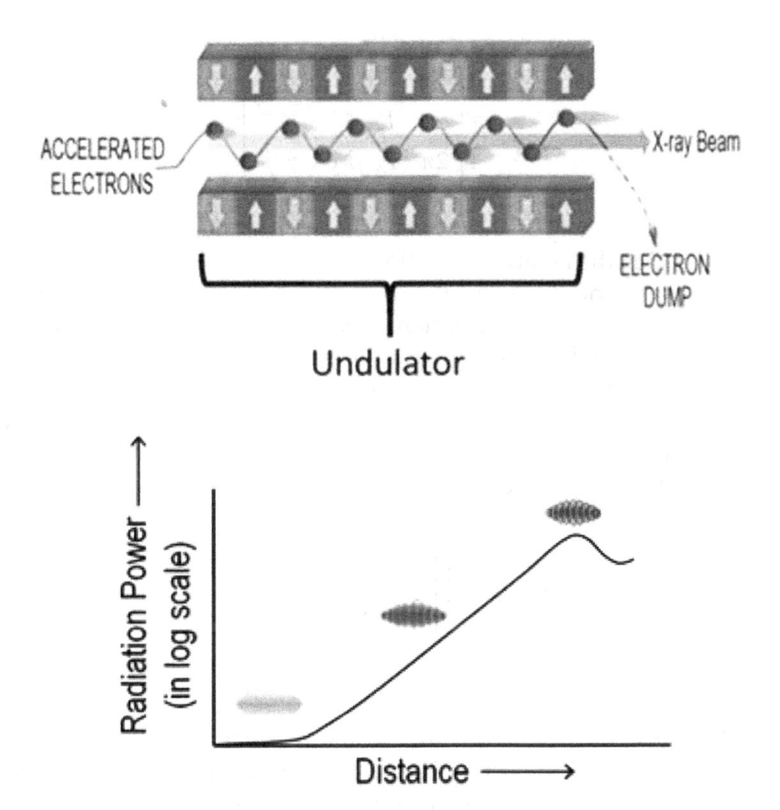

FIGURE 9.2
Schematic of self-amplified spontaneous emission (SASE) in an undulator. The density modulation (micro-bunching) of the electron bunch is represented in the lower part of the figure along with the resulting exponential amplification of the electromagnetic field along the undulator.

and decelerate, whereas the electrons that are out of phase with the radiations will gain energy and hence accelerate. Because of such interactions, a longitudinal fine structure called micro-bunching is formed, which in turn, amplifies the electromagnetic field. The modulated electron bunch loses its kinetic energy to amplify certain photon energies until the time the system undergoes saturation. The energy spectra of SASE has a noise-like distribution with intense spikes. The phase space volume, which is available to the photons, is decreased by the micro-bunching; therefore, most likely, they preferred to have a similar phase and the emission of the beam is quasi-coherent. The schematic representation of the SASE mechanism of X-ray FEL is depicted in the upper part of Figure 9.2.

A good estimate of the radiation wavelength is given based on the resonance condition, as

$$\lambda_r = \frac{\lambda_u}{2\gamma^2}\left(1+\frac{K^2}{2}\right) \tag{9.1}$$

where λ_r, γ, and λ_u, respectively, refer to the wavelength of radiation, the relativistic factor, and magnetic period length (wavelength) of the undulator. The undulator parameter, K, is given by $K = eB_u\lambda_u / 2\pi mc$, where e, c, B_u, and m correspond to the electronic charge, speed of light, magnetic field of the undulator, and electron mass, respectively.

On the other hand, FEL parameter, ρ, is defined as

$$\rho \approx \frac{1}{4}\left[\frac{1}{2\pi^2}\frac{I_p}{I_A}\frac{\lambda_u^2}{\beta\epsilon_N}\left(\frac{K}{\gamma}\right)^2\right]^{\frac{1}{3}} \tag{9.2}$$

where I_p is the peak current, I_A is the Alfven current, ϵ_N is the normalized emittance of the electron beam, and β is the betatron function.

In an undulator, some additional criteria must be required to amplify the radiation power exponentially in the spontaneous emission. One criterion is to have an electron beam of superior quality. There should be enough of an overlap between electron bunch and radiated pulse along the undulator. For enough overlapping, the electron beam should have low emittance, low divergence, and high charge density along with a specific magnetic field. Also, the beam steering should be perfect along the long undulator. During the amplification process the electron beam energy spread, σ_δ, must follow the following condition

$$\sigma_\delta < \rho \approx \frac{1}{4}\left[\frac{1}{2\pi^2}\frac{I_{pk}}{I_A}\frac{\lambda_u^2}{\beta\epsilon_N}\left(\frac{K}{\gamma}\right)^2\right]^{\frac{1}{3}} \tag{9.3}$$

The modulation in the electron density is due to the arrangement of electrons in longitudinal direction separated by equidistant spaces of λ_r, where λ_r is the wavelength of the emitted radiation. The radiations emitted from the micro-bunched electrons are superimposed coherently. The coherent superposition increases because a greater number of electrons start to radiate in phase. However, the density modulation of the electron bunch increases with an increase in intensity of electromagnetic field and vice versa.

Initially, all the electrons $N_e \geq 10^9$ in a bunch can be assumed as individually radiating charges in the case when there is no micro-bunching and power of the spontaneous emission ($P(z)$) is directly proportional to N_e. When there is micro-bunching, the electrons in the bunch radiate in phase with each other and the power of the spontaneous emission varies as $P(z) \propto N_e^2$. Hence, there is an amplification of spontaneous emission of the undulator by several orders of magnitude. Due to the progress in micro-bunching, there is an exponential growth in the radiation power $P(z)$ of such SASE FEL along with the distance z of the undulator, as per the relation

$$P(z) = AP_{in}\exp\left(\frac{2z}{L_g}\right) \tag{9.4}$$

where P_{in}, A, and L_g correspond to the 'effective' input power, input coupling factor, and field gain length, respectively. For an ideal electron beam, A is equal to $\frac{1}{9}$ in 1D FEL theory. The estimate of the effective input power of the shot noise P_{in} can be known by using the radiation power of spontaneous emission on the first gain length within the FEL bandwidth and within a coherence angle. At the entrance of the undulator, the input power is given by

$$P_{in} = 6\sqrt{\pi}\rho^2\frac{P_b}{N\sqrt{\log(N/\rho)}} \tag{9.5}$$

where ρ, N, and P_b, respectively, are the FEL parameter, number of electrons per radiation wavelength, and electron beam power, and $N = I_p\lambda/ce$ with I_p as the peak current and e

as the charge of the electron. From Figure 9.2, we can see that until the electron beam is completely bunched, the exponential growth takes place, after this it is over modulated, resulting in saturation.

9.3.2 Development of First Hard X-Ray Free-Electron Laser (XFEL)

In 2009, the world's first hard X-ray FEL (XFEL), Linac Coherent Light Source (LCLS), began lasing. This has been made possible by the availability of very high-quality electron beams, which typically have energy 2.4–15.4 GeV, beam charge 250–300 pC, energy spread $\sigma_E/E < 10^{-3}$ and an emittance close to 1π mm-mrad (Ishikawa et al. 2012; Loos 2013). This produces an extremely bright source of hard X-rays: the peak brightness of LCLS is 10^{33} photons/s/mm^2/mrad^2/0.1%BW. Since the beam grows from noise, it has poor temporal coherence and a relatively 'large' bandwidth (~20 eV). The bandwidth can be improved to 0.4 eV, along with an order of magnitude increase in brightness, by a self-seeding setup (Amann et al. 2012). The peak X-ray brightness for a selection of third- and fourth-generation light sources has shown a huge increase (~10^9) in brightness due to lasing.

LCLS X-ray pulses can be tuned to have energies of 0.25–10.5 keV and durations of 5–500 fs, with more than 10^{12} photons / pulse (Loos 2013). Studies carried out at LCLS have advanced multiple fields of science including X-ray absorption and emission of materials on a femtosecond timescale (Vinko et al. 2015), femtosecond crystallography (Martin-Garcia et al. 2016), dynamics of superfluid (Gomez et al. 2014), studies of high-temperature superconductors (Gerber et al. 2015), and femtosecond chemistry (Minitti et al. 2015). A few to a few tens of femtosecond pulses have been produced by all XFEL facilities.

9.4 Attosecond X-Ray Laser

Recently attosecond and nanoscale physics are two areas of research that came together. The interaction of ultrashort laser pulses on the timescale of femtosecond and sub-femtosecond duration, with the atoms, molecules, or solids has been dealt with in attosecond physics. Therefore, the measurement, control, and finally the manipulation of electron dynamics on the timescale of attoseconds can be made. These electron dynamics gives the information about the chemical and physical changes that occur at fundamental levels. A single isolated pulse of attosecond duration is possible to generate, as confirmed by the several researchers (Hentschel et al. 2001; Kienberger et al. 2004; Sansone et al. 2006; Schultze et al. 2007; Goulielmakis et al. 2008; Mashiko et al. 2008) and it finds applications (Goulielmakis et al. 2004; Uiberacker et al. 2007) in investigation of faster processes such as on the atomic scale. The aim of the attosecond science is to probe the electronic motion on the atomic length scale. More specifically, it belongs to the ultra-fast motion of the charges (including holes, electrons, and also protons in some cases) and interactions among themselves.

Sources based on SASE schemes such as X-ray free-electron laser (XFEL) present some special characteristics such as high peak intensity, extraordinary power, and coherence characteristics. In addition to this, in the hard and soft X-ray regimes, the higher photon energy of the order of the energy of core electrons inside the atoms and molecules (see Figure 9.3) gives access to previously unexplored phenomena (Krausz and Ivanov 2009).

To get spatial resolution on the atomic scale, the radiation wavelength should be of the order of 0.1 nm or shorter; therefore, it is exceedingly beneficial to have the attosecond

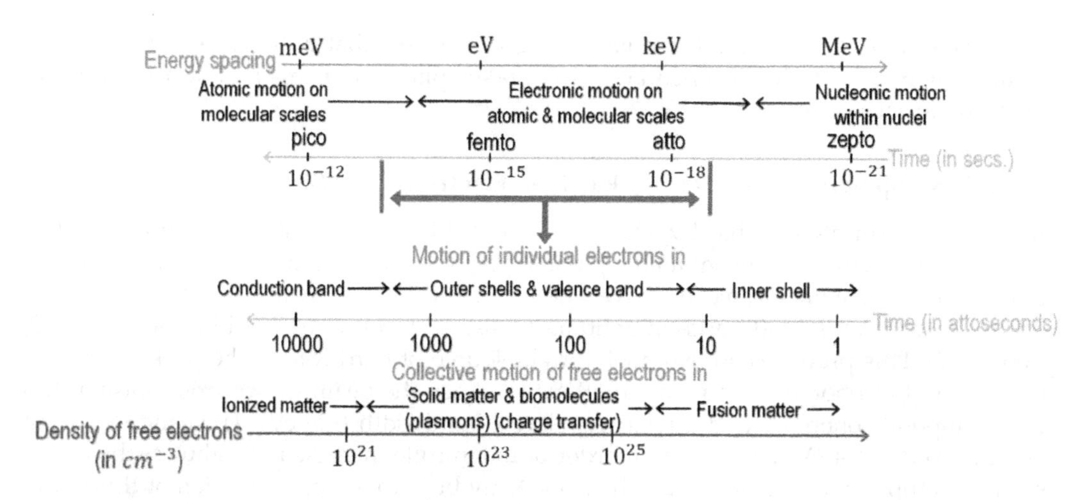

FIGURE 9.3
Characteristic timescales of microscopic motion of the electrons. The upper panel shows an energy gap between stationary states. The lower panel shows the time-scales (characteristic) for the motion of the individual electrons and collective motion of the free electrons.

pulse at 0.1 nm or shorter to get temporal resolution on the scale of attoseconds and spatial resolution on the scale of nanometres. Hence, to achieve the duration of XFEL pulse to the level of sub-femtosecond, some theoretical efforts have been made. The experimental effort has started for the realization of an attosecond XFEL and, in 2017, the LCLS a project for the X-ray laser-enhanced attosecond pulses was started, leading to the formation of a new field of science that needed to be explored.

9.5 Methods for Attosecond X-Ray Pulse Generation in XFEL

In this section, some methods for attosecond pulse generation in XFEL are described. The assessment and the possibility of each scheme to understand the terawatt-attosecond (TW-as) XFEL are discussed in detail. These methods involve electron beam current modulation, beam emittance modulation, and beam tilting for the selection of good electron trajectory in XFEL.

9.5.1 Methods Based on Energy and Density Modulation

There are six methods that fall within this category: current enhanced SASE (E-SASE) scheme, electron beam delayed scheme in XFEL, current modulation by frequency-chirped laser in XFEL, optical beam delayed scheme in XFEL, current enhanced scheme with undulator tapering in XFEL, and mode-locked technique in XFEL. These will be discussed one by one.

9.5.1.1 Current E-SASE

This scheme provides a single isolated attosecond pulse in XFEL with slight modification in between linear accelerator end and the undulator front (Figure 9.4).

In this scheme, there is a wiggler magnet with two periods, and a bunch of electrons from a linear accelerator is made to inject on it. Inside the wiggler, there is few-cycle laser pulse propagating along with the Ti:Sapphire laser pulse, and energy modulation of the electron

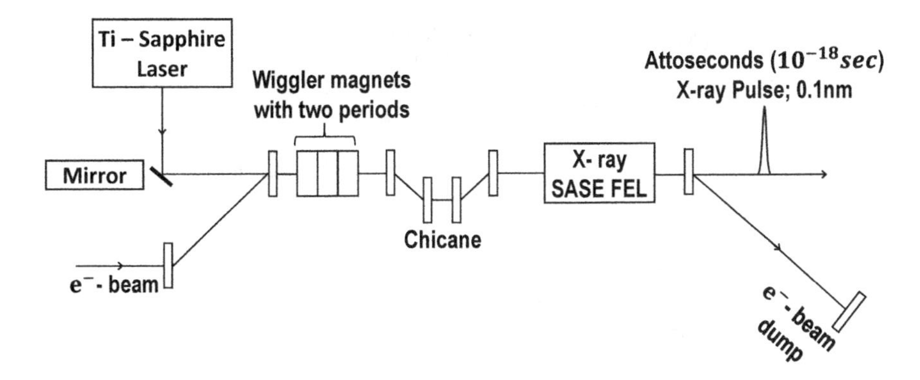

FIGURE 9.4
Current enhanced scheme for attosecond pulse generation. Here the Sapphire laser with 5-fs pulse and wavelength of 800 nm is adopted.

bunch takes place. The laser interacts through a small longitudinal section of the bunch of electrons. To get a better modulation, random noise in electron energy should be smaller than the energy modulation amplitude, and for that laser peak power is selected accordingly. The electron beam is now made to fall on the chicane, which causes dispersion and leads to a density modulation. The density modulation creates micro-bunching of the electrons with a separation of periodic current peaks and laser wavelength. An increment in the energy spread of electrons leads to an increase in the peak current. At last, to improve the micro-bunching, the electron bunch is made to enter a long undulator. At a resonant wavelength, the current spikes radiate strongly through the standard process of SASE (Figure 9.4). From the simulation study, it is reported that the production of an isolated attosecond hard X-ray results from the manipulation of the energy beam profile of the electrons.

9.5.1.2 Electron Beam Delayed Scheme in XFEL

In the previous method, the number of current spikes can be more than one depending on the chosen laser pulse duration used for energy modulation. Therefore, they will radiate together inside the undulator and at the end of the undulator, one cannot obtain single radiation pulse. In this method, multiple current spikes are generated by using a many-cycle laser pulse. However, by seeding via repeated delay of electron beam before entering the undulator section, all these current spikes can be made to contribute to a single radiation pulse. At the first stage, a large delay is introduced between the current spikes and radiation pulse (Tanaka 2013). For delaying the X-ray pulses, a series of mirrors are used. Now the pulse of delayed radiation is aligned with the target current spike [see Figure 9.5(A)]. The delayed radiation pulse is termed as target radiation pulse. By seeding the tale current spike, amplification of target radiation pulse is achieved. However, the amplification of other radiation pulses is achieved, but the degree of amplification is relatively smaller than the target radiation pulse. For third undulator stage amplification, the target radiation pulse is aligned with the second to last current spike. This procedure is repeated at several undulator stages and all current spikes are utilized for seeding, and the target radiation pulse gets amplified only.

At the current peak positions, various X-ray pulses are created at the end of the fourth segment [Figure 9.5(B)]. With the help of suitable electron and optical delays, the target pulse is amplified in further segments. In figure, an arrow has neen inserted to indicate

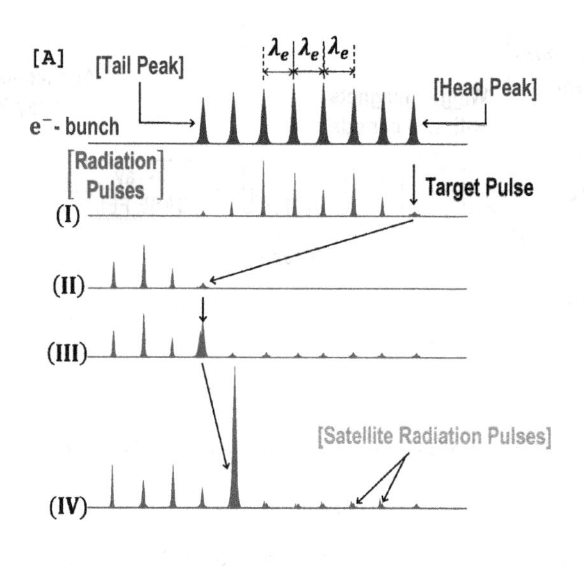

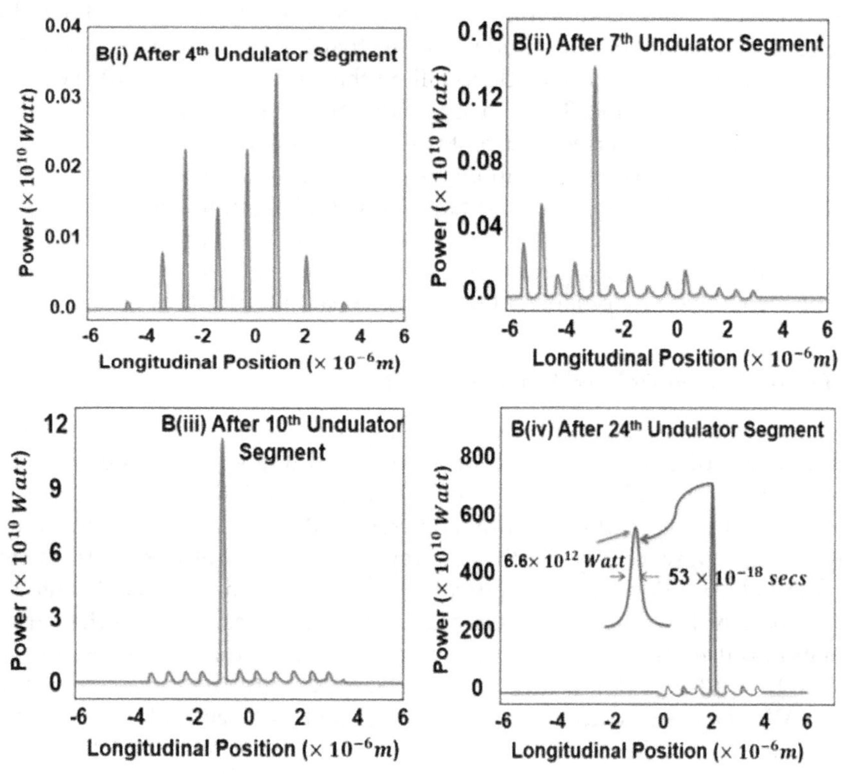

FIGURE 9.5
(A) Target pulse amplification process by using regularly spaced current spikes. The purple diagram at the top shows the electron bunch with equal spacing of multiple current spikes. (I) The first undulator unit generates many radiation pulses. (II) A delay is introduced in the radiation pulse so that it matches the tail peak of the target pulse. (III) The amplification of the target pulse is done in the second undulator. (IV) Again, delay is introduced in the electron bunch so that the second to last current spike matches the amplified target pulse. A repetition of this process is made. (B) Behavior of power in gigawatts as a function of longitudinal position in micrometres after B(i) 4th, B(ii) 7th, B(iii) 10th, and B(iv) 24th (final) segments.

the target pulse. At the exit of the tenth segments, a solitary pulse is formed. From this method, Tanaka (2013) reported X-ray pulse from the SPRing-8 Angstrom Compact Free-Electron Laser (SACLA) XFEL. The pulse obtained has a 53-attosecond pulse width and 6.6 TW radiation power at a wavelength of 0.1 nm after final 24th segment. The same is shown in Figure 9.5(B).

There is also a number of drawbacks of this process. First is the generation of many radiation pulses by all the current spikes because they radiate independently. Because of equal spacing of current spikes and radiation pulses, unwanted radiation pulses are amplified due to other current spikes. This is the second disadvantage. Since the spread in the energy of the matched current spike is large because of the amplification mechanism of the target radiation pulse, other radiation pulses amplify slowly compared with the fresh current spike. The creation of pure isolated X-ray pulse is becoming challenging through the use of regularly spaced current spikes because of its capability to intensify the radiation pulses. Figure 9.5(B) shows that at the exit of the 24th segment, beside the main target radiation pulse, there are several satellite radiation pulses.

9.5.1.3 Current Modulation by Frequency-Chirped Laser in XFEL

Continuing the efforts to obtain the isolated attosecond X-ray pulse not having satellite pulses, irregular spacing among current spikes is created in the E-SASE section. By doing so, the satellite radiation pulse, in principle, can be removed completely. This idea was proposed by Tanaka et al. (2016), as shown in Figure 9.6(A). Indeed, three different concepts were proposed to produce the current spikes with irregular spacing, the details of which can be found in Tanaka et al. (2016). Among these three concepts, the most real and simple idea was the frequency-chirped laser method. Here, the current spikes with irregular spacing would be generated if a frequency chirped laser is introduced in the E-SASE section for the energy modulation inside the wiggler. After that, there is a change in the wavelength of the chirped laser pulse (Tanaka et al. 2016; Wang et al. 2017; Kumar et al. 2018).

To elaborate this method, let us consider unequally spaced current spikes / peaks of the electron beam [Figure 9.6(A), purple]. The gap between ith and $(i + 1)$th current spikes is not equal and is denoted by τ_i. In the normal SASE process, X-rays are generated (orange) independently by the current spikes / peaks at the position (i) of Figure 9.6(A). In this figure, the X-ray at the end of the tail of the current spikes of electron beam is denoted by blue and is served as the target (main) pulse in the next levels. In the chicane, when there is a delay in electron beam by the factor of τ_1, the target (main) pulse corresponds to the second to last current peak / spike. Here, other pulses do not coincide; therefore, there is an amplification of the main pulse only. After this, there is a delay by a factor of τ_2 ($\tau_1 \neq \tau_2$) in the electron beam. Now, the target pulse coincides with the third to last current peak / spike and further amplification of this target pulse is reported [Figure 9.6(A), (iii)]. By repeating this process several times, we have a target pulse corresponding to the head of the current spikes of the electron beam with a sufficient amplification of the target pulse. Thus, in this method, there is a selective amplification of the target pulse only.

Figure 9.6(B) shows the simulation results for five different cases. The highest power obtained for XFEL with the chirped laser method is 1.8 ± 0.45 TW with the pulse duration of 62 ± 3 attoseconds. However, the preferred configuration strongly relies on conditions like electron beam parameter, specifications of E-SASE laser, and number of undulator series. With a significant change in these conditions, the output of various cases would be modified considerably.

Therefore, to remove satellite radiation pulses that were present in the work of Tanaka (2013), the idea suggested by Tanaka et al. (2016) seems to be more feasible. The good thing

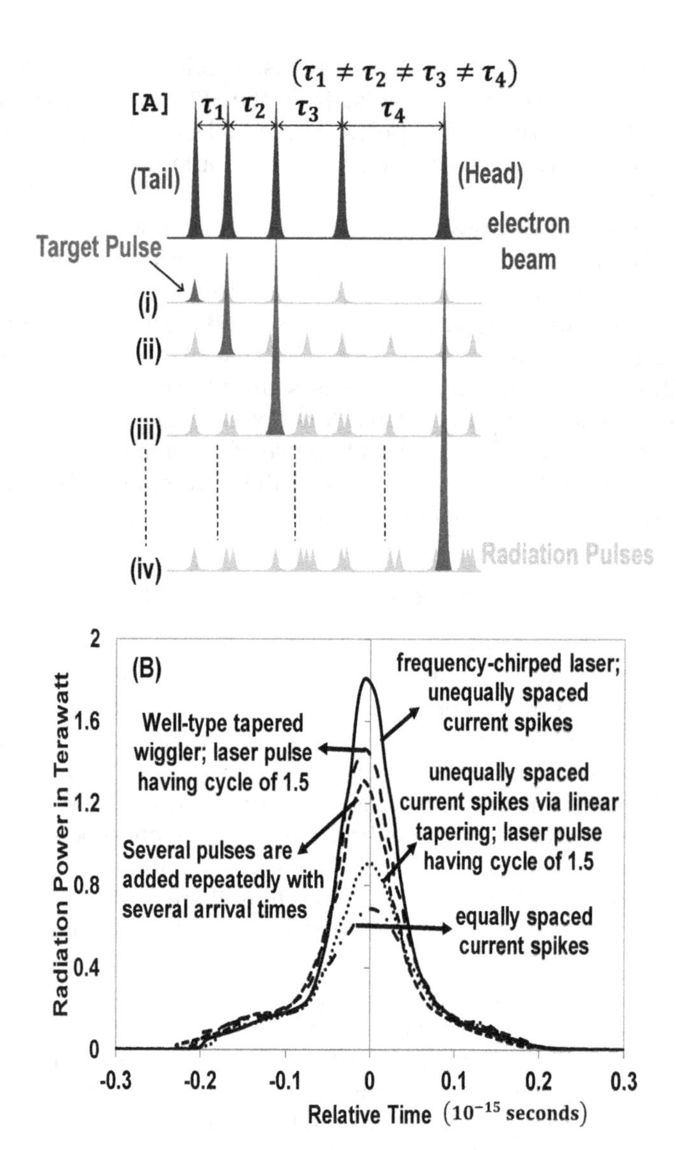

FIGURE 9.6
(A) Unequally spaced current spikes of electron beam (top, purple). (i) In the first undulator section, the corresponding radiation pulses generated are orange spikes. In the same row, the blue spike is the target spike (TP) of the radiation. (ii) TP is aligned with second to last current spike. (iii) TP is aligned with third to last current spike and further amplified. (iv) This process is repeated for all current spikes. (B) Five different simulations of radiation pulse power using the original method (equally spaced current spikes) in Tanaka (2013) including unequally spaced current spikes via linear tapering of the wiggler with a laser pulse having cycle of 1.5; well-type tapered wiggler with a laser pulse having a cycle of 1.5; frequency-chirped laser used for unequally spaced current spikes; and several pulses formed by splitting a laser pulse and then added repeatedly with several arrival times, like τ_1, τ_2, ... as shown in the top part of (A), are compared.

in this method is that no optical delay unit, i.e. no X-ray mirror, is used. Hence, the implementation of this method is easy compared with the one suggested by Tanaka (2013). However, there is a challenge to resolve the stability issue of electron beam delay. Moreover, in an electron delay unit, the fluctuations in the field of magnets may be caused by the fluctuation of a power supply. Hence, there would be a deviation in the field amplitude, which may result in mismatching of the electron current spikes and radiation pulse.

9.5.1.4 Optical Beam Delayed Scheme in XFEL

Various methodologies have been adopted for this scheme. For TW-attosecond (TW-as) X-ray generation, a single current spike with slightly higher current is utilized instead of adopting several current spikes. In the previous section, in the multi-current spike method, a peak of current ~10 kA was considered. However, a peak of current of 30 kA is employed for undulator radiation in the optical beam delayed scheme in XFEL. The idea is to amplify an X-ray pulse by a large number of times by using the single current spike several times (Kumar et al. 2016). In the radiation generation process, an X-ray pulse slips ahead of the current spike. However, the current spike is still able to be reused as it shows low divergence and good emittance. Hence, such a high current spike is used several times for the amplification process. There is further contribution of electrons in the amplification process if we delay the X-ray pulse to match the same current spike again. This process is depicted in Figure 9.7 (Kumar et al. 2016).

The radiation travels ahead of the electrons in an undulator by one λ_r (wavelength of the radiations) per λ_u (undulator period) and slips ahead by N times of λ_r, where N refers to the total number of passed undulator periods. A chicane mirror unit, which is comprised of a set of plane mirrors and four dipole magnets [Figure 9.7(iii)], is included among UMs to compensate for the radiation slippage. During SASE radiation, the micro-bunching generated in the UMs gets diluted and the electron bunch gets delayed by the chicane. However, a temporal delay is provided by the reflective mirror unit to the radiation compared with the bunch of the electrons. To minimize the pulse broadening and radiation slippage, a

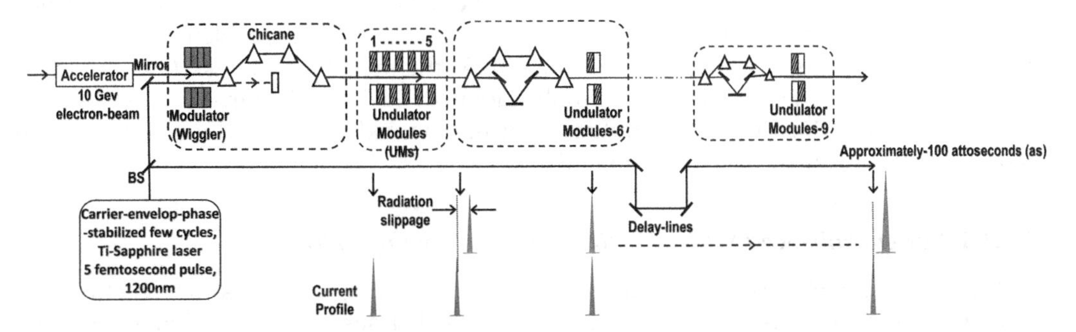

FIGURE 9.7
This is the single current spike method with the modulation of the electron beam of 10 GeV in the density and energy by (i) chicane and modulator system. Carrier envelop phase (CEP) stabilizes a few cycles, and Ti:Sapphire laser of 5-attosecond pulse duration and wavelength of 1200 nm is adopted for the modulation of energy of the electron beam. (ii) SASE undulator is composed of five undulator modules (UMs) and adopted for seed radiation production. (iii) Combination of radiation alignment and 1 UM is used for the amplification of radiation. This is the chicane mirror system. (iv) The process identical to (iii) is repeated except that the size of the chicane mirror system is compatibly tinier. The position of current spikes and the analogue's radiation spike are depicted by blue and red, respectively.

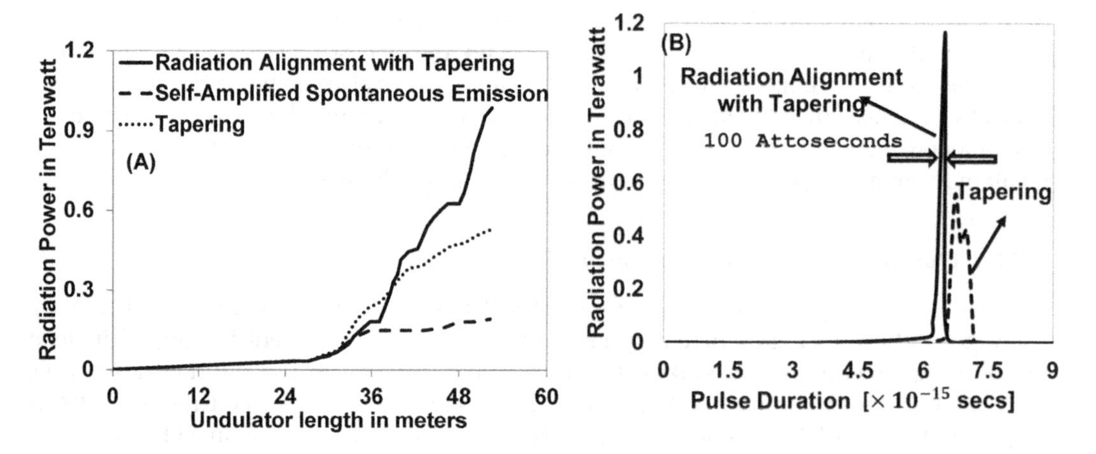

FIGURE 9.8
Simulation results using the recycle method with a single current spike. (A) Behaviour of radiation power with undulator length. Dashed line depicts the SASE process without any tapering in the undulator line. For this, the recorded output power is 0.17 ± 0.1 TW. In the undulator line, the results for the optimal tapering without the recycle method is shown by the dotted line. The result shows a 0.5 ± 0.11 TW radiation power and 300-attosecond pulse duration. The solid line shows the tapering case along with the alignment of radiation, for which it is possible to obtain a radiation power greater than 1 TW, i.e. 1.01 ± 0.36 TW using the recycle of a single current spike. (B) FEL radiation's temporal profile.

small UM is adopted in the next level of amplification. To receive a clean-single amplified pulse, this UM can be included many times [Figure 9.7(iv)]. In the study conducted by Kumar et al. (2016), a total of nine UMs were included to have an amplified pulse of terawatt level.

Figure 9.8 shows that the terawatt power obtained for tapering and tampering with radiation alignment is 0.55 and 1.2 TW, respectively, whereas the pulse duration for these cases is 300 and 100 attoseconds, respectively. The delay of both the X-ray and electrons is necessary in this method. X-ray mirrors suffer from a vibration problem in addition to the stability of the power supply (Tanaka 2013). At the first stage, Tanaka (2013) used an X-ray delay unit only; whereas for the amplification, the delay unit in every step was used by Kumar et al. (2016). The overall radiated power obtained decreases due to the multiple reflections that resulted from the X-ray mirrors.

9.5.1.5 Current Enhanced Scheme with Undulator Tapering in XFEL

In the previous methods, for the attosecond XFEL, the major issue is the synchronization between X-ray pulse and optical laser pulse. This issue is fundamentally beneficial because of its application in performing well-defined pump-probe experiments. For the realization of these schemes, in addition to the inherent synchronization, synchronization at the chicane-mirror setup is also required because the performance of the scheme could be affected by the electron beam energy jitter and the timing jitter between modulation laser and electron beam. In pump-probe experiments, the jitter in the optical path lengths is due to the mechanical vibration. For the stable operation of these schemes, one should take care of all these things. Another alternative for the stable operation is the use of a single current spike exclusive of an optical delay unit (Figure 9.9). The single current spike is used with a peak current of 37 kA and no extra element was used [electron beam delay

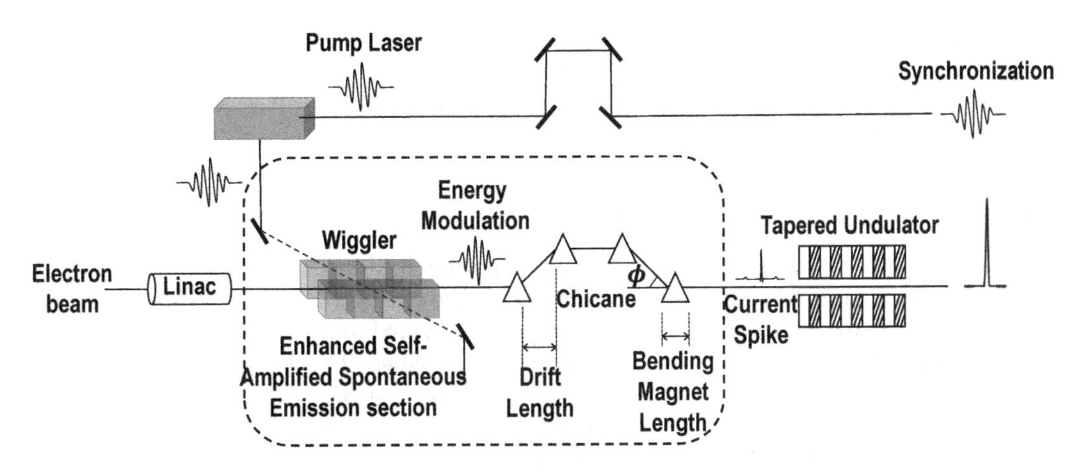

FIGURE 9.9
Single current spike process for stable operation without optical delay and electron beam delay.

(magnetic chicane) and optical beam delay (X-ray mirrors)] in this case between the undulator units to generate an X-ray pulse of terawatt level with Pohang Accelerator Laboratory X-ray Free-Electron Laser (PAL-XFEL) parameters.

The undulator line is tapered optimally. The power of FEL radiation grows very quickly in the initial level of the undulator line because of high peak current. The radiation pulse is separated from the current spike in the middle part of the undulator line. However, a background current of a few kiloamperes still remains that amplifies the radiation pulse slowly. Figure 9.10(A) depicts the power growth of X-ray free-electron laser radiation and Figure 9.10(B) depicts their profile (Kumar et al. 2017; Shim et al. 2018). The average power generated in this method is 1.0 ± 0.3 TW. The profile of radiations showed a single spike with full-width half-maximum (FWHM) of around 36 attoseconds. This is the shortest pulse among the schemes investigated in this section. This scheme seems to be more realistic and the method is simple. These outcomes are important since, in the undulator line,

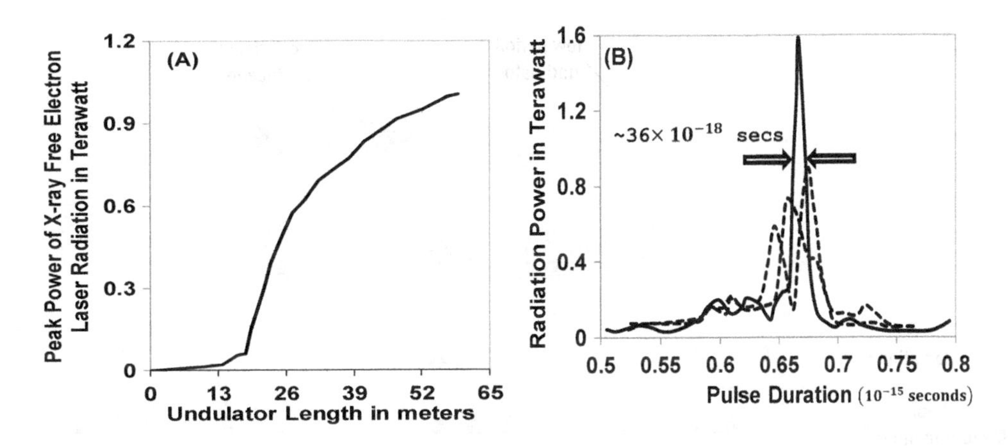

FIGURE 9.10
(A) Radiation power growth along the length of the undulator, and at the end of the undulator line, amplified radiation pulse profile (solid line) and (B) the fluctuations of the radiation pulse (dashed lines). Here, a current spike of 60 nm width, peak current of 37 kA, and laser of 2000 nm wavelength and 0.7 mJ energy was used.

no magnetic chicane or X-ray mirrors have been used. Due to a higher space charge effect, the generation of a high peak current of 37 kA may be of great concern. In the case of high energy beam, for example, 10 GeV, the impact of space charge effect on the performance of the FEL is miniature, as shown by the literature.

9.5.1.6 Mode-Locked Technique in XFEL

In the previous methods, there is a trade-off between minimizing the pulse duration that requires a short interaction and maximizing the emitted power that requires a long interaction due to the movement of the radiation forward to the electrons. In the mode-locked FEL amplifier concept, this problem is diminished (Dunning et al. 2014). In this technique, the long FEL interactions result in the formation of series of short interactions. The separation of short interactions is made by the longitudinal realignment of electron beam and radiation. To achieve the realignment, magnetic chicanes are used. A delay in electron beam with respect to radiation is introduced by the chicanes. This is portrayed in Figure 9.11.

This case is not beneficial for the amplification of an isolated ultrashort pulse because the radiation pulse propagates ahead of the electrons. However, the amplification of a train of ultrashort pulses can be made using this method. If there is one pulse in the train, the micro-bunching of the electron beam is increased because the pulse interacts with the electron beam through several undulator periods, resulting in an increment of the radiation intensity. Now it shifts forward with respect to the electrons and interacts with the electron beam, which is aligned previously to the radiation pulse preceding it in the train. The minimization of the pulse duration and maximization of the power at the same time is allowed by a series of short interactions. An external pulse train source can be amplified using such a technique (Thompson and McNeil 2008). Since at hard X-ray wavelengths there is unavailability of such sources, noise is always in the FEL. In this case, support is provided by multiple interleaved pulses, and a comb structure variation must be employed to the properties of the electron beam to choose a single clean pulse train structure (Dunning et al. 2013, 2014; Prat and Reiche 2015). In this method, the minimal

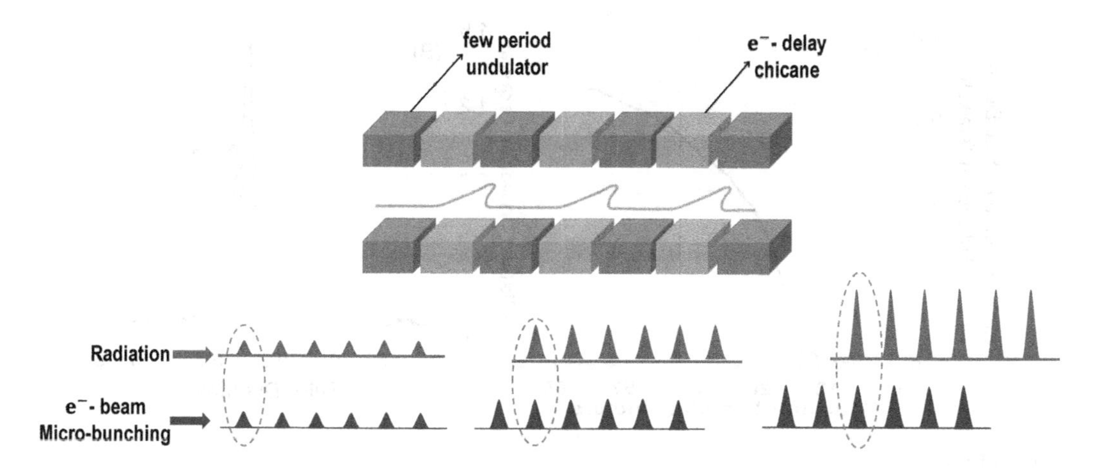

FIGURE 9.11
Schematic of mode-locked FEL technique in XFEL. The chicanes are used to introduce the delay in the electrons beam.

number of optical cycles per pulse is given by the number of undulator periods per section; therefore, there are few-cycle pulses. However, in this method, a considerable modification in the existing FELs is required, which are generally segregated into modules of various hundred periods.

In another method, through the addition of a short afterburner extension, the existing facilities of X-ray FEL generated the trains of few-cycle radiation pulses. This is relatively easy to add in the existing facilities. In this technique, an electron beam of high quality with periodic regions is formed before it is injected into a normal FEL amplifier. The energy modulation of the electron beam is done to achieve this. This can also be done by using a current or emittance modulation. The strong FEL interaction occurs through the high-quality electron beam regions while any short-scale structure in the radiation is washed out by the slippage, which occurs in the amplifier. This way leads to the generation of a periodic comb structure in the FEL-induced micro-bunching. The electron beam is made to inject on the 'mode-locked afterburner' before FEL saturation after the micro-bunching comb is developed properly. The injected electron beams map the comb of radiation intensity to identical combs of the electron micro-bunching. The 'afterburner' is made up of a series of a few period UMs. The electron delay chicanes separate the modules. This arrangement is identical to the one adopted in mode-lock amplifier FEL (Dunning et al. 2014). The undulator-chicane modules maintained an overlap between the developing radiation comb and the comb of bunching electrons. Therefore, it allows the exponential growth in the power towards FEL saturation power levels, say gigawatt. Alternatively, for the sake of simplicity, a single short undulator with lower power can be adopted (Parc et al. 2018). For minimal pulse duration, i.e. few-period (short) undulators, there is no limitation in taking the parameters of the afterburner because there is no modification in the main undulator in this method.

9.5.2 Other Methods for Attosecond Pulse in an XFEL

There are two methods in this category. The first one is based on the use of multi-slotted foil for emittance spoil and the second one is based on the tilting electron beam inside the undulator method in an XFEL. These will be discussed one by one.

9.5.2.1 Use of Multi-Slotted Foil for Emittance Spoil

It is clear that a low-emittance electron beam is assumed to be a better candidate for the maximum amplification inside the undulator than the electron beam of larger emittance. If both low- and high-emittance regions are present within an electron beam, the amplification of radiation will be faster in the region of low emittance compared with the region of high emittance. This implies that it is possible to produce an FEL pulse whose duration is smaller than the electron pulse. This process was used to generate a TW-as XFEL pulse for the modulation in the energy of the electron beam without a laser (Prat and Reiche 2015). If a slotted foil is used in bunch compressor, then there is local deterioration in the emittance of the electron beam. If the foil has many slots with unequal spacing, then there is a modulation in the emittance of electrons after they pass through the foil. This is depicted in Figure 9.12(A). Good emittance regime electrons contribute to the radiation process. The several radiation pulses having similar spacing as in the good emittance region can be obtained in the first undulator unit.

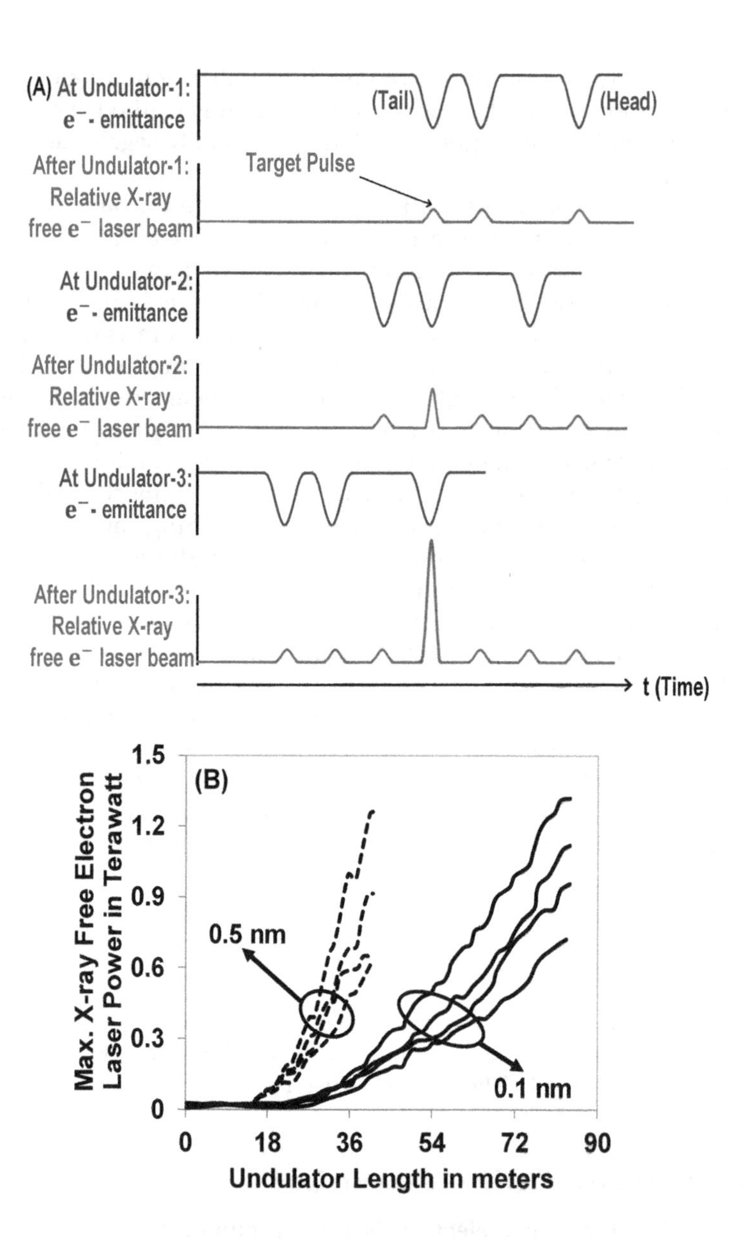

FIGURE 9.12
(A) Good emittance region method mentioned in Prat and Reiche (2015). Good emittance regions having different spaces relative to the neighborhood good emittance region are shown by the purple first line at the top. In the first undulator unit, the generated radiation pulses from the electron beam are shown by the first blue line. The target radiation pulse is represented by the last blue peak in the 'tail' part. The delay introduced in the electron beam to match the second good emittance region to target radiation pulse is represented by the second purple line. In the second undulator unit, the amplified target radiation pulse is represented by the second blue line. Further delay is introduced in the electron beam to match the third good emittance region to the target radiation pulse (third purple line). In the third undulator unit, further amplification of the target radiation pulse by the third good emittance region is represented by the third blue line. Time is in arbitrary units. (B) Simulation results of wavelength of 0.1 nm (1 Å) and 0.5 nm (5 Å). Based on the various seeds for shot noise production of the electron beam, four simulations are performed.

The delay in the electron beam is introduced to match the target radiation pulse to the second to last good emittance region. In this delay process, due to irregular spacing between the other radiation pulses, they are unable to access a good emittance region. The target radiation is amplified to the end by the repetition of this process. However, there is no chance for the amplification of the other radiation pulses. Using this method for Swiss-FEL parameters, it is possible to obtain ~1 TW power, in which the energy of the beam is 5.8 GeV for both the X-ray cases of 0.1 and 0.5 nm [Figure 9.12(B)]. For the cases of 0.5 nm, the required length of the undulator to attain this power level is only half compared with that of 0.1 nm.

Since in the emittance modulation process no laser is used, there is no inherent synchronization of the target X-ray pulse with any other external laser. In the pump-probe experiment, the probe or pump laser may be the external laser source. The scheme is appropriate for those experiments in which there is a requirement of only high power and short X-ray pulses.

9.5.2.2 Tilting Electron Beam Inside Undulator Method in XFEL

In the FEL radiation process, only the electrons that remain in a straight path in an undulator, i.e. in the good trajectory region, can participate. This implies that the duration of the FEL can be smaller than that of electron pulse itself. A new criterion was introduced to identify this fact for the TW-as XFEL pulse, which was tilted in an electron beam at the undulator entrance (Prat et al. 2015). The delay can be introduced, if a part of the electron beam is controlled to remain in the straight path and the other parts are controlled not to do so. Figure 9.13(A) explains the same. The photon pulse is amplified by several times with the help of electron-delay units in between undulators. In this figure, the good emittance region of the undulator is represented by an orange solid line. The electron beam distribution in space is represented by the purple line. In the good emittance region, the longitudinal beam distribution is tilted. The tail part can be controlled to remain in the good emittance region at the initial stage of the undulator line. The electrons, which are overlapped with the good emittance region, radiate during the first undulator section. This radiation is represented by the blue line and can be regarded as the target radiation pulse. In the second undulator unit, the purpose of delay in the tilted electron beam is to match the middle part of the electron beam to target radiation pulse. This is shown in the middle in Figure 9.13(A). At the third undulator unit, further delay is introduced for the final amplification through the matching of head part of the electron beam to target radiations. To tilt an electron beam, various methods have been employed and transverse deflector structure is one of them (Prat et al. 2015).

In the horizontal direction, two initial tilts are considered: one with strong and the other with modest tilt amplitude in the offset to demonstrate the effect of tilting of the electron beam. The amplitude of the tilt in the offset for these cases is 1 and 0.5 mm, respectively, along the complete bunch. Power generation for these cases have been compared in Figure 9.13(B). Attosecond X-ray pulses can be obtained as for Swiss-FEL parameters. The pulse has a power of 2.38 ± 0.94 TW, the duration of pulse is 363 ± 6 attoseconds, and energy of the pulse is 732 ± 41 μJ for the stronger tilt and 20 undulator sections [Figure 9.13(BI)]. However, the pulse has a power of 0.91 ± 0.21 TW, duration of pulse is 563 ± 17 attoseconds, and energy of the pulse is 492 ± 24 μJ for the modest tilt and 10 undulator sections [Figure 9.13(BII)].

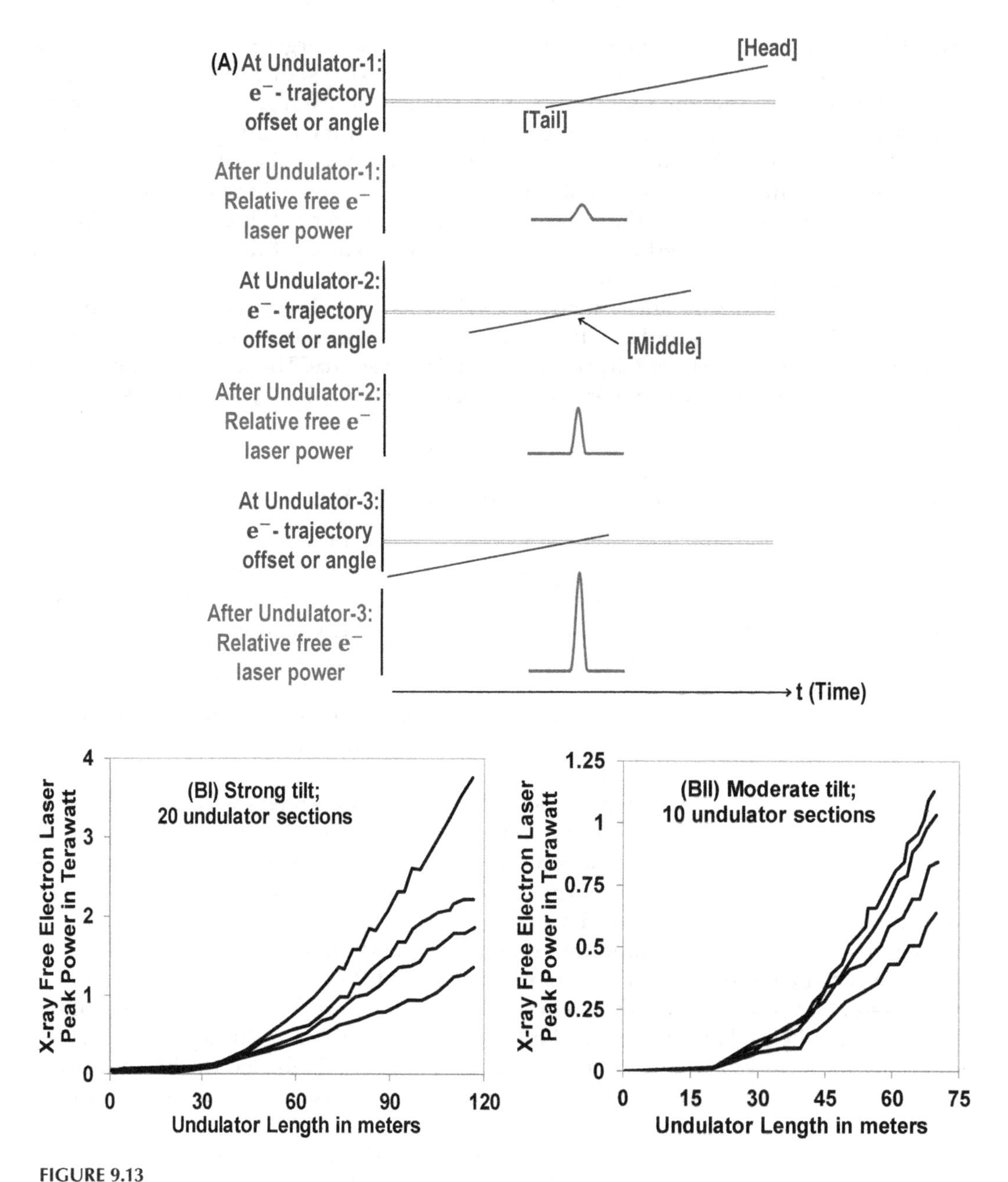

FIGURE 9.13
(A) Tilting of the electron-beam method to amplify the radiation. The good trajectory region is represented by the orange solid line. The tilted direction of the electron beam is represented by the purple solid line. The blue line pulses represent the amplified target radiation pulse in different undulator stages. Time is in arbitrary units. (B) XFEL peak power variation along the undulator length. (BI) Simulation results for strong initial tilt in offset of 100 with 20 undulator sections. (BII) Simulation results for moderate initial tilt in offset of 50 with 10 undulator sections. Based on the various seeds for shot noise production of the electron beam, four simulations are performed.

9.6 Applications of Attosecond XFEL Pulse

Attosecond X-ray pulses offer unlimited applications. Attosecond pulses in the X-ray regime would immediately enable the pursuit of mainly three topics of study among several applications in biology, chemistry, and physics. These are discussed below.

9.6.1 To Observe the Electron Clouds in Atoms and Molecules

Most of the X-ray photons are scattered by electrons in molecules and atoms. For the dynamics of electrons in atoms, an attosecond X-ray pulse enables us to follow the electron motion on the timescale of a few femtoseconds or less. Therefore, in quantum mechanics, we are able to measure the time-dependent analysis of the probability density distribution. With the help of real-time imaging of the quantum motion of electrons with real-space resolution, it is possible to understand various chemical processes like bond formation, bond breaking, and migration of charges in the molecules. It is reported from recent theoretical works that a proper quantum electron dynamical (QED) treatment is required for X-ray scattering, not only the simple expansion of the classical picture. This primary issue is definitely dealt with attosecond XFELs and offers a new area to manipulate the electrons in their real space and time.

9.6.2 To Observe the Dynamics of Hollow Atoms and Warm Dense Matter

In the highly excited level, a hollow atom is unstable; therefore, it will make a transition to the stable level through several ways. Attosecond pulses are required to investigate the dynamical changes occurring through various paths because these occur within a timescale of a few femtoseconds. Similarly, WDM refers to the state of matter in which ions' densities are so immense that ion-ion coupling becomes considerable and the temperature of electrons become analogous to Fermi energy. In this non-equilibrium state, it is necessary to investigate the electron-ion coupling dynamics to understand the energy relaxation. The ultrafast electron-ion energy relaxation in WDM is difficult to answer in spite of the theoretical progress. For the future progress, using only the theoretical efforts are not enough; it also requires a broad range of independent experiments for the detailed measurements of the dynamics of electron-ion coupling. There is a requirement of attosecond X-ray pulses to perform such experiments.

9.6.3 Single-Molecule Imaging

The direct structural information without crystallization can be obtained through the single-molecule imaging method, and this has become an eventual goal for the imaging of the biological molecule. To get a better signal-to-noise ratio or a reasonably strong scattering signal from a single molecule, a considerable number of photons ($\sim 10^{12}$) per pulse is desirable. The single fragile molecule can be destroyed using the strong X-ray pulse. Therefore, an ultrashort pulse is required to avoid structural changes during the extraction of structural information from a molecule. Since the femtosecond X-ray pulse is even too long, the attosecond X-ray pulses become necessary to obtain the structural information. These pulses open a new area of research and development in engineering and biomedical science. Thus, for single-molecule imaging, the TW XFEL pulse is essential with a pulse duration of several 100 attoseconds or less.

9.7 Alternative Improved X-Ray: γ-Ray Light Source

Both third- and fourth-generation light sources have had a huge impact on multiple scientific fields. However, these devices are both large and very expensive and have limited acceleration gradient and synchrotron power losses. In a way these highlight the perceived importance of bright sources of X-rays to the scientific community and indeed citizens in general. There are applications that require all the features of these sources: hard, short-pulse X-ray beams of extremely high brightness with high repetition rate. However, if the constraint of the high repetition rate can be relaxed one might consider the use of a different source of bright X-rays.

Betatron radiation from the laser wakefield accelerators (LWFAs) has become a useful and viable source in this regard where repetition rates of 0.1–1 Hz can be tolerated. As Figure 9.3 showed, it has a brightness higher than modern synchrotron sources, especially in the 10s-keV region of the spectrum. The figure also demonstrates that this is a relatively hard source of synchrotron-like radiation. In addition, betatron sources have a significant cost advantage, with a laser of several hundred terawatts. However, at present such lasers have a significantly lower repetition rate than sources driven by conventional accelerator technology. Progress is being made in this area, and it seems likely that within the next few years lasers with a power of several hundred terawatts will be operating at 10-Hz repetition rates.

In the range of 10–100 keV, highest brightness is obtained by betatron radiation among the LWFA-driven light sources. However, at higher energies Compton scattering sources become more efficient. Using Compton scattering, with the relatively modest energies of the electron beam, it is possible to have high-energy scattered photons (beyond 100 keV) compared with betatron X-ray radiation obtained in similar region (Krausz and Ivanov 2009; Phuoc et al. 2012). In the inverse Compton scattering (Compton 1923) process, the energy of the scattered electrons is upshifted.

Hence, an LWFA-based betatron source and a Compton scattering–based light source can be treated as future generation light sources. One of the beneficial methods to accelerate the charged particles and to produce a compact radiation source is laser-driven plasma waves. Compared with conventional sources, they are hundreds to thousands times smaller. These waves have various advantageous properties, such as high peak current (up to 35 kA), normalized emittance $\epsilon_n < 1\pi p$ mm-mrad, sub-10-fs electron bunches that are potentially shorter (<1 fs), and the energy spread is ~ $\Delta\gamma/\gamma < 1\%$. Higher charge and shorter bunch length, stability, repetition rate, beam transport, energy spread, and emittance are some technical issues that can be overcome with the rapid advancement of technology and research. The e-bunch slice values are essential for FELs and are potentially ten times superior. They demonstrate broad wavelength and energy range such as THz–X-ray. They are favorable applicants for an FEL in terms of coherence of the radiation and wavelength tunability. Betatron light source will produce radiation towards femtosecond duration gamma rays. However, it still requires a few years to show potential. They are well-synchronized with the laser and can combine particles (protons, electrons, ions), radiation, and intrinsic synchronization. In conclusion, the γ-ray light source is a compact light source and a fifth-generation light source for the paradigm shift science.

9.8 Conclusions

In this chapter several methods have been explored for attosecond X-ray pulse generation in XEFL. Using the electron beam–delayed scheme in XFELs, the X-ray pulse obtained has a 53 attosecond pulse width and 6.6 TW radiation power at a wavelength of 0.1 nm. The highest power obtained for XFELs with the chirped laser method is 1.8 ± 0.45 TW with the pulse duration of 62 ± 3 attoseconds. In the optical beam–delayed scheme in XEFLs, it is possible to obtain a radiation power greater than 1 TW, i.e. 1.01 ± 0.36 TW, using the recycle of a single current spike. A radiation power larger than 1 TW and a pulse duration of 36 attoseconds were achieved by a current enhanced scheme with undulator tapering in XFELs. All these methods are based on energy and density modulations. In the method of multi-slotted foil for emittance spoil, it is possible to obtain ~1 TW power, in which energy of the beam is 5.8 GeV for both the X-ray cases of 0.1 and 0.5 nm. For the cases of 0.5 nm, the required length of the undulator to attain the power level of ~1 TW is only half compared with that of the 0.1 nm cases. Attosecond X-ray pulses with a power of 2.38 ± 0.94 TW and wavelength of 0.1 nm with duration of pulse of 363 ± 6 attoseconds are obtained from tilting the electron beam inside the undulator method in XFELs. One can choose any of these methods as per the specific application.

9.9 Selected Problems and Solutions

PROBLEM 9.1

Wave 1 (intensity I_1) and wave 2 (intensity I_2) are incident on a nonlinear crystal at its front at $z = 0$. This process is used to generate a sum frequency wave with the use of d_{eff} as the effective nonlinearity of the crystal. What will be the dependencies of I_1 and I_3 (sum frequency) on the propagation distance z, if the intensity of wave 1 is such that it does not deplete in the process?

SOLUTION

For sufficiently small propagation distance, wave 1 does not deplete. Then the intensity of the sum frequency wave can be found by solving the wave equation as

$$I_3(z) = I_1(0)\frac{z^2}{L_{NL}^2}\sin\left(\frac{\Delta kz}{2}\right)$$

where $L_{NL}^2 = \frac{\varepsilon_0 c^3 n_1^2 n_2}{2\omega^2 I_1 d_{\text{eff}}^2}$ is called nonlinear length and $I_1(0)$ is the intensity of the wave at $z = 0$.

PROBLEM 9.2

In the above problem, what would be the intensity of the sum frequency wave for the phase-matching condition $\Delta k = 0$?

SOLUTION

The intensity I_3 of the sum frequency wave, when the phase-matching wave number Δk is zero, is given by

$$I_3(z) = I_1(0)\frac{\omega_3}{\omega_1}\sin\left(\frac{z}{L_{NL}}\right)$$

PROBLEM 9.3

Find the expression for sum frequency intensity I_3 if wave 1 with intensity I_1 does not get depleted and the phase-matching wave number (Δk) is not zero.

SOLUTION

When the intensity I_1 does not get depleted, then the expression is given by

$$I_1(z) = I_1(0)\left[\frac{\Delta k^2}{4g^2} + \frac{1}{L_{NL}^2}\frac{\cos^2 gz}{g^2}\right]$$

As per the problem $k = 0$ is not met. Then sum frequency intensity is given by

$$I_3(z) = I_1(0)\frac{\omega_3}{\omega_1}\frac{1}{L_{NL}^2}\frac{\sin^2 gz}{g^2}$$

where $g = \sqrt{\frac{1}{L_{NL}^2} + \frac{\Delta k^2}{4}}$.

PROBLEM 9.4

Calculate the undulator parameter and period of undulator having a magnetic field period length of 20 mm, magnetic field of 0.64 T, and length 1 m.

SOLUTION

The period of undulator is

$$N = \frac{L}{\lambda_u}$$

where L is the length of undulator and λ_u is the magnetic field period length. Hence

$$N = \frac{1}{20\times10^{-3}}$$
$$= 50$$

The undulator parameter K is given by

$$K = \frac{eB_u\lambda_u}{2\pi mc}$$

where B_u is the magnetic field. Putting in the values, one gets

$$K = \frac{1.6 \times 10^{-19} \times 0.64 \times 20 \times 10^{-3}}{2 \times 3.14 \times 9.1 \times 10^{-31} \times 3 \times 10^{8}} = 1.195$$

PROBLEM 9.5

For a BBO crystal, what will be the sum frequency intensity if the phase-matching condition $\Delta k = 0$ and when $\lambda_3 = 0.9\ \mu$m and $\lambda_1 = 0.8\ \mu$m?

SOLUTION

For phase-matching condition $\Delta k = 0$. The sum frequency intensity is given by

$$I_3(z) = I_1(0)\frac{\omega_3}{\omega_1}\sin\left(\frac{z}{L_{NL}}\right)$$

where $\omega_3 = \frac{2\pi c}{\lambda_3} = \frac{2\times 3.14\times 3\times 10^8}{0.9\times 10^{-6}} = 2.09\times 10^{15}\,\text{rad.s}^{-1}$

Similarly, $\lambda_1 = 0.8\ \mu\text{m} \Rightarrow \omega_1 = 2.35\times 10^{15}\,\text{rad.s}^{-1}$

Therefore,

$$I_3(z) = 10^{17} \times \frac{2.09\times 10^{15}}{2.35\times 10^{15}} sin\left(\frac{0.215}{0.0894}\right)$$

$$I_3(z) = 3.73\times 10^{15}\,\text{Wm}^{-2}$$

PROBLEM 9.6

For a BBO crystal, what will be the sum frequency intensity I_3 if wave 1 having intensity I_1 does not get depleted and $\Delta k = 0.5 \times 10^6$ m^{-1} when $z = 57.29\ \mu$m? Given that, when the intensity I_1 does not get depleted, the expression is given by

$$I_1(z) = I_1(0)\left[\frac{\Delta k^2}{4g^2} + \frac{1}{L_{NL}^2}\frac{\cos^2 gz}{g^2}\right]$$

$k = 0$ is not met. Then sum frequency intensity is given by

$$I_3(z) = I_1(0)\frac{\omega_3}{\omega_1}\frac{1}{L_{NL}^2}\frac{\sin^2 gz}{g^2}$$

where $g = \sqrt{\frac{1}{L_{NL}^2} + \frac{\Delta k^2}{4}}$.

SOLUTION

The sum frequency intensity I_3 is calculated by using the given formula

$$I_3(z) = I_1(0)\frac{\omega_3}{\omega_1}\frac{1}{L_{NL}^2}\frac{\sin^2 gz}{g^2}$$

Calculating g: $g = \sqrt{\frac{1}{L_{NL}^2} + \frac{\Delta k^2}{4}} = \sqrt{\frac{1}{0.008} + \frac{\left(0.5\times10^6\right)^2}{4}} = 2.5\times10^5\,\text{m}^{-1}$

Hence,

$$I_3(z) = 10^{17} \times \frac{2.09\times10^{15}}{2.35\times10^{15}} \times \frac{1}{0.008} \frac{\sin^2\left(2.5\times10^5\times57.29\times10^{-6}\right)}{\left(2.5\times10^5\right)^2}$$

$$I_3(z) = 1.088\times10^7\ \text{Wm}^{-2}$$

Suggested Reading Material

Amann, J., Berg, W., Blank, V., ... Hastings, J. (2012). Demonstration of self-seeding in a hard-X-ray free-electron laser. *Nature Photonics, 6*(10), 693–698.

Compton, A. H. (1923). A quantum theory of the scattering of x-rays by light elements. *Physical Review, 21*(5), 483.

Dunning, D. J., McNeil, B. W. J., & Thompson, N. R. (2013). Few-cycle pulse generation in an x-ray free-electron laser. *Physical Review Letters, 110*(10), 104801.

Dunning, D. J., McNeil, B. W. J., & Thompson, N. R. (2014). Towards zeptosecond-scale pulses from x-ray free-electron lasers. *Physics Procedia, 52*, 62.

Ferrari, F., Calegari, F., Lucchini, M., ... Nisoli, M. (2010). High-energy isolated attosecond pulses generated by above-saturation few-cycle fields. *Nature Photonics, 4*(12), 875–879.

Gerber, S., Jang, H., Nojiri, H., Matsuzawa, S., Yasumura, H., Bonn, D. A., ... & Song, S. (2015). Three-dimensional charge density wave order in YBa2Cu3O6. 67 at high magnetic fields. *Science, 350*(6263), 949–952.

Gomez, L. F., Ferguson, K. R., Cryan, J. P., ... Bernando, C. (2014). Shapes and vorticities of superfluid helium nanodroplets. *Science, 345*(6199), 906–909.

Goulielmakis, E., Loh, Z.-H., Wirth, A., ... Kling, M. F. (2010). Real-time observation of valence electron motion. *Nature, 466*(7307), 739–743.

Goulielmakis, E., Schultze, M., Hofstetter, M., ... Kienberger, R. (2008). Single-cycle nonlinear optics. *Science, 320*(5883), 1614–1617.

Goulielmakis, E., Uiberacker, M., Kienberger, R., ... Drescher, M. (2004). Direct measurement of light waves. *Science, 305*(5688), 1267–1269.

Hentschel, M., Kienberger, R., Spielmann, C., ... Krausz, F. (2001). Attosecond metrology. *Nature, 414*(6863), 509–513.

Ishikawa, T., Aoyagi, H., Asaka, T., ... Furukawa, Y. (2012). A compact X-ray free-electron laser emitting in the sub-Ångström region. *Nature Photonics, 6*(8), 540–544.

Kienberger, R., Goulielmakis, E., Uiberacker, M., ... Heinzmann, U. (2004). Atomic transient recorder. *Nature, 427*(6977), 817–821.

Krausz, F., & Ivanov, M. (2009). Attosecond physics. *Reviews of Modern Physics, 81*(1), 163.

Kumar, S., Landsman, A. S., & Kim, D. E. (2017). Terawatt-isolated attosecond X-ray pulse using a tapered X-ray Free electron laser. *Applied Sciences, 7*(6), 614.

Kumar, S., Lee, J., Hur, M. S., & Chung, M. (2018). Generation of an isolated attosecond pulse via current modulation induced by a chirped laser pulse in an x-ray free-electron laser. *JOSA B, 35*(4), A75–A83.

Kumar, S., Parc, Y. W., Landsman, A. S., & Kim, D. E. (2016). Temporally-coherent terawatt attosecond XFEL synchronized with a few cycle laser. *Scientific Reports, 6*(1), 1–9.

Li, N., Bai, Y., Miao, T., Liu, P., Li, R., & Xu, Z. (2016). Revealing plasma oscillation in THz spectrum from laser plasma of molecular jet. *Optics Express, 24*(20), 23009–23017.

Loos, H., (2013). Advances in X-ray free-electron lasers II: instrumentation, Proceedings of SPIE 15422, SPIE Optics/Optoelectronics p. 5.

Martin-Garcia, J. M., Conrad, C. E., Coe, J., Roy-Chowdhury, S., & Fromme, P. (2016). Serial femtosecond crystallography: A revolution in structural biology. *Archives of Biochemistry and Biophysics, 602*, 32–47.

Mashiko, H., Gilbertson, S., Li, C., & ... Chang, Z. (2008). Double optical gating of high-order harmonic generation with carrier-envelope phase stabilized lasers. *Physical Review Letters, 100*(10), 103906.

Minitti, M. P., Budarz, J. M., Kirrander, A., ... Lemke, H. T. (2015). Imaging molecular motion: Femtosecond x-ray scattering of an electrocyclic chemical reaction. *Physical Review Letters, 114*(25), 255501.

Parc, Y. W., Shim, C. H., & Kim, D. E. (2018). Toward the generation of an isolated TW-attosecond X-ray pulse in XFEL. *Applied Sciences, 8*(9), 1588.

Phuoc, K. T., Corde, S., Thaury, C., ... Rousse, A. (2012). All-optical Compton gamma-ray source. *Nature Photonics, 6*(5), 308.

Prat, E., Löhl, F., & Reiche, S. (2015). Efficient generation of short and high-power x-ray free-electron-laser pulses based on superradiance with a transversely tilted beam. *Physical Review Special Topics-Accelerators and Beams, 18*(10), 100701.

Prat, E., & Reiche, S. (2015). Simple method to generate terawatt-attosecond X-ray free-electron-laser pulses. *Physical Review Letters, 114*(24), 244801.

Sansone, G., Benedetti, E., Calegari, F., ... Velotta, R. (2006). Isolated single-cycle attosecond pulses. *Science, 314*(5798), 443–446.

Schultze, M., Goulielmakis, E., Uiberacker, M., & ... Kleineberg, U. (2007). Powerful 170-attosecond XUV pulses generated with few-cycle laser pulses and broadband multilayer optics. *New Journal of Physics, 9*(7), 243.

Shim, C. H., Parc, Y. W., Kumar, S., Ko, I. S., & Kim, D. E. (2018). Isolated terawatt attosecond hard X-ray pulse generated from single current spike. *Scientific Reports, 8*(1), 1–10.

Tanaka, T. (2013). Proposal for a pulse-compression scheme in X-ray free-electron lasers to generate a multiterawatt, attosecond X-ray pulse. *Physical Review Letters, 110*(8), 84801.

Tanaka, T., Parc, Y. W., Kida, Y., ... Prat, E. (2016). Using irregularly spaced current peaks to generate an isolated attosecond X-ray pulse in free-electron lasers. *Journal of Synchrotron Radiation, 23*(6), 1273–1281.

Thompson, N. R., & McNeil, B. W. J. (2008). Mode locking in a free-electron laser amplifier. *Physical Review Letters, 100*(20), 203901.

Uiberacker, M., Uphues, T., Schultze, M., ... Lezius, M. (2007). Attosecond real-time observation of electron tunnelling in atoms. *Nature, 446*(7136), 627–632.

Vinko, S. M., Ciricosta, O., Preston, T. R., & ... Engelhorn, K. (2015). Investigation of femtosecond collisional ionization rates in a solid-density aluminium plasma. *Nature Communications, 6*(1), 1–7.

Wang, Z., Feng, C., & Zhao, Z. (2017). Generating isolated terawatt-attosecond x-ray pulses via a chirped-laser-enhanced high-gain free-electron laser. *Physical Review Accelerators and Beams, 20*(4), 40701.

10

Lasers for Thermonuclear Fusion

10.1 Introduction

Controlled and guided thermonuclear fusion is undoubtedly one of the most challenging issues in modern physics. The effective utilization of unlimited resources of thermonuclear energy, to be released after the fusion of light atomic nuclei and forming heavier nuclei, is always an attraction for the scientific community. Protium, deuterium, and tritium are the three isotopes of hydrogen atom in which the deuterium and tritium are mainly the nuclear fuels for the production of thermonuclear energy. Enough deuterium reserves are available even after the excessive consumption of humankind for the next millions of years. However, there are exceptional obstacles in achieving the controlled fusion that need to be solved by the researchers in the present scenario. The nuclear fusion process can only occur if the two nuclei proceed towards each other at a minimum distance of $\sim 10^{-13}$ cm, which is the order of the nuclear size. This may happen only if positively charged nuclei attain sufficiently high kinetic energy by some mechanism such as heating to a very high temperature to overcome the mutual electrostatic repulsion. In general, such phenomena occur in the interiors of stars (like the sun), which are natural thermonuclear reactors, and we have been harvesting thermonuclear energy for a long time by radiation energy or electromagnetic radiation. The researchers have already investigated the source of such energy and reproduced this process on Earth to make the most powerful weapon, which is nothing but a thermonuclear bomb. However, attempts to achieve a guided thermonuclear fusion reaction have made it essential to carry out extensive theoretical research to build unique equipment, and to perform the exceptionally and extensively fundamental research in nuclear and plasma physics.

The energy produced by nuclear fusion reactions of hydrogen into helium within the sun is the most fundamental ingredient to fostering life on the Earth. If a similar nuclear fusion reaction could be simulated, then we definitely could solve two major issues, namely energy production and the environment problems of our Earth. Ocean is a rich source of deuterium, and approximately 20 litres of seawater comprises about 600 mg of deuterium. The amount of energy that can be drawn from ~5000 litres of fuel (oil) is equivalent to the energy from a nuclear fusion of 600 mg of deuterium. Figure 10.1 shows the schematic representation of equivalent energy production from deuterium and oil. The other energy fuel is tritium, which can be produced in industrial plants. If the practical laser-induced fusion technique (discussed later) comes into the picture, and if we can produce the power at the cheap rate, then we will enter into an era of everlasting industrial growth.

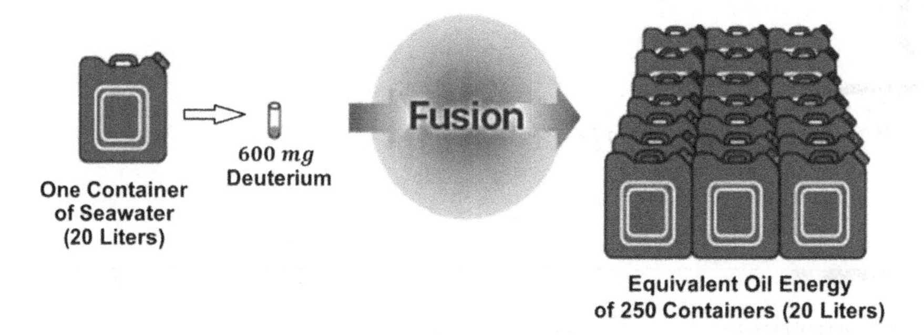

FIGURE 10.1
Schematic representation of equivalent energy production from two sources, i.e. deuterium and oil.

10.2 Physics of Fusion

Chemical scientists have already explored the possibility of harnessing huge amounts of energy by rearranging atoms in the atomic structure; indeed this is the way that one can extract energy from coal, oil, and natural gases. But by the end of first quarter of twentieth century with the knowledge of Einstein's mass-energy equivalence, we have found another way of rearranging the protons and neutrons in the atomic nucleus. It was found that nuclear reactions could release a million times more energy than any chemical process. The energy released by a nuclear reaction is governed by the nuclear binding energy (B.E.) curve (Figure 10.2).

Figure 10.2 shows that a huge amount of energy can be released by fusion of lighter elements into heavier elements. Actually, this is the source of energy of the sun and stars. For most of the fusion-related experiments, deuterium ($D = {}^2H$) and tritium ($T = {}^3H$) are

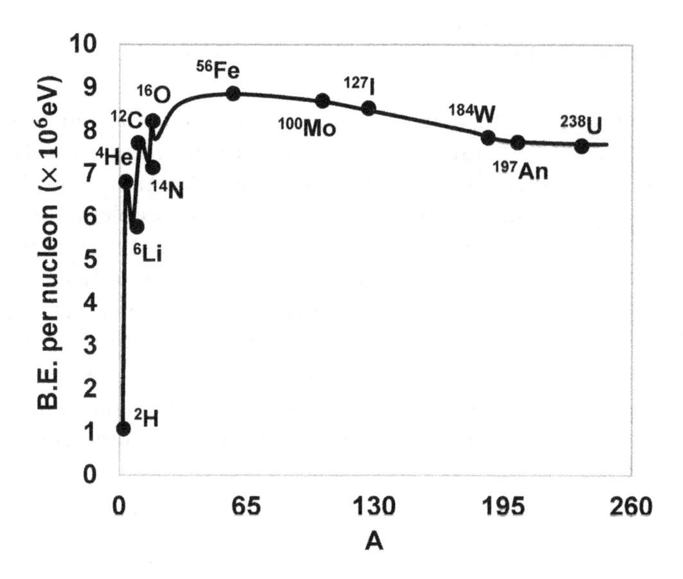

FIGURE 10.2
Binding energy (B.E.) per nucleon as a function of mass number (A).

considered to be the lighter nuclei to fuse together and the corresponding fusion reaction, known as the D-T reaction is

$$D + T \rightarrow {}^{4}He + n + 17.6 \text{ MeV}$$

Deuterium exists in nature (0.016% abundant in hydrogen), tritium is not readily available and is produced via ${}^{6}Li + n \rightarrow T + {}^{4}He$, and lithium is naturally abundant. Each D-T fusion reaction yields a helium nucleus (alpha particle), a neutron, and 17.6 MeV energy out of which 3.5 MeV is carried by the alpha particle and 14.1 MeV by the neutron. The energy of the neutron can be captured and used to drive a turbine and generate electricity. The energy released in fusion is so high that 1 gm of DT fuel is equivalent to 80 tons of trinitrotoluene (TNT), which is a high explosive derived from toluene. Fusion energy releases up to 360 billion J/g of fusion fuel, whereas coal burning gives only 20–30 thousand J/g. There is a factor of 18 million between fusion and chemical reaction.

It is very difficult to fuse the two nuclei as they are positively charged and need to overcome very strong Coulomb repulsion to fuse together. Basic electrostatics calculations show that hundreds of kiloelectron volts of energy or equivalently a temperature of 10^{9}–10^{10} K are required to overcome the electrical repulsion between the nuclei. Such a huge requirement of temperature would rule out any possibility to build a practical fusion reactor on Earth. However, quantum tunneling allows the reaction to proceed even if the barrier is not fully overcome, and this reduces the required energy to 20 keV with a corresponding temperature of about 2×10^{9} K. The most important aspect of fusion is to have the breakeven; a fusion reaction must release more energy than the energy put into heating up the fuel. This breakeven condition in the most general form is known as Lawson criterion (Lawson 1957).

10.3 Condition for Fusion and Lawson Criterion

By putting the high density of deuterium and tritium ions at critical ignition temperature for a long time, a net yield of energy can be obtained. The neutrons and remaining helium nuclei carried away about 80% and 20% energy, respectively, in the plasma. The resultant helium ricochets around the vessel hit the unburnt fuel target nuclei, enhancing its temperature. This may result in fewer requirements of the external heating systems. This process needs enough time to happen depending on the plasma density and temperature. In 1957, J. D. Lawson revealed that the optimum conditions for productive fusion can be calculated by multiplication of ion density and confinement time. The said product and condition is basically known as Lawson criterion. A similar approach to Lawson's criterion has been applied to the Tokamak Fusion Test Reactor (TFTR) at the Princeton Plasma Physics Laboratory (PPPL) (Meade 1987). TFTR was in operation from 1982 to 1997. TFTR has set a number of world records by attaining a very high plasma temperature of 510 million degrees. This was the highest temperature ever generated in a human-made lab and was over the 100 million degrees required for a commercial fusion reaction. The required ignition temperature has been achieved and found very close to Lawson's criterion, but this was possible only at different time.

In the late 1990s, TFTR was the world's first magnetic fusion device to carry out dedicated experiments with plasmas comprised of 50-50 D-T fuel. The 50-50 mixture was needed for real reactors to generate power. Concurrently, TFTR also set a world record when it generated controlled fusion power of 10.7 million Watts in 1994, which is more than enough to fulfill the requirement of more than 3000 homes. Such experiments have also been performed to focus on the study of alpha particles produced in the D-T reactions. Though the higher temperature may be required to overcome the Coulomb barrier in nuclear fusion, a critical density of ions may also be required to achieve higher energy yield due to higher chances of collisions. In the case of hot plasma, the confinement time may be derived from the density to calculate the total energy yield. The minimum criterion for fusion reaction is called Lawson criterion, and it is a result of the multiplication of ion density and confinement time. Plasma can be confined in the inertial confinement method by using Lawson criterion based on time and density parameters of the plasma.

10.3.1 Confinement Time for Fusion

The criterion that Lawson derived is the limit for the product of fuel density and confinement time. The confinement time for the thermonuclear fusion reaction is described as the time the plasma is confined above the critical ignition temperature. In some sense, this is a measure of the rate at which the system loses energy to its environment. This criterion is also called the breakeven criterion, and it is defined as the condition for which the total generated fusion power is equal to the power that heats the plasma. To yield more energy from the fusion than its input energy to heat up the plasma, the plasma should be confined at this critical temperature for a certain amount of time. The minimum confinement time can be calculated using the following formulae

$$\tau = \frac{2\times 10^{14}}{n} s \text{ for deuterium-tritium (D-T) fusion}$$

$$\tau = \frac{5\times 10^{15}}{n} s \text{ for deuterium-deuterium (D-D) fusion}$$

where n is the ion (or electron) density (per cm^3) in the plasma. Lawson estimated the minimum temperature for the D-T reaction to take place as 30 million degrees, which is equivalent to 2.6 keV. For the D-D reaction/fusion this temperature was estimated as 150 million degrees, which is equivalent to 12.9 keV. Lawson had calculated this by equating the radiation losses and the volumetric fusion rates.

10.3.2 Lawson Criterion

In principle, a combination of three parameters, i.e. temperature, density, and time, is required for fusion. The criterion for self-sustaining fusion and the generation of sufficient heat in a reaction to keep the plasma hot enough after removal of the external heating systems is known as the Lawson criterion. This is given by

$$n\tau \geq \frac{3kT}{\frac{\eta}{4(1-\eta)}\langle\sigma v\rangle \Delta E - \alpha T^{1/2}} \tag{10.1}$$

where n is the particle density of the plasma (either of ions or electrons), τ is the confinement time, T is the temperature, η is the efficiency parameter, $\langle\sigma v\rangle$ is the average reaction

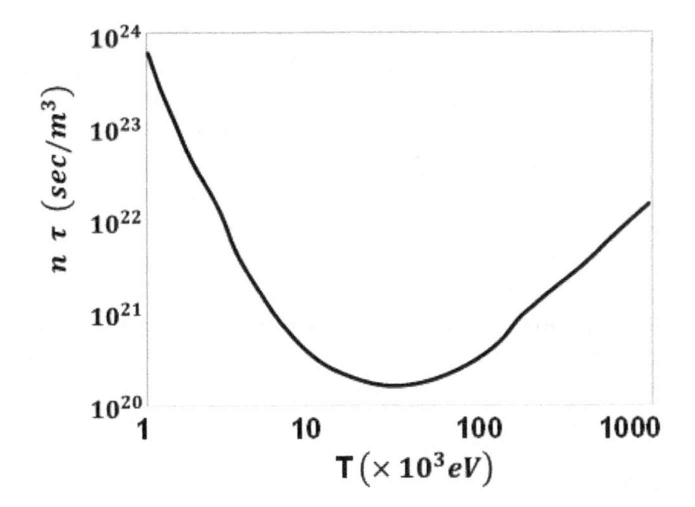

FIGURE 10.3
The Lawson criterion, i.e. the minimum required value of product of electron density (n) and energy confinement time (τ) for the concept of self-heating as a function of temperature (in keV). Ignition will take place for the values greater than this curve.

rate, k is Boltzmann constant, and α is the Bremsstrahlung coefficient corresponding to the energy lost in radiation. The curve that shows the variation of the minimum value of the product of the particle density and the confinement time with the temperature is called a Lawson curve. The Lawson curve is shown in Figure 10.3.

All efforts during the last six to seven decades for nuclear fusion by both magnetic confinement or inertial confinement have focussed on attempting to achieve conditions that satisfy the Lawson criterion.

10.4 Principle of Laser-Driven Thermonuclear Fusion

For several decades, scientists have been working on the world's most powerful lasers to reestablish the conditions for thermonuclear fusion on Earth that usually occur in the stars on a regular basis. Now, researchers found that the laser-induced thermonuclear fusion reactions are possible. In the late 1940s, scientists performed the confinement of hot plasmas using strong magnetic fields. Plasma in this state consists of turbulent mixtures of free electrons and ions, and they were heated to very high temperatures of the order of $\sim 1 \times 10^8$ to 3×10^8 K under the strong magnetic field.

For laser-induced fusion reactions, hydrogen works as a fuel and it consists of deuterium and tritium, which are packed in a golden tube under high pressure and low temperatures. The positively charged deuterium nuclei and tritium nuclei may overcome their repulsive electrostatic force, which helps in the fusion reactions, and produce a helium nucleus with two protons and two neutrons. The newly produced helium nucleus has a slightly smaller mass than the sum of two hydrogen nuclei's masses. The difference in masses leads to the generation of kinetic energy as per the mass-energy equivalence ($E = mc^2$) given by Einstein. The products of the thermonuclear fusion, i.e. alpha particle and neutron, carry the kinetic energy, and when they react with the other matter, their energy is transformed into heat energy. In the early 1970s, researchers had started using intense laser beams to compress the fuel of deuterium and tritium and increase their temperature to achieve fusion. Fusion achieved by this kind of technique is known as laser-driven fusion.

In this process, a small pea-size pellet, containing deuterium and tritium (D-T fuel), is irradiated symmetrically by several intense and powerful laser beams to ignite the thermonuclear fusion. The laser is used for rapid heating of target fuel and exploding its outer layer. The inner layer of the target is forced in an inward direction due to Newton's law, which leads to a rocket-like implosion. This results in the D-T fuel squeezing inside the capsule and leads to the production of shock waves, which help in further heating and a self-sustaining burn of fuel. The thermonuclear fusion burning process explores in the outward direction through the comparatively cooler outer part of the pellet more rapidly than the explosion of the capsule. Here, the confinement of plasma occurs due to its own mass of inertia instead of magnetic fields. Therefore, such an approach to fusion is known as inertial confinement fusion (ICF).

There are three major processes involved in laser-driven fusion reactions: implosions, ignition, and burning of the fuel. In the implosions, powerful lasers are focussed on the fuel through all directions and form a hot plasma with a density of about 10^3 times the solid density. In the second step, the temperature of the imposed plasma is increased to approximately 10^8 °C by using intense lasers of 1 quadrillion Watts or more. This process is called the ignition. The central ignition (CI) is a way to ignite and burn the fuel by using lasers just by compression. However, heating lasers are used in the fast ignition (FI) to focus on the plasma, which is already compressed by lasers, and causes the combustion process. This may create the possibility of higher power gain at a lower input energy, which may lead to building compact and modular thermonuclear power reactors in the future. Finally, there is a burning process, in which ignition of the target fuel leads to the combustion of the thermonuclear fusion fuel. This process iterates around 10 times per second and the energy produced is transformed into electricity. The schematic of laser-driven thermonuclear fusion is represented in Figure 10.4. We will discuss all these processes in detail in the following sections.

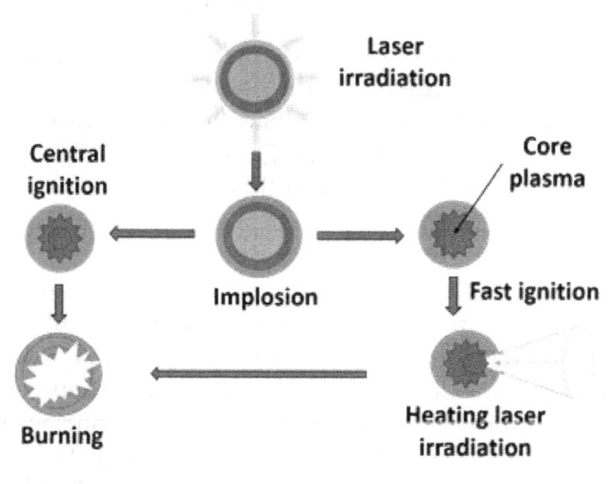

FIGURE 10.4
Schematic of laser-driven thermonuclear fusion.

10.5 Inertial Confinement Fusion (ICF) and Magnetic Confinement Fusion (MCF)

ICF and magnetic confinement fusion (MCF) are the two main areas of fusion research. In contrast to MCF, where the plasma is confined by strong magnetic fields at lower densities (about 10^{14} to 10^{15} cm^{-3}) for a longer time, ICF starts thermonuclear fusion by heating and squeezing the nuclear fuel. The target fuel, in a pellet format, comprises deuterium (D) and tritium (T) elements. For fusion of the D-T fuel, the Lawson criterion must be satisfied by both the MCF and ICF approaches according to which the fuel temperature is in the excess of 10^8 K and the product of its density (n) and the confinement time (τ) are such that $n\ \tau = 10^{14}$ to 10^{15} sec/cm^3. Both MCF and ICF aim to satisfy the Lawson criterion with parameters in different regimes. In MCF density is restricted by the requirement of a very high magnetic field and is of the order of $n \sim 10^{14}$ cm^{-3} with the corresponding confinement time $\tau \sim 1$ sec. On the other hand, in the ICF approach, a very high plasma density is attained by shock-wave compression after the D-T pellet is imploded by ultraintense laser beams. The particle density at the order of 10^{25} cm^{-3} can be achieved, which reduces the requirement of confinement time to a very low value as $\tau = 10^{-10}$ sec. The inertia of the imploding fuel is used to hold the pellet material together during this confinement time for considerable thermonuclear burn to occur. This also helps to release more energy than the deposited energy by the source.

The source in the ICF is used as a driver and it may be the high-energy laser source. The radiation pulses or X-rays of high-energy lasers are used to focus directly on the target fuel. They interact with the internal walls of a hollow metallic cylinder containing the fusion target. The continuous radiation pressure of a laser light or X-rays leads to the subsequent rapid heating and ablation of some of the outer layers of the pellet, which results in a force in the opposite direction. This force is now responsible to create a set of powerful shock waves travelling very fast in the inward direction and compression of the remaining material as a consequence of the momentum conservation. The energy released during compression and burning of the fuel will significantly exceed the energy used to implode the capsule under proper conditions. A laser having an intense and high energy (~10^5 J) or beam of light, electrons, or ions is required to compress the fuel up to the desired high density and heat it to reach a very high fusion-ignition temperature (~10^8 K) at the centre. The heating process leads to a chain reaction among the fuel and burns a significant portion of the fuel. This process is designed in such a way that it heats the interior of the fuels (D and T) and the heating is strong enough to result in the thermonuclear fusion of D-T isotopes. The ICF approach reveals a long-term and unlimited energy source by thermonuclear fusion reactions based on a secure and limitless fuel. Typically, the fuel pellets with the size of a pinhead contain about 10 mg of fuel, and a small amount of fuel goes into the fusion reaction for huge energy production. There will be a production of enormous energy if the total amount of fuel is utilized for thermonuclear fusion.

A laser with a single beam is used as a driver in the ICF process. One needs to supply the energy externally for the confinement of fuel, and this is possible by two ways, either by laser beam or particle beam. The single beam is divided into many beams, which are further amplified of the order of a trillion times. The amplified beams are sent to the target chamber, where the fuel is placed, with the help of a number of mirrors. The mirrors are placed in such a way that the beam illuminates the target uniformly on its complete surface. The radiation pressure of the laser driver results in the heating of the outer layer of the fuel target and the target is exploded. This process is similar to the reaction of the

H-bombs in which the outer layer of the bomb's target fuel cylinder is irradiated by X-rays from the fission device. The material, which is exploding off from the surface, leads the rest of the material to be driven inward with a larger force. This results in collapsing the material into a tiny spherical ball. Currently, the density of the resulting fuel mixture is as much as 100 times the density of lead, i.e. around 10^3 g/cm^3 in ICF devices.

In the late 1970s and early 1980s, laser-driven ICF rapidly evolved and started delivering from only a few Joules of laser energy per pulse to ~10^4 J to a target. To perceive the interactions between intense laser light and plasma, many experiments have been performed. Experiments held during this time demonstrated that the energy efficiencies of ICF and MCF were much lower than expected, and it was difficult to achieve the required condition of ignition. It was always an objective to generate the enormous energy from thermonuclear fusion to overcome the losses by ICF and MCF. Basically, when a laser light falls on the surface, a critical surface density is formed at the target surface depending on the light wavelength, intensity, and pulse length, whereas the beams (electron or ions) penetrate up to a certain range into the matter on the basis of its energy and surface density. It is therefore the interaction process that depends on the laser driver used in the process.

Presently, a number of laser facilities around the world are approaching the required conditions of ignition by inertial confinement. Scientists were able to design the latest and larger machines, which could attain the required condition for ignition energies. National Ignition Facility (NIF) located at the Lawrence Livermore National Laboratory (LLNL) in California is the largest ICF experimental facility and has been operational since 2008, whereas the ignition campaign started in 2010. This was a major step forward. The NIF was designed after decades long experience and experimental results. This was the biggest laser system ever built. NIF failed initially to attain ignition condition many times, but it produced about one-third of the energy levels required as of 2015 and set a milestone for the first commercialization of fusion. It was possible to generate more energy (~2 MJ) from a fuel capsule than the input energy applied to it. NIF also has reached many other records like production of numbers of neutrons and alpha particles. Now, there is a need for phenomenally large scientific instruments for experimentation. The 10-beam LLNL Nova laser and the 20-beam neodymium doped glass Shiva laser (its predecessor) have entered the realm of 'big science'.

10.5.1 Drivers for ICF

These are the following drivers for ICF in the United States, France, and Japan:

1. Heavy ion driver at Lawrence Berkeley National Laboratory (LBNL).
2. Diode pumped solid-state neodymium lasers at LLNL.
3. Krypton fluoride (KrF) lasers at the U.S. Naval Research Laboratory (NRL).
4. Laser capsule irradiation at the University of Rochester.

10.6 Conditions for Fusion Control

In principle, thermonuclear fusion is a process of the combination of two lower mass nuclei (hydrogen) and forming higher mass nuclei (helium). Deuterium and tritium are the most energetically fusion reactants. The differences in masses of resultant nuclei transform in the high-energy particles during the fusion process. Suppose we require a controlled fusion reaction having a positive energy balance, then the plasma temperature–dependent Lawson criterion should be satisfied, and D-T plasma needs to be heated to a very high temperature (about 10^8 degrees). This temperature is typically six times the sun's internal temperature. Under this condition, the D-T plasma is confined in a limited space and for a longer time. This is required to compensate for the energy losses and input energy (required for plasma production) to the energy released through fusion processes. The combination of deuterium and tritium is the most suitable and readily available ingredients for fusion reaction; deuterium and tritium can simply be extracted by hydrogen and lithium, respectively. When the neutron flux penetrates the lithium, tritium can be produced, and the lithium may be applicable for the purpose of blankets for the reactor core. The product of the plasma particle density and the confinement time should be higher than 3×10^{20} m^{-3} s to satisfy the Lawson criterion, while the plasma is present in the form of the D-T mixture with a lower content of impurity and at the temperature of about 10^8 degrees.

It is a difficult task to confine the plasma in a vessel at such high temperatures. There are two ways for plasma containment: inertial and magnetic. Basically, an inertial confinement is a virtual phenomenon; such confinement does not occur in actuality. Theoretically, lasers or particle beams can be used to increase the temperature of dense plasma very fast. The temperature would increase first using a smaller portion of D-T pellet surface, which results in blowing off of the material on the surface; hence, the fusion reaction takes place due to implosion of the remaining material. There is always a general question of the possibility of a big explosion during fusion in the reactor, but practically it is completely ruled out due to the presence of a very small amount of fuel in the reactor vessel. This is one of the advantages of the fusion reactor used for society. Now, the question is about the radioactivity generated by waste products during the fusion. The waste material is completely safe and is no longer radioactive. Of course, the inner part of the reactor may be radioactive, but the waste is not. The handling of waste product disposal is common and safe for all operational reactors. There is the possibility that the electricity can be generated using plasma and the conversion efficiency can be increased up to 90% in the limited fuel cycles of the fusion power plant.

In another kind of process, a magnetic field helps to confine the plasma. When there is no field, the particles move randomly and interact with the reactor vessel's wall. This may lead to cooling down the plasma making it difficult for fusion reactions. When magnetic fields are applied, the particles move in the spiral trajectory around the fields' lines, avoiding the vessel walls. The plasma confinement by magnetic field is classified into two systems, namely closed and open. The magnetic fields' lines start from the plasma forming a loop in the open in the closed system. The open system may work on the principle of mirror reflection (which is an open tube system with a magnetic field), magnetic well, or theta pinch theory and it may leak plasma more than the closed system. The plasma is trapped in the centre due to the weaker magnetic field in the middle and is stronger at the ends in this type of mirror reflection device.

10.7 Laser Plasma Interaction and Physics of Fusion

In laser-driven thermonuclear fusion, the hydrogen fuel consisting of D-T is packed in a golden tube under high pressure and low temperatures. An ultraviolet laser is irradiated at the fuel centre to start the thermonuclear fusion and a large amount of X-rays are produced in this process. The generated X-rays start heating the outer layer of stored hydrogen fuel. This heating explodes and compresses the internal portion enough to ignite the nuclear fusion reaction.

In the conventional technique of laser-driven thermonuclear fusion, a nanosecond laser beam or X-rays are illuminated symmetrically on the spherical target (D-T) fuel. This leads to fast expansion of plasma on the target fuel surface. The implosion of the target fuel and its compression can be controlled by the momentum of the internal part of the target fuel where the said momentum could be balanced by the momenta of expanding plasma. There is a decrement in the target volume of imploding fuel, and this is escorted through the heating of the central part of the fuel and increases the target density. After ample reduction in the target volume, the temperature of the central part of the fuel increases to a very high value (~10 keV). The temperature is so high that there is a development of a hot spot in the central region, and this helps to start the self-ignition of the target fuel. This kind of fuel ignition process is called central ignition or central hot-spot ignition (CHSI).

If E_{fus} is the energy output of the fusion reaction and E_{las} is the laser energy input for the reaction to occur, then the energy gain is defined as $G = \frac{E_{fus}}{E_{las}}$. The goal of fusion is to achieve $G > 1$.

The basic approach is to compress a D-T fuel target exceeding a critical value to very high density and very high temperatures. This is achieved by using a nanosecond laser (~10 ns) with extremely high energy ($E_{las} \geq 1\text{MJ} = 10^6\,\text{J}$), which eventually creates a central hot spot where ignition occurs. This CHSI approach needs to satisfy the high symmetry of the target's illumination and various other technical conditions.

10.8 Central Hot-Spot Ignition (CHSI)

To meet the Lawson criterion for nuclear fusion we need to create a very high temperature and a very high density plasma for a very short time. This is achieved by compressing a spherical tiny fuel target by illuminating it by laser (direct drive) or X-rays (indirect drive). Inertial confinement is based on imploding the small D-T fuel pellets by intense pulses of laser light illuminated on its surface. The laser light uniformly heats the outer layer and this rapid heating causes the outer layer to ablate. By rocket effect, that is conservation of momentum, there will be inward motion of the first layer inside the ablated material. Pressure from the rocket effect is generated by inward propagation of the shock wave produced at the surface, if the process is sufficiently symmetric. This results in the formation of a central hot spot at the core of the pellet where the extreme density and temperature required for fusion are achieved for a billionth of a second. The ignition occurs at the central hot spot and burn propagates with the help of alpha heating.

The various stages of the CHSI fusion have been represented in Figure 10.5. (i) Laser beams or laser-produced X-rays instantly heat the surface of the fusion pellet, forming plasma expanding outwards with a high velocity ($10^8\,\text{cm/s}$). (ii) Strong and focussed shock waves are produced close to the plasma ablation surface, compressing the D-T fuel into the pellet. This phenomenon occurs due to the blowing off of hot plasma following the principle of rocket. (iii) The density of D-T fuel becomes highest at the final stage of implosion, and the temperature in the central portion of pellet near the hot spot attains a very high temperature of ~10 keV; hence, it leads to ignition of the hot spot. (iv) Thermonuclear burn

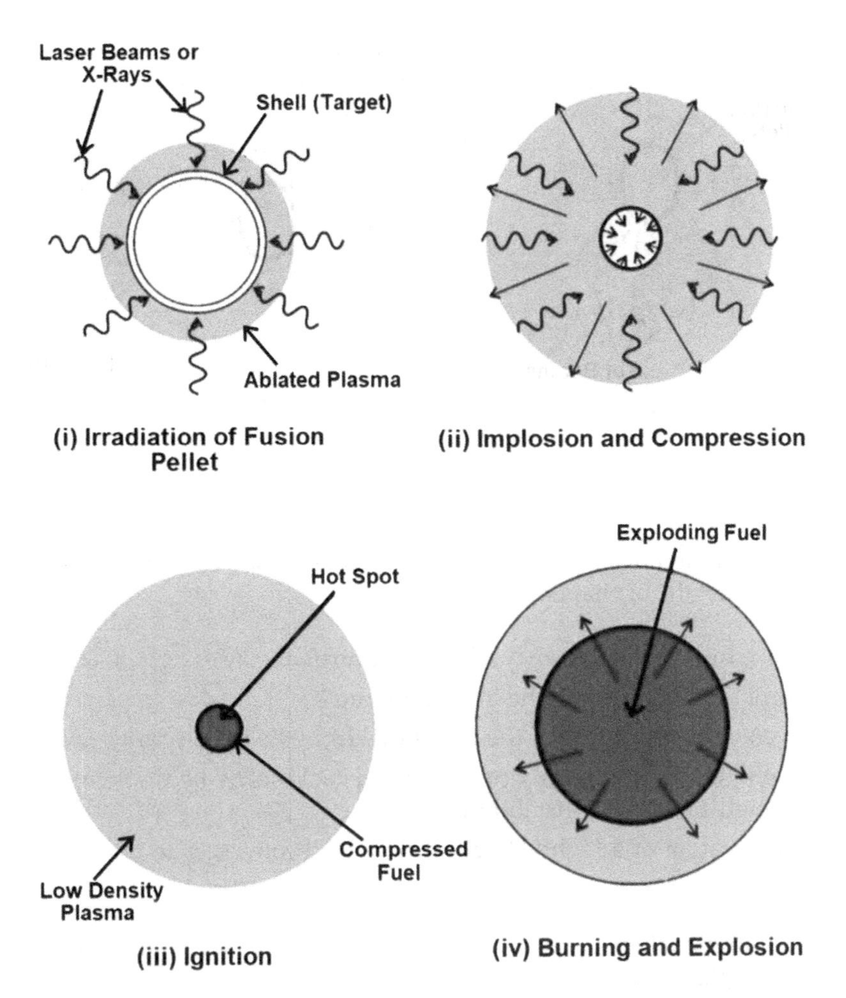

FIGURE 10.5
CHSI processes: (i) irradiation of fusion pellet, (ii) implosion and compression, (iii) ignition, and (iv) burning of fuel and explosion.

expands out very quickly with the help of compressed high-density fuel, which results in higher yields than the input laser energy.

The CHSI process can be followed by using the direct drive or the indirect drive (Figure 10.6). In the case of direct drive, the fusion pellets are irradiated directly and symmetrically using many (~10) ultraviolet nanosecond laser beams. However, in the indirect drive, the pellets are kept in a small (~1 cm) cylinder, which is known as the hohlraum, and it is made of high-Z metal (Lindl 1995). The laser beams are used to focus the internal portion of the hohlraum and heating enough to produce the hot plasma, which starts radiating mainly the thermal soft X-rays. The soft X-rays are absorbed from the target fuel surface and it starts imploding them. This process is similar as it happens when the surface is focussed and interacts directly with the laser beams. The efficiency of the absorption of thermal X-rays from target fuel is found to be higher than that in the case of direct absorption. In general, the hohlraum also needs sufficient energy to get heated on its own; therefore, it might reduce significantly the overall efficiency of the energy transfer from laser to target.

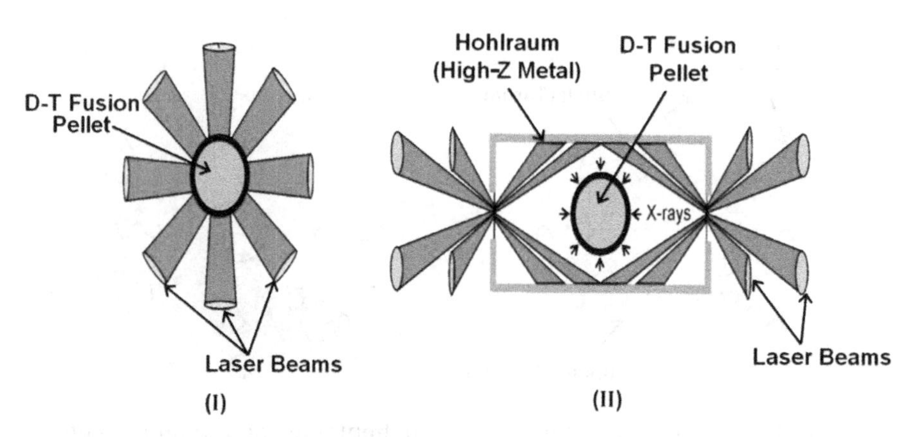

FIGURE 10.6
(I) Direct-drive and (II) indirect drive.

There are still numerous problems to resolve in these cases, i.e. in the direct and indirect drive CHSI. The details of challenges are as follows:

1. To achieve the higher efficiency of energy transfer by laser light to target fuel.
2. To control the symmetry of the imploding fuel.
3. To prevent the pre-heating of the target fuel from the hot electrons and X-rays.
4. To prevent the premature mixing of hot and cool fuel by hydrodynamics [mainly the Rayleigh-Taylor (RT) instabilities].
5. In the generation of a 'tight' shock wave, which converges to the centre of compressed fuel.

To overcome these challenges, we need to have a laser that comprises the radiation of shorter wavelengths so that they can be used to focus in the centre to compress the fuel. The fuel pellets must be made of high-Z material and shaped with extremely high precision. The pellets must follow the spherical symmetry without any kind of aberrations (not more than a few microns on its surface, if any). The laser (X-ray) beams might be very accurate and coherent so that the beams arrive on the pellet surface at the same time. The indirect drive technique is found to be more stable and it has been observed that its symmetric implosion and higher ablation pressure are attained in this case. The interaction of the pellets by thermal radiation in the hohlraum is found to be more homogeneous than by many laser beams; therefore, it results in more symmetric implosions (Figure 10.7). In such cases, less energy is absorbed by the pellets due to the shorter wavelength of X-rays; therefore, the energy requirement to compress the target fuel will become less in magnitude.

However, we require a higher input energy from the laser for ignition due to lower energy transfer efficiency in the indirect drive than in the case of direct drive, but the energy requirement is about 1 MJ in both the cases. Currently, NIF is the only laser facility in the world that produces energy more than 1 MJ, and it fulfills all the conditions of fusion reaction, ignition, and energy gain. There is another megajoule laser facility in France which took 15 years in construction and was declared operational on October 23, 2014. The construction of the Laser Mégajoule (LMJ) was launched in a single set of eight beamlines. This was called the Laser Integration Line (LIL) and it was powered by a 450-MJ energy bank. LIL was completed in 2002 and was predicted to be launched in May 2014, but the

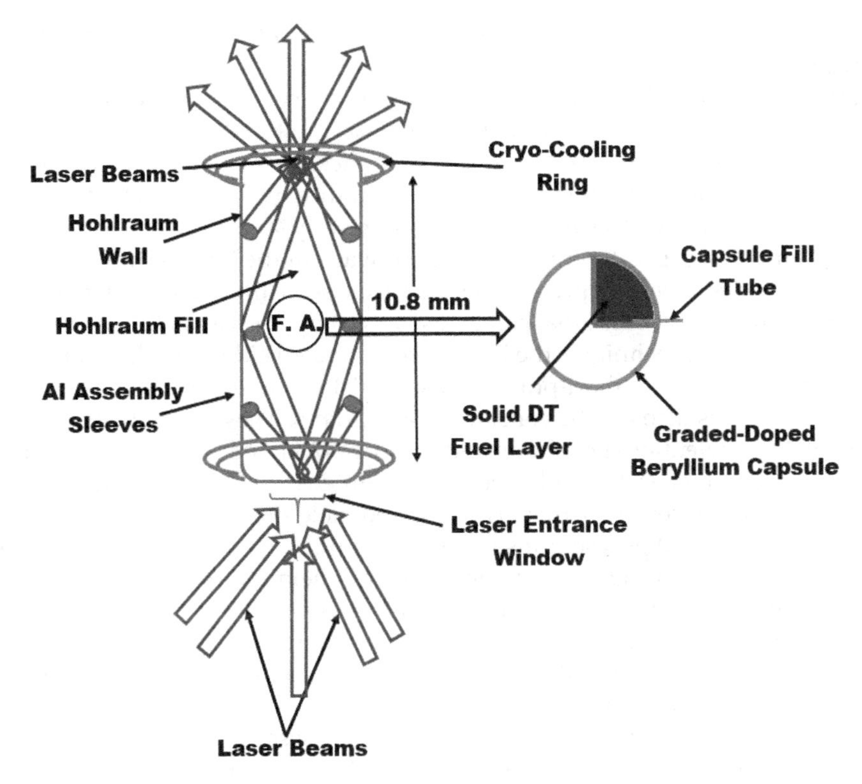

FIGURE 10.7
Hohlraum target fuel applied for indirect drive CHSI at NIF. F. A., fuel assembly.

operations were postponed until December 2014. Both NIF and LIL are designed to satisfy the required condition for ignition and energy gain of the order of $G \sim 10$ by using the indirect drive CHSI process.

Many ways to start the ignition of fuel were proposed to reduce the previously mentioned conditions and requirements and generally include fast ignition (FI), shock ignition (SI), and impact ignition (II).

10.8.1 Fast Ignition (FI)

FI is a technique for power generation, and it is used for experiments using the ICF process to achieve the ignition and burn of D-T fuel for thermonuclear fusion reaction. This process is unlike the conventional CHSI fusion where compression and ignition of the D-T are done by separate drivers. FI works on the principle of simultaneous ignition and compression of D-T fuel during an implosion. In FI, the long-pulse (nanoseconds) drivers/laser beams or X-rays are used to compress the D-T target fuel to a very high density. Then the ignition of the compressed fuel core is performed by using a short-pulse (picoseconds) ultraintense ($\sim 10^{20}\,\mathrm{W/cm^2}$) lasers/particle beams. An advantage of FI over the conventional laser fusion technique is that there are fewer requirements of density, pressure, and spherical symmetry of the fuel capsule for the implosion. It requires less input energy to provide the significantly higher energy gain, which is about 10 to 20 times than the central hot-spot technique due to the smaller mass ignition area used in the process. FI could provide a simple process with fewer requirements of laser driver energy and about 300 times fusion energy gain.

Though the FI approach is not much more in the advance stage than the central hot spot, it could give the solution to the next stage's physics problems. It needs a smaller region ($\sim 10^{-5}$ g) of D-T fuel compressed to about 10^3 times the solid density ($\rho_f > 200\text{g/cm}^3$), which is heated by an external ignitor to a temperature of about 10 keV. If the volumetric amount of the compressed fuel is found to be sufficiently large (the confinement parameter $\rho_f r_f > 2\ \text{g/cm}^2$), an ignited thermonuclear burn wave in the hot spot propagates through the fuel and results in energy production. As discussed earlier, there are two techniques, direct drive and indirect drive that can be used to compress the fuel.

In the indirect drive technique, the X-rays are generated either by a laser or by a Z-pinch or by a heavy ion beam. A short-pulse (picoseconds) PW laser-driven light ion beam, fast electrons, or protons also can be considered as ignitors. They might be used by the currently available conventional methods and could be obtained from the accelerated heavy ion beams. Tabak et. al. (1990) had used the ultraintense (~10 ps) laser pulse to penetrate up to the core of dense target fuel. This was possible only with the help of a channel bored in plasma by ~100 ps laser pulse used as pre-pulse of the ~10 ps ignition pulse. The relativistic electron beams (about MeV energy) are generally used to ignite the target fuel. These electrons are produced when ultraintense pulses counter on the critical surface of the dense plasma core. A critical stage in this process is the efficient formation of the channel and transportation of the ultraintense pulses through it. Recently, it was proposed that a hollow high-Z cone, packed in a spherical target, can be used as an alternative to FI (Figure 10.8). The high-Z cone helps an ultraintense laser beam focus into the cone and to produce fast electrons at tip point close by the dense plasma generated from the cone-guided implosion. A thin foil target is used to generate the proton/ion beam under the cone scheme after igniting the compressed fuel where the foil was placed near the tip. This concept makes the transportation of ignited laser beam easier to the dense core of the target fuel, but it creates complications in shaping the target and it disturbs the spherical symmetry of the implosion process.

The first integrated FI-related experiment was carried out by the scientist team of Japan and the UK. They have performed experiments with the PW laser (300 J/0.5 ps) for heating the plasma very fast and the Gekko XII nanosecond laser (9 beams/2.5 kJ/0.53 μm) for CD shell implosion in the cone-shape structure. There is an increment of about 10^3 times in the neutron yield (from 10^4 to 10^7) due to the heating of the PW laser that results in the enhancement of the PW laser-thermal plasma coupling efficiency by 20–30%.

The cone could not degrade the target implosion very efficiently. The results of the path-breaking experiment favor the FI technique, which has exploded the research work on FI worldwide. There are various 2D numerical hydrodynamic simulations that are used to determine the importance of FI. The ignitor beam energy E_{ig} and intensity I_{ig} to be delivered to the fuel and corresponding optimal pulse duration τ_{ig} and beam radius r_{ig} can be derived from the fuel density as per the relations:

$$E_{ig} = E_{\text{opt}} = 18\left(\frac{\rho_f}{300\ \text{g/cm}^3}\right)^{-1.85} \text{kJ} \tag{10.2}$$

$$I_{ig} = I_{\text{opt}} = 6.8 \times 10^{19}\left(\frac{\rho_f}{300\ \text{g/cm}^3}\right)^{0.95} \text{W/cm}^2 \tag{10.3}$$

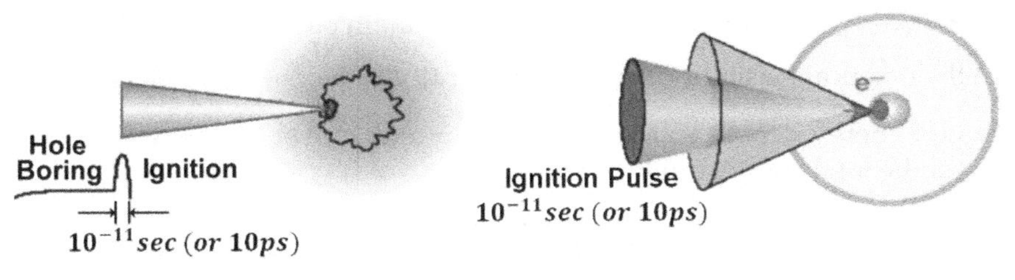

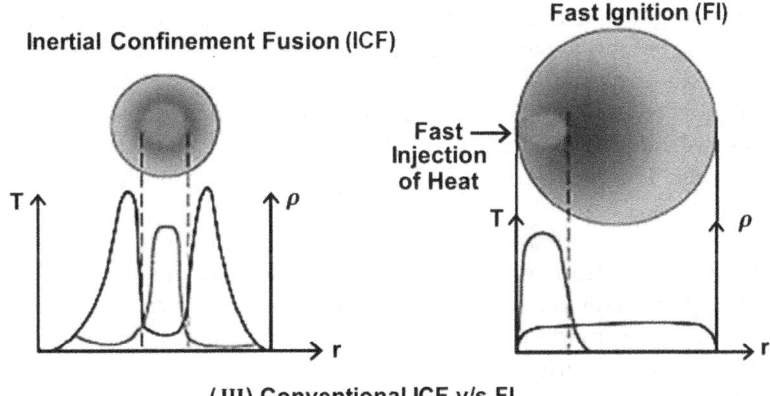

FIGURE 10.8
The fundamental fast ignition concepts: (I) the concept explained by Tabak et al. (1990) wherein ignition pulses and hole boring are used, (II) the cone-in-shell concept where, through a hollow cone, a picosecond pulse penetrating near the dense fuel core was used, and (III) conventional ICF versus FI.

$$\tau_{ig} = \tau_{\text{opt}} = 21\left(\frac{\rho_f}{300\ \text{g/cm}^3}\right)^{-0.85}\ \text{ps} \tag{10.4}$$

$$r_{ig} = r_{\text{opt}} = 20\left(\frac{\rho_f}{300\ \text{g/cm}^3}\right)^{-0.97}\ \mu\text{m} \tag{10.5}$$

The formula (10.2) tells that the energy of the laser beam required for compression, and ignition decreases fast when the fuel density enhances. However, to achieve high fuel density high-energy input from the compression driver is required. There was a need for reconciliation of the magnitude of fuel density ρ, and the value of $\rho \approx 300\ \text{g/cm}^3$ was finally accepted. For such kinds of densities, the ignitor parameters are calculated as

$$E_{ig} \approx 17\ \text{kJ},\ I_{ig} \approx 7\times10^{19}\ \text{W/cm}^2,\ \tau_{ig} \approx 20\ \text{ps, and}\ r_{ig} \approx 20\ \mu\text{m} \tag{10.6}$$

The abovementioned parameters are exceptionally demanding and they may not be possible through a conventional particle accelerator. One way to generate such ultraintense particle beams is by using the laser acceleration. Eqs. (10.2)–(10.6) refer to the required parameters of a particle beam, which need to be focussed to the target fuel. We need to represent some coupling factors for the determination of some important parameters of the ignitor drivers, e.g. laser and so forth. Here, the main parameter, which will be used to measure the practical feasibility of FI, is the quantity $\eta_E = \frac{E_{ig}}{E_L^{ig}}$. This is elucidated as a ratio of the ignitor energy deposited to the fuel and ignitor driver E_L^{ig} energy. The conversion efficiency of the total energy can be expressed as

$$\eta_E = \eta_{prod} \times \eta_{\text{transp}} \times \eta_{\text{dep}} \tag{10.7}$$

which refers to mainly the three phases of the ignitor driver and fuel interaction. Here, η_{prod} is the efficiency of the production of energetic particles, η_{transp} is the efficiency of the transport of the energetic particles to the dense fuel, and η_{dep} is the efficiency of deposition of energy of these energetic particles on the dense target fuel. The particular efficiencies and η_E are defined substantially, which depend on particle type evolved by the driver. The value of η_{prod} is found usually higher for electrons compared with protons or ions in the case of the laser driver, whereas it was found that the transportation of heavier particles is always easier. The energy deposition may be higher and it is done in a good manner for heavier particles.

Currently, both the electrons and protons/ions are considered separately as possible energetic particles to transfer and deposit their energy to the fuel core for FI, and research and efficiency analyses related to both of them have been carried out by physicists. The first effort has been made to control the two main parameters of laser-produced electron beams such as energy spectrum and angular divergence. There is a requirement to control the beam energy deposition to compress the fuel and beam transportation of the ultrahigh current (~ GA) electrons in the dense plasma. Significant achievements have been made so far in the progress of laser-driven ultraintense proton/ion sources for FI. Techniques to produce highly efficient ($\eta_{\text{prod}} > 10 - 20\%$) proton or ion beams for FI have been proposed, investigated, and analyzed experimentally as well as by using the advanced computer codes.

FI is an ingenious technique for laser-induced fusion reaction, ensuring high energy gain at the expanses of low energy and at a minimum cost of a driver. Such an approach is supported by the first integrated FI-related experiments, but great efforts are required around the world for the complete validation of the FI process and to understand the risk factors. The ignition and energy gain are found feasible as per the numerical modelling with total driver energy found to be ≤300 kJ and the ignition laser energy ≤100 kJ. Recently, laser facilities are being constructed using multi-kJ PW lasers to benchmark experiments of a sub-ignition scale and make the proof-of-principle. Now, it seems that the full-scale FI experiments will be technologically possible in the coming decade. This would explore the FI research in a broad way and proceed to the more advance stage where the FI energy could be determined similar to ICF.

10.8.2 Shock Ignition (SI)

In this case, converging strong shock waves are focussed on the D-T target fuel just after compression by a highly intense laser pulse (Figure 10.9). SI is assisted by CI proposed by Betti et al. (2007) as an alternative to FI. Here the target is irradiated first symmetrically but driven at lower velocity than for CHSI. The temperature of the hot spot produced is found below the ignition threshold at the end of the process of implosion. In this technique, the fuel pellets have been irradiated by using intense sub-nanosecond

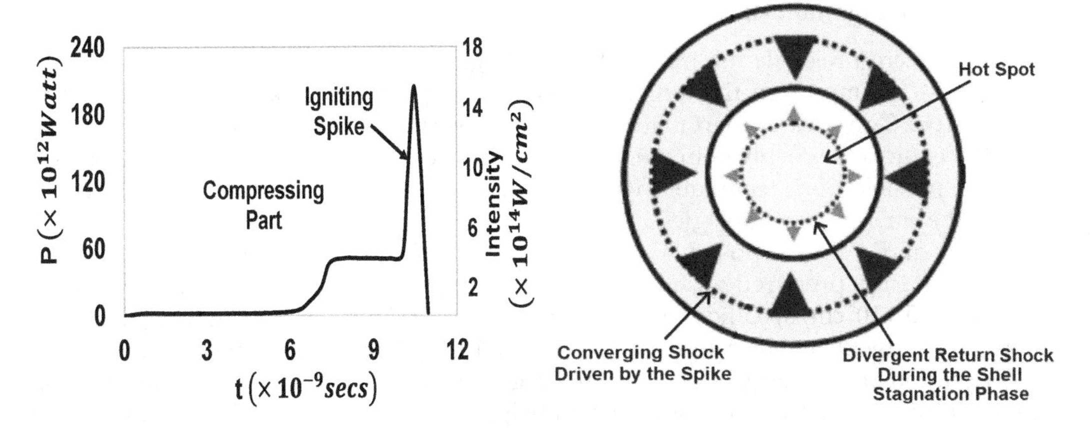

FIGURE 10.9
A strong and focussed shock wave controlled through the high-intensity sub-nanosecond laser spike, which ignites fuel in its final stage of compression in the SI technique.

laser spikes (Figure 10.9). The irradiation takes place for an appropriate time, which drives strong converging shock waves towards the end of the implosion. The initial pressure was observed to be 300 Mbar, which was amplified by convergence, and again it was further amplified to converging shock waves. Such types of shock waves collide with the return shock waves and provide the necessary amount of energy to trigger ignition from a central hot spot. The theoretical study of the SI concept has been validated by using the advanced computer simulations and by a 'proof-of-principle' experiment. SI experiments with the OMEGA Laser at The Laboratory for Laser Energetics (LLE) has produced high areal densities (defined later) and demonstrated the significant yield enhancement (Theobald et al. 2008).

The main advantages of SI concepts are as follows:

1. It requires a low-power (≈200 TW) ignition pulse compared with FI, which requires very high power (7000 TW ignition pulse). Energy of the ignition pulse is low (200–300 kJ).
2. High energy gain (G ≈80–100) is expected.
3. Simple hydrodynamics is required. Due to low implosion speed, it is less effected by RT instability.
4. Conventional laser technology can be used.
5. It requires non-isobaric fuel assembly, so target fabrication is simple.

To produce highly symmetric and strong shock waves by few laser beams, a number of problems are faced. There are numerous problems unsolved so far, such as the efficient and precise shock timing during the interactions of the shock waves with the compressed shell, bouncing and divergent shock, and degradation of laser-plasma coupling due to parametric plasma instabilities at high-intensity spike. SI is considered to be tested at full scale both at NIF and LMJ. However, some significant changes are required in the target geometry in both the facilities for irradiation to accomplish such experiments. However, the SI technique can be realized only in the direct drive scheme.

10.8.3 Impact Ignition (II)

When a micro-projectile of mass $\sim 10^{-6} - 10^{-4}$ gm is accelerated with high velocity (>10^3 km/s), its impact on the target fuel leads to a quick temperature rise of the compressed fuel. This idea of thermonuclear fusion was started in the early 1960s. However, only II technique has evolved in the past a few decades and the feasibility of this process could be possible. Such techniques were required to be accomplished with the present as well as the emerging technologies.

Caruso and Pais proposed a scheme in 1996 [Figure 10.10(a)], where a micro-projectile (~1 μg) prepared with high-Z material (e.g. Au) and accelerated to very high velocity ($\sim 5\times 10^8$ cm/s) collides with the compressed D-T fuel ($\rho_f \geq 200$ g/cm^3). Here, the projectile is quickly evolved to high densities (> 1000 g/cm^3) due to the collisions and a larger portion of its energy is given to the target fuel in a very short time ($\sim 10^{-11}$ s). This results in the creation of a hot spot, which helps in the ignition of the fuel. In this case, the projectile particles are required to carry only ~10–20 kJ of kinetic energy, which in comparison is much less than the laser energy, E_{las}, required to compress the fuel. On the other hand, Murakami and Nagatomo proposed a more advanced II scheme in 2005 [Figure 10.10(b)]. In this approach, the ignition of compressed D-T fuel has been carried out by the impact of collisions of another portion of imploded D-T fuel, which has been accelerated in the hollow conical target with hyper-velocity ~10^8 cm/s. Here, the kinetic energy is transformed into thermal energy with respect to its temperatures >5 keV after the interaction with the main portion of the target fuel. Here, the self-heating of the target is instrumental for the ignition process.

The irradiation of the ignitor shell is carried out mostly using a nanosecond laser pulse having higher beam intensities $> 10^{15}$ W/cm^2 and a shorter wavelength (e.g. 0.35 μm). This is performed to increase the ablation pressures above 100 Mbar on D-T target fuel by accelerating projectiles. According to Azechi et al. (2009), the neutron yield was found to be enhanced by two orders of magnitude using the CD shell target by this technique instead of the conventional CHSI scheme. Here, various significant achievements have been revealed using this scheme over the CHSI, mainly the higher energy gain (G ~100) at the cost of lower laser driver energy. On the other hand, the total laser energy required for the compression process and ignition of D-T fuel is calculated to be ~200–300 kJ in this

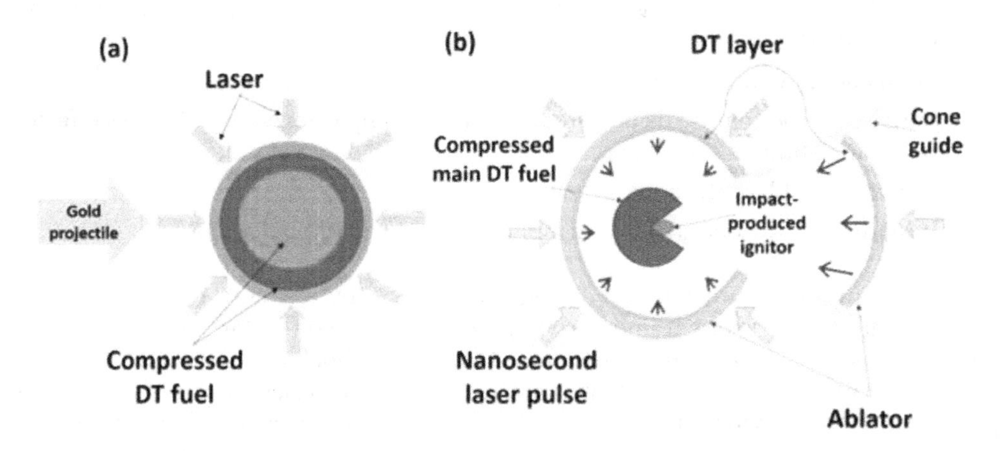

FIGURE 10.10
(a) II phenomenon using a high-Z projectile (Caruso and Pais 1996). (b) II phenomenon using a D-T plasma projectile (Murakami and Nagatomo 2005) accelerated by laser plasma ablation inside a guiding cone.

process, which is much lower in magnitude than in the CHSI scheme. The importance of the II process for fusion energy production has been given in detail during the experiment performed in Osaka with the multi-kJ Gekko XII laser (Shiraga et al. 2011). If we compare this to FI, we find that it has its own significant merits, which are simpler, and a short-pulse multi-PW laser may not be required. It is a cost-effective and limited required laser technology widely used worldwide. The major challenge in this process is the acceleration of a multi-μg, high-density ($\rho \geq 50$ g/cm^3) projectile to a very high velocity $> 10^8 \text{cm/s}$.

The new achievements and advancements made in this scheme provide the assurance of getting higher energy gain G at the cost of less input energy of the laser than the conventional CHSI. However, there are a few limitations for the compressed D-T fuel that need to be overcome to achieve better ignition and higher energy gain. They are mostly related to the confinement of target parameters such as ion temperature T, density ρ_f, and areal mass density $\rho_f r_f$. Here, r_f is the radius of compressed target fuel. The temperature should be estimated to be in the range of ~10–100 keV. However, the rate of fusion reaction of D-T fuel is found much lower for this temperature range than the rate at the optimum temperature T ~30 keV. Minimum values for ρ_f and $\rho_f r_f$ could be obtained by assuming the acceptable mass of D-T fuel be limited to m_0 ~10 mg, as

$$m_f = (4\pi/3)\rho_f r_f^3 < m_0 \tag{10.8}$$

and the burn fraction ϕ (which refers to the consumed portion of fuel) is greater than some real value of ϕ_0:

$$\phi = \frac{\rho_f r_f}{\rho_f r_f + H_B} > \phi_0 \tag{10.9}$$

where $H_B(T)$ is known as the burn parameter (the function that varies very slowly and is a function of the ion temperature) equal to ~7 g/cm^2 at the temperature T ~30 keV.

If we assume $m_f = 10 \text{ mg}, H_B = 7 \text{ g/cm}^2$ and $\phi_0 = 0.2$, the parameters will be as follows:

$$\rho_f > 200 \frac{\text{g}}{\text{cm}^3} \tag{10.10}$$

$$\rho_f r_f > 2 \ g/cm^2 \tag{10.11}$$

However, it is understood that the hot-spot areal density ($\rho_{hs} r_{hs} \approx 0.2 - 0.4 \text{ g/cm}^2$) could be much lower than that calculated by the relation (10.11). The hot-spot density could be greater and temperature T_{hs} should be around 10 keV.

10.9 Implosion and Burn of ICF Targets

Inertial fusion relies on extracting fusion energy from a freely expanding plasma in a series of micro-explosions. The confinement time is determined by inertia of the fuel as per the Lawson criterion and after this time cooling occurs due to expansion, which leads to a rapid drop in the fusion reaction rate. To achieve the required ignition in ICF, the target mass assembly determining the inertia and duration of the converging shock waves determining the confinement time are fundamentally important.

For ICF, the Lawson criterion is expressed in different units, which are density of the compressed fuel (ρ: g/cm^3) and the radius of the compressed fuel (r: cm). The quantity ρr is known as 'areal density' and considered as a confinement parameter. Assuming constant density and temperature during the confinement time, the burn fraction (ϕ) at 20 keV is given by

$$\phi = \frac{\rho r}{\rho r + 6\ \text{g/cm}^2} \tag{10.12}$$

Usually, meeting the Lawson criterion poses a challenge for researchers working on MCF, whereas keeping the burn fraction as given by relation (10.12) to a high value is the basic requirement to be fulfilled for ICF.

The proper target fabrication and point design of laser specifications are very important to achieve the required burn. This requires very high precision in physics and systems. The target capsules consist of a shell with a total mass of 0.58 mg composed of an ablator and an inner region of D-T, which forms the main fuel (Figure 10.11). In the figure, R is the shell radius and ΔR is its thickness during the implosion. The D-T fuel has to be compressed to a very high density, about thousands of times the solid density using intense laser beams. The laser beam inside the plasma is subject to various instabilities of which the hydrodynamic RT instabilities, which usually grow at the ablation front of the pellet, are very important. The exponential growth factor of even the fastest mode of RT instabilities should be limited to 6 or 7. This imposes a limit on in-flight aspect ratio $R/\Delta R$ requiring it to be about 25 to 35. For this, the peak intensity of the compressing laser beam should be 10^{15} W/cm^2, causing peak ablation pressure of 120 – 142 Mbar. With these parameters, the implosion velocity would be about 3.8×10^5 m/s (Glenzer et al. 2010). The implosion velocities must be uniform to maintain spherical implosion, which along with controlling RT instabilities to avoid fuel mixing are the most crucial factors for the creation of a central hot spot, where the fusion and a thermonuclear burn front propagate rapidly outward into the main fuel, producing high gain.

The ICF capsule is compressed by heating the outside layer through ultraintense laser beams to produce an expanding plasma that acts like a rocket propelling the shell inwards. As a result, the D-T fuel collapses inward with the ablation of its outer layers. If the implosion velocity exceeds 10^5 m/s, then the collision of the walls at the centre leads to temperatures sufficient to start fusion reactions. The fusion burn propagates out and heats the rest

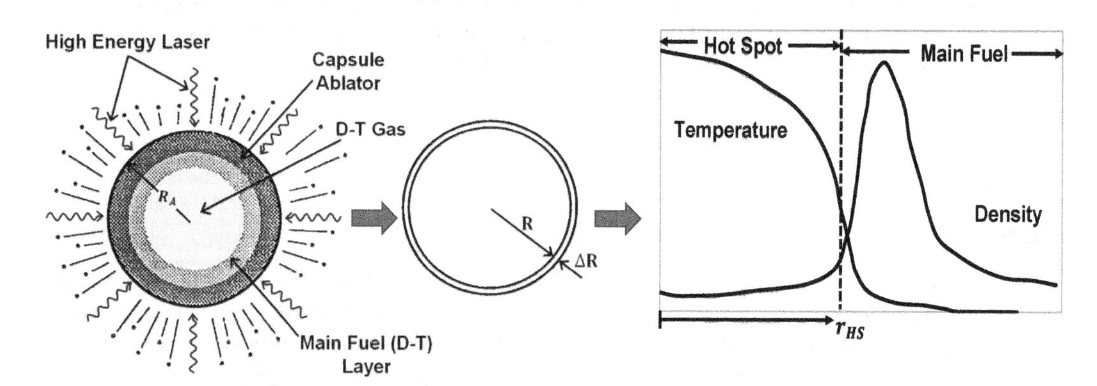

FIGURE 10.11
Laser-driven implosions and burning of target (D-T Fuel).

of the plasma. A large fraction of the D-T fuel is required to burn before it expands. For ignition to occur, the energy deposited to the core during one confinement time should be equal to the energy required to heat the plasma up to thermonuclear burn temperatures.

The confinement time is determined by inertia according to the formula

$$\tau \sim \frac{r}{C_s} = r\sqrt{\frac{3m_i}{5kT_e}} \tag{10.13}$$

where, C_s is the expansion velocity.

Considering equal temperatures of the ions and electrons (20 keV), the specific energy required to heat a D-T plasma is

$$E_{\text{heat}} = 0.1152 \times 10^9 T = 2.3 \times 10^9 \text{ J/g} \tag{10.14}$$

For D-T fusion specific energy of fusion (energy per unit mass) is given as $\varepsilon_f = 3.54 \times 10^{11}$ J/g.

A single D-T fusion yield of 17.58 MeV energy of which 3.52 MeV is carried by the alpha particle and the remaining 14.04 MeV is carried by a neutron is realized as

$$D + T \rightarrow n(14.06 \text{ MeV}) + \alpha(3.52 \text{ MeV}) \tag{10.15}$$

If we assume that the energy of all of the alpha particles is deposited efficiently, then the energy deposited in the fuel per gram is

$$E_\alpha = \frac{\varepsilon_f}{5}\phi = \frac{6.68 \times 10^{10} \rho r}{\rho r + 6 \text{ g/cm}^2} \text{ J/g} \tag{10.16}$$

Equating E_{heat} and E_α from relations (10.14) and (10.16) shows that at 20 keV, ignition occurs for areal density of $\rho r > 0.21$ g/cm^2 ($nr > 1.7 \times 10^{14}$ sec/cm^3), which corresponds to a burn efficiency of about 3.4%.

However, the above calculation for $\rho r > 0.21$ g/cm^2 is based on ideal implosion and driver efficiencies. This value increases 10–20 times due to the actual implosion efficiency and 3–20 times depending on the actual driver efficiency, whereas increasing the burn efficiency ϕ can limit this value. The calculation shows that for fusion to occur in the laboratory, the confinement parameter $\rho r \approx 3$ g/cm^2 should be achieved. However, Atzeni and Ciampi (1997) have reported that a larger value of ρr leads to substantial neutron scattering in the target.

For a spherical D-T pellet, mass M, radius r, and density ρ are related as $M = 4\pi/3(\rho r)^3/\rho^2$. For a pellet of 5 mg mass, to achieve $\rho r = 3$ g/cm^2 required for $\phi = 1/3$, the pellet density should be $\rho = 400$ g/cm^3, which corresponds to a plasma density of $n = 10^{26}$ cm^{-3}, and is about 1000 times solid density. The need for such astronomical densities is the major challenge of inertial fusion (Atzeni and Meyer-ter-Vehn 2004). If a very high density pallet of 5 mg mass is imploded by an ultraintense laser for 10 ns and fusion is achieved, it would have a yield of about 6×10^8 J. With the repetition rate of about 5–6 per second, such targets could drive a 1-GW electric reactor for the power production.

The fuel compression is energetically efficient if the D-T remains nearly Fermi degenerate (FD). For an efficient compression it is necessary to produce a sequence of shock waves that follow the adiabatic compression curve for the fuel as closely as possible. To achieve this, the laser pulse is properly tailored such that the pressure on the target surface increases gradually, so that each shock generated rises in strength. The energy required to compress 5 mg of D-T fuel to 400 g/cm^3 would be only 6.5×10^4 J, and that will not be sufficient to initiate thermonuclear burn for which the entire 5 mg mass needs to be heated to 5 keV. This would require 3.0×10^6 J. As the implosion efficiency is found to be as low as 5%, to meet these requirements the driver would need to deliver 6×10^7 J.

The creation of a hot spot inside the main cold D-T fuel is the key scientific challenge for the feasibility of the scheme. Another key issue is the efficient transfer of laser beam energy to the hot spot. The high precision in target fabrication and point design of compression beam specification with high implosion efficiency would result in target gains of 40 or more, which is required for fusion reactors based on ICF.

10.10 Direct and Indirect Drive Implosions

There are two main classifications of ICF implosions, direct drive and indirect drive. For the direct drive approach, a laser light with powerful beams is focussed on a tiny spherical pellet comprising micrograms of deuterium and tritium. The steep rise in the temperature of the pellet caused by the laser driver leads the target to explode its outer layer and its other part is forced to a rocket-like implosion in the inner direction as per the principle of Isaac Newton's third law. This leads the capsule fuel compression and the development of shock waves that result in the central fuel heating and hence a self-sustaining burn. The fusion burn exits outward cooler but much faster than the capsule can expand fully. Here, the confinement of plasma is done by its own mass of inertia rather than its own magnetic fields.

There is another technique known as the indirect drive method and it was tried first at NIF. In this case, the lasers are used to heat the inner walls of a gold cavity, called a hohlraum, which contains the pellets. Glenzer et al. (2010) have reported that the hohlraum can absorb more than 90% of the laser energy incident on it and gets heated. The heating makes the plasma superhot and it starts radiating the soft X-rays uniformly. This results in the fast heating of the pellet's outer surface, which causes a high-speed ablation of the pellets and imploding the capsule in the same way as lasers directly. The fusion takes place due to the plasma ignition in the central hot-spot area when the radiation compresses the capsule symmetrically and fuel burns. This was a major step for energy gain in the NIF in the direction of production of fusion energy for commercial power plants. Indirect drive research has been explored and carried out more than the direct drive for two decades, but it does not mean that the targets used in indirect drives are a better choice for the upcoming ICF reactor. Seventy to 80% of the absorbed energy is used in the generation of X-rays for planar targets. X-ray production efficiency depends on the optimum irradiances of 10^{14} to 10^{15} W/cm^2 for ions with energy ranging from 30 MeV to 10 GeV depending on their masses. There is still an open question and challenge for scientists to work out a more suitable scheme of energy-producing ICF reactor.

Researchers are still working hard to achieve the focussed intensity of the ion beam. The advantage of the direct drive is that it is more efficient in transporting driver energy to the fusion capsule. Here, we need to compromise with the spatial quality of the illuminated radiation. The main concern about direct drive targets is the nonuniformity of radiation. The direct drive implosions are much more susceptible to nonuniform radiation due to the hydrodynamic instability than in the indirect drive approach. The indirect drive approach is less sensitive to the irradiating beams. In addition to this, the implosions due

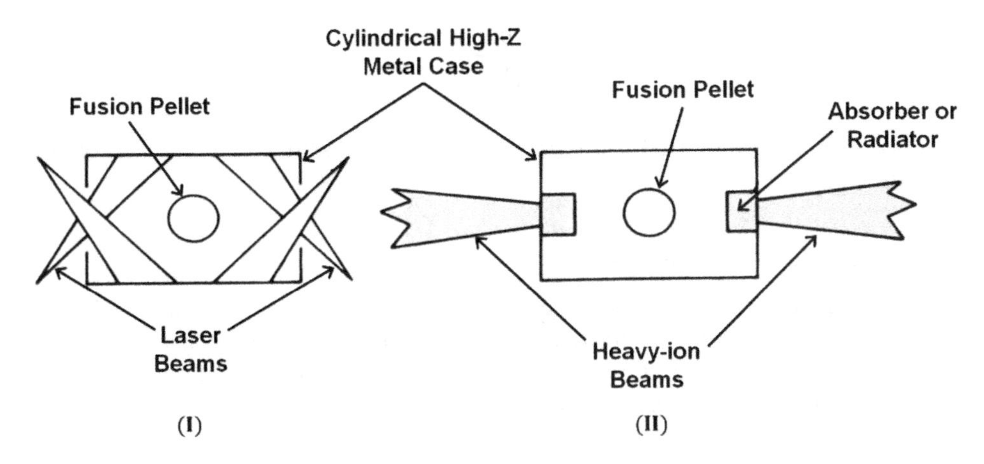

FIGURE 10.12
Hohlraums for indirect drive ICF using (I) laser beams and (II) heavy-ion beams.

to X-ray-driven ablation are hydrodynamically more stable. The laser propagates through a large volume of plasma generated in the indirect drive approach and it is more susceptible to laser-driven parametric instabilities. Scientist have achieved significant progress in the direction of ignition and higher energy gain using both indirectly and directly driven ICF targets. Figure 10.12 shows the schematic representation of hohlraums for indirect drive ICF, which uses lasers and heavy ion beams.

10.11 The Implosion Process

The compression of fuel towards the centre of the target and producing higher densities is referred to as the implosion process. It follows the reverse rocket process. There are three types of compressions during the laser-driven fusion: (1) linear compression, (2) cylindrical compression, and (3) spherical compression.

10.11.1 Linear Compression

A linear compression is generally a 1D compression of the mixture of fuel and air contained in the cylinder of an internal combustion engine. Suppose the original length of the gas column is r_0 in the cylinder of cross-sectional area A, which becomes r after the compression. The volume and density before compression are V_0 and ρ_0, respectively, which change to V and ρ after compression. As per the conservation of mass of the gas in the cylinder, the density after compression may be expressed as

$$V_0\rho_0 = V\rho = \text{mass} = \text{constant}$$

$$Ar_0\rho_0 = Ar\rho = \text{constant}$$

$$r_0\rho_0 = r\rho = \text{constant}$$

$$\frac{\rho}{\rho_0} = \frac{r_0}{\mathrm{r}} \tag{10.17}$$

The relation (10.17) states that the compressed density is inversely proportional to the change in length.

10.11.2 Cylindrical Compression

The cylindrical compression is nothing but squeezing a tube. This is also called the 2D compression where the tube radius decreases uniformly. Suppose the original radius is r_0, V is the volume, and L is the length of the cylinder. Let us assume that r_1 is the radius after compression. In this situation, the following equations can be derived:

$$V_0\rho_0 = V\rho = \text{mass} = \text{constant}$$

$$L\pi r_0^2 \rho_0 = L\pi r^2 \rho = \text{constant}$$

$$r_0^2 \rho_0 = r^2 \rho = \text{constant}$$

$$\frac{\rho}{\rho_0} = \left(\frac{r_0}{\mathrm{r}}\right)^2$$

$$\rho = \rho_0 \left(\frac{r_0}{\mathrm{r}}\right)^2 \tag{10.18}$$

The relation (10.18) shows that the compressed density is now inversely proportional to the square of the change in radius.

10.11.3 Spherical Compression

Now consider the situation where squeezing of a sphere takes place in such a way that its radius decreases uniformly in the three dimensions. This kind of 3D compression is known as the spherical compression. For the spherical compression, the following can be derived:

$$V_0\rho_0 = V\rho = \text{mass} = \text{constant}$$

$$\frac{4}{3}\pi r_0^3 \rho_0 = \frac{4}{3}\pi r^3 \rho = \text{constant}$$

$$r_0^3 \rho_0 = r^3 \rho = \text{constant}$$

$$\frac{\rho}{\rho_0} = \left(\frac{r_0}{\mathrm{r}}\right)^3$$

$$\rho = \rho_0 \left(\frac{r_0}{\mathrm{r}}\right)^3 \tag{10.19}$$

Here, the relation (10.19) shows that the compression density is inversely proportional to the cube of the change in radius.

Based on the above calculations, we can reveal the following general expression:

$$\rho = \rho_0 \left(\frac{r_0}{\mathrm{r}} \right)^{\mathrm{n}}, \text{ where } n = \text{compression dimension} \tag{10.20}$$

For 3D compression, $n = 3$. If there is a reduction in the radius by a factor of ½, i.e. when $r = \frac{r_0}{2}$, then the density increases by a factor of 8 on the compression, and the final density will become

$$\rho = \rho_0 2^3 = 8\rho_0$$

On the other hand, if the radius is reduced by a factor of 1/10, i.e. when $r = \frac{r_0}{10}$, then the density increases by a phenomenal factor of 1000 on the compression. It means

$$\rho = \rho_0 10^3 = 1{,}000\rho_0$$

Accordingly, a higher order of implosion results in a higher degree of compression for the same change in radius. A rapid compression is possible in the 3D implosion processes. This kind of compression is symmetrical in all the directions and it can be easily analyzed theoretically.

10.12 Physics of Implosion

The implosion process, considering the situation of directly driven laser fusion reactions, has been demonstrated here to explain the physics scenario. The process of laser fusion is shown in Figure 10.13, right from the laser absorption to energy productions from fusion reaction. The essential scenario of capsule implosion is the hydrodynamic process and its sequential flow at the centre. The energy transport and preheating processes can be analyzed from the elements on the left-hand side, whereas the hydrodynamic stability of the implosion is discussed on the right-hand side. Almost all physics scenarios shown on the left-hand side and in the centre are mainly 1D, while multi-dimensional treatment in the space is required for the elements shown on the right-hand side. Here, we will discuss the implosion process in detail from top to the bottom, as shown in Figure 10.13.

Numerous studies have been done on the laser-plasma interactions to analyze the absorption and electron heating. The nonlinear laser-plasma interactions lead to the production of hot electrons. Today, the absorption takes place mainly due to the classical process because the laser intensity can be controlled in such a way that there will be no undesirable nonlinear interactions. The ablation is maintained when the energy of the electrons is transferred to the high-density plasma regime. The heating of electrons is achieved by laser absorption. Here, the ablation front is separated from the laser absorption region using the conduction region. This separation is not as wide as the mean free path of the electrons. This results in ion wave turbulence and magnetic field generation. Hence, the non-local transport and transport inhibition become essential here. The development of an

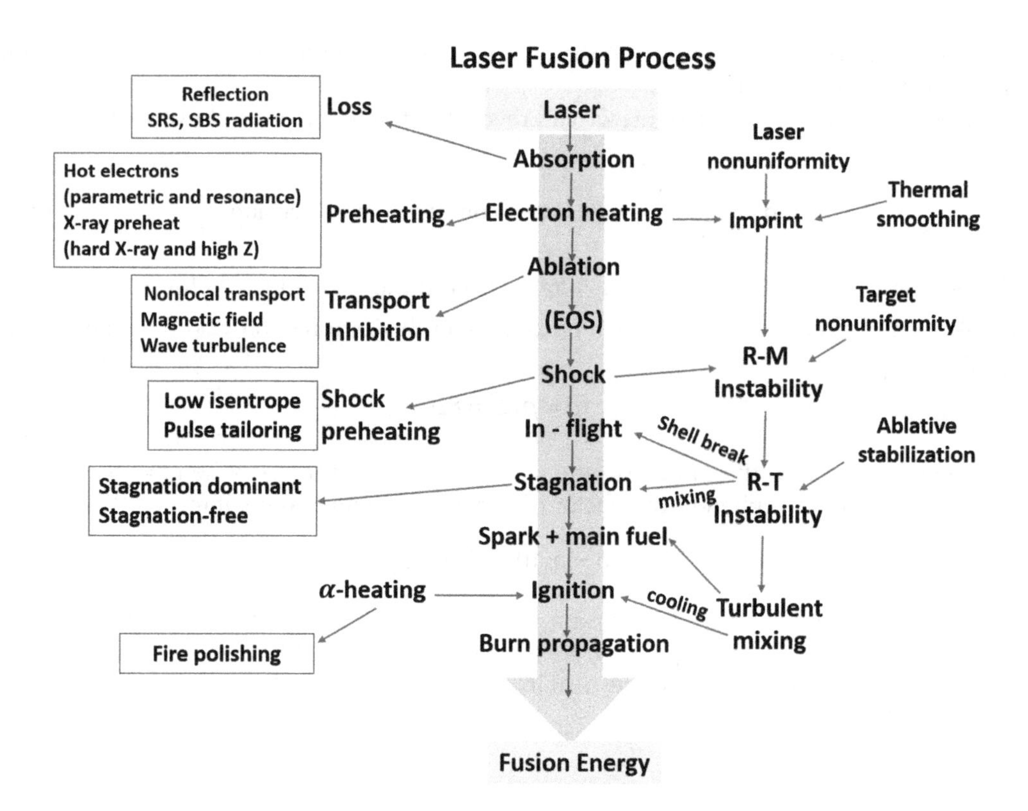

FIGURE 10.13
Laser-driven implosion from laser irradiation to energy production in the fusion reaction.

immense pressure of the order of 10–100 Mbar is achieved at the ablation front, which produces the shock waves. The material equation of the state (EOS) governs these parameters and their characteristics. The target entropy increases to a very high level before compression for the too strong shock waves. Therefore, there is a requirement of laser pulse engineering to suppress pre-heating due to shock waves. Initially, pre-compression of the fuel target is attained and then, using main laser pulses, it is accelerated towards the centre. This process is referred to as the in-flight phase. Once the in-flight phase is over, the collisions of the shell target at the target centre begin and then the stagnation dynamics starts. The main fuel is produced in the stagnation process. Basically, the triggering of ignition in spark is created and the propagation of the burned wave is started. All such processes could be visualized in 1D implosion codes.

On the target, a non-uniform ablation is generated by a non-uniform laser irradiation. The work of Aglitskiy et al. (2010) has shown that the Richtmyer-Meshkov (R-M) instability may be introduced due to the non-uniformity in the intensity of the laser. During the transport of energy from the regime of absorption to ablation surface, the non-uniformity in ablation pressure may be relaxed after this stage because of the thermal smoothing. The growth of RT instability occurs during the main pulse acceleration. However, the stabilization of the shorter wavelength perturbation is possible owing to ablative stabilization. Whenever the aspect ratio of the in-flight phase is enormous, it is possible to break the accelerating shell very easily. The mixing of the main fuel and spark is also possible for the symmetric acceleration. In the stagnation phase, if the mixing occurs because of an explosive growth of RT instability, then the spark is cooled and it is not possible to have any kind of fusion ignition. Basically the RT instability may not be generated in a single mode; rather, several modes are grown simultaneously, which result eventually in inducing the turbulent mixing. Even with the presence of a few states of uneven structure of the main fuel, the smoothing of the burning wave can be achieved through fire polishing by the heating of the alpha particle provided spherically symmetric implosion is maintained.

A key to laser ICF is to keep spherically symmetric implosions to the final stage. To achieve this there are several different zones of plasmas at the target to be studied in detail. The laser-induced fusion can be studied in different fields of physics such as laser plasma interaction, dynamics of implosion, electron energy transport, atomic physics and radiation transport, hydrodynamic instability, and target design. Let us briefly discuss the various processes.

10.12.1 Laser Plasma Interaction

Researchers have proved that the high-power lasers of pulse lengths >1 ns play a key role in the study of high-energy density physics. There are numerous important applications in the field of ICF and astrophysics. Plasma physics covers a variety of topics, namely parametric instability, inverse Bremsstrahlung absorption, wave breaking, and so forth. In the early 1970s, physicists tried to find the answer to a fundamental question about electron heating by intense laser light. Scientists extensively studied hot electron generation in the resonance absorption condition in the 1980s. This was done considering the relation with hot electron transport. At the NIF, various challenges have been answered for the interactions of plasma with long-pulse high-energy lasers by considering the nominal hohlraum target for achieving ignition. The research related to parametric instability suppression has been performed with a broadband and random phase laser beam in the last few decades. Recently, focus is oriented on stimulated Raman scattering (SRS) and the associated generation of high-energy electrons in reactor-sized plasma, the coupling of hydrodynamic instability with filamentation instability, and so forth. The physics of plasmas and laser plasma interactions attract many physicists to work towards the indirect drive fusion scheme.

10.12.2 Electron Energy Transport

In the case of electron energy transport, a self-consistency of laser absorption is essential. We cannot neglect the effect of the magnetic field in this case where it mainly consists of the study of non-local transport of electrons in laser-induced plasma. For coupling of electrons with hydrodynamic code, some research was performed in the 1980s to work out the Fokker-Planck equation (Bell et al. 1981; Matte and Virmont 1982). In energy transport, non-local coupling becomes essential due to a shorter length of temperature scale compared with the mean free path of the electrons, which dominantly transports the heat flux. The electron distribution function is not Maxwellian in nature and it should be treated in the corrected manner. The magnetic field generation during the intense laser irradiation is a very interesting part for the basic physics study apart from laser-induced fusion reactions. Using multi-dimensional hydrodynamic code, scientific interest is now focussed towards the analysis of non-uniform relaxation of the conduction regime through the solution of multi-dimensional Fokker-Plank equation.

10.12.3 Dynamics of Implosion

The dynamics of implosion is concerned with the fluid dynamics. The basic concept behind the laser-induced fusion is based on the rocket model where, due to conservation of momentum, the outward motion of ablative plasma causes the compression of the spherical pellet and creation of central hot spot. There were plenty of discussions about many modes of implosions such as explosive and ablative implosions, as discussed by Bodner (1981). Because of the suppression of hot electrons and the conversion of laser wavelength to second-order/third-order harmonics, a change in the implosion mode has been seen from the explosive mode of hot electrons, which was dominant, to a comparatively low isentropic ablative mode. If there is an enhancement in the ablation pressure, it helps to make a path in achieving the required performance of cryogenic implosion on the lasers that would scale hydrodynamically to the ignition process. The higher ablation pressure permits a more massive shell and a greater adiabatic to attain the required hot-spot pressures (>100 Gbar), areal densities (>300 mg/cm^2), and ignition velocities (>3.5×10^7 cm/s). In general, a micro-balloon made of plastic shells and glass has been imploded using a laser beam having 10^4 J of energy at multi-beam systems with a large laser facility. The scientists have reported achieving a high density of the order of 600 times solid density (Nakai 1989; Nakai et al. 1990; Azechi et al. 1991) and a high temperature, i.e. around 10^4 eV (Yamanaka and Nakai 1986; Takabe et al. 1988) in a multi-beam facility. The theoretical and experimental data, like compressed core density and neutron yields, have been analyzed and compared with 1D simulations. In the converging implosion regime there are still some disparities in the compared data. This may be discussed and analyzed quantitatively to uncover the physics regarding the discrepancies.

10.12.4 Hydrodynamic Instability

This kind of instability is the main concept behind the ICF research. The RT instability occurs due to the density gradient, and in an ablative acceleration its excitation occurs where the high-density ablator is triggered by low-density hot plasma. In the acceleration phase, the impact of ablation on the growth of instabilities has been studied analytically by several physicists (Takabe et al. 1983, 1985; Kull and Anisimov 1986; Kull 1989). The numerical investigations have been performed by Tabak et al. (1990) and Gardner et al. (1991), whereas the experimental studies have been done by Glendinning et al. (1992) and Remington et al. (1993). These examinations have helped to resolve the problems related to the ablation effect. Now, stabilization and reduction in the growth rate, respectively, are possible for shorter wavelength perturbations and longer wavelength perturbations. The key factor for the establishment of ICF as an appropriate approach is the ablative stabilization effect. At the ablation front, further study is required to confirm whether the effect of ablation is also helpful in the stabilization or control of R-M instability growth. This also may be helpful in the formulation of stabilization. It is imperative to portray the time dependence of the instability growth in accordance with the evolution of implosion dynamics.

In the basic physics of fluids, the concept of nonlinear effect is very attractive. It can be studied in the mode-mode coupling among different wave numbers and RT modes. We can formulate the evolutions of various instabilities arising from the linear to turbulent and nonlinear levels, where the turbulent mixing and mode-mode coupling have attracted the interest of researchers.

The development of the turbulent mixing layer with the width of the mixing zone of about $0.7\alpha_A g t^2$ in the classical fluid is made, where α_A is the Atwood number. Here, we need to find out why at the ablation front the growth of the turbulent mixing layer can be controlled and reduced due to the ablative stabilization effect (Takabe and Yamamoto 1991). Also, the type of generated turbulence spectrum should be depicted. The energy transported by radiation, alpha particles, and electrons in laser fusion is strongly connected with the fluid motion, and evolution of hydrodynamics is affected. Ideally, one wants to gently heat the plasma by collisional absorption of the intense laser beam. Of course, such energy transports strongly affected the turbulent mixing, and heat conduction strongly influenced the small-scale vortices specifically because they are distinguished from classical fluid turbulence. The phenomena of turbulence and instability, in ICF, can be discussed through the self-consistent treatment of such energy transports.

10.12.5 Atomic Physics and Radiation Transport

For this, a wide range of basic physics needs to be covered. Here a non-local thermodynamic equilibrium (non-LTE) atomic process, a database of rate coefficients, detail configuration accounting (DCA), an atomic model of a multi-electron system, solving rate equations, and so on have to be dealt with. Moreover, through the radiation transport, a coupling of these issues with hydrodynamics has appeared. The dominance of line radiation for the energy transport has appeared in higher-Z plasmas. However, transition arrays and so many lines should be evaluated. One way to carry out such a challenging job is to use super transition array (STA) (Bar-Shalom et al. 1989) and an unresolved transition array (UTA) (Bauche-Arnoult et al. 1979). One of the imperative works for investigating laser-produced plasmas is to extend such models to non-LTE situations.

10.12.6 Target Design

Depending on the physical models described above, an implosion code is required to achieve a desirable target design. Initially, a target gain must be optimized with the 1D implosion code. After this, using similar physics for multi-dimensional code as that of the 1D code, the sensitivity analysis of the optimized target to hydrodynamic instability should be entertained.

10.13 Energy Gain Systematic: Propagation From a Hot Spot

Imploded hot spot's ion temperature is an imperative parameter for ICF. Here, a small fraction of target mass is heated to attain the energy up to 10 keV. This is a natural implosion process that occurs in the center of the target. At the center, the developed shock converges, whereas, off the center, it reflects. This leads to a further increase in temperature of the available low-density D-T gases. The energy level of the hot spot is reflected by the ion temperature, and it has a critical dependence on the implosion speed and implosion symmetry. The density and temperature of the hot spot are raised to approximately 50 g/cm^3 and 10 keV, respectively. The hot spot acquires an area nearly half of the radius of the overall assembled fuel. Basically, there is a layer of cold fuel with high density of the order of 500 g/cm^3 around the central hot-spot region. There is about 1–2% of the mass of the main fuel, which has been heated to 10^4 eV. The order of this is $\rho_{HS} R_{HS}^3 = \left(\frac{\rho_F}{10}\right)\left(\frac{R_F}{2}\right)^3$, where ρ_F and ρ_{HS} are fuel and hot-spot densities, respectively and R_F and R_{HS} are fuel and hot-spot radii, respectively. Within the hot-spot area, the alpha particles, which are generated through the D-T fusion, can be stopped if the magnitude of $\rho_{HS}R_{HS}$ is of the order of 0.3 g/cm^2 for a hot spot that is about a plasma of 5–10 keV. This may result in the thermally self-heating of the energy to about 30–40 keV. Accordingly, we observe that the magnitude of $\rho_{HS}R_{HS}$ of the order of 0.3 g/cm^2 supplies enough burn fraction f_b to generate sufficient alpha particles to stop in the high-density cold matter's adjacent shell (AS).

Here, we will try to solve the energetics by considering that the mass of the AS is three times that of the hot-spot (HS):

$$M_{HS} = \left(\frac{1}{3}\right) 4\pi R_{HS}^2 \left(\rho_{HS} R_{HS}\right) \tag{10.21}$$

whereas, $M_{AS} = 4\pi R_{HS}^2 \left(\rho_{HS} R_{HS}\right)$ is the mass of thin shell. Here, $\rho_{AS}R_{AS}$ and $\rho_{HS}R_{HS}$ have been taken of the order of 0.3g/cm^2; therefore, $M_{HS} = \left(\frac{1}{3}\right) M_{AS}$ or $M_{AS} = 3M_{HS}$. One fused D-T pair showed that the hot spot produced alpha particles of energy 3.6 MeV. There are many $\left(\frac{M_{HS}}{5} \text{ amu}\right)$ potential pairs available to produce the alpha particles. The number fused is f_b times that potential, and f_b is about 5%. Thus, the hot plasma can produce the alpha particles with energy about $(18 \times 10^4 \text{ eV}) \times \left(\frac{M_{HS}}{5} \text{ amu}\right)$. The alpha particles, those stopped in three times massive adjacent shells ($\rho_{AS}R_{AS}$ = 0.3 g/cm^2), can be expressed as $3 \times \left(\frac{M_{HS}}{5} \text{ amu}\right)$ D-T pairs. In general, 60 keV per D-T pair is required to heat all of them. The hot spot supplies the total energy about $(60 \text{ keV}) \times 3 \times \left(\frac{M_{HS}}{5} \text{ amu}\right) = (180 \text{ keV}) \times \left(\frac{M_{HS}}{5} \text{ amu}\right)$. When the AS, adjacent to the hot-spot shell, runs away, it generates enough alpha particles creating sufficient energy. This energy is used to supply the next shell (NS) of thickness $\rho_{AS}R_{AS} = 0.3$ g/cm^2, which is equal to an alpha range. This energy will be used to heat them up to 10^4 eV; therefore, the propagation of thermonuclear burn wave is started. The energy production of the AS is enough and is about three times that of the HS, since the mass of the NS with high density and thin shell is almost equal to mass of the AS and both thin shells have the same ρR. The HS produces as much energy to heat the AS because the mass of the AS is almost three times the HS mass. In this process, the neutrons of energy 14 MeV are streamed to the chamber walls. The range of the neutrons of the order of 5 g/cm^2 has been reported. This reported range is slightly greater than that of the alpha range, whereas it is even greater than the usual total target ρR of 3 g/cm^2.

To summarize this process, we say that the fusion process helps to start the burning of high-density cold fuel that surrounds it. The fusion takes place only in 1% of the mass of the hot spot. The process continues, but it does not mean that the gains are infinite. The energy is used to compress the cold fuel surrounded by the hot spot. Since the ignition temperature is about 5 keV, it is concluded that the hot-spot cooling by mixed-enhanced conduction loss plays a key role in setting the operational boundary for the ICF ignition process. A minimum amount of energy will be required for the case if the fuel is preserved on the lowest acceptable isentrope. Such an isentrope investment is designated by the FD isentrope. Due to the effective quantum pressure, an optimum energy is needed to compress the cold fuel. On compressing the cold fuel, the electrons squeeze to inter-particle distances, which are shorter compared with their de Broglie wavelengths. As the wavelength of electrons is squeezed smaller and smaller, the momentum of each electron will be enhanced by the uncertainty principle. Such an increment in the electron's momentum behaves as a pressure. In centimetre-gram-second (CGS) units, the formula to determine the pressure for cold main fuel is

$$P_F = \alpha_{\mathrm{FD}} P_{\mathrm{FD}} = \alpha_{\mathrm{FD}} (2 \times 10^{12}) \rho^{5/3} \tag{10.22}$$

and for cold compression, the formula to determine the required specific energy is

$$\varepsilon_F = \alpha_{\mathrm{FD}} \varepsilon_{\mathrm{FD}} = \alpha_{\mathrm{FD}} (3 \times 10^5) \rho^{\frac{2}{3}} (\mathrm{J/g}) \tag{10.23}$$

We designate α_{FD} in both the cases by a multiplier, whose value is ≥1. Actually, this is the value that confirms the minimum required FD isentrope. Thus, to obtain the $\varepsilon_F = 3 \times 10^5$ (J/g), we require the densities of the order 10^3 g/cm^3. It also may be compared with the fusion output per gram of 3×10^{11} J/g. We can obtain an inherent gain of $\varepsilon_{DT} / \varepsilon_{FD}$ of the order 10^4 times f_b* times the coupling efficiencies $\eta_T = \eta_C \eta_H$ from hot-spot ignition and subsequent burn of the dense surrounding fuel. This is the gain value required to achieve the goal of gains of the order of 10^2.

10.14 Energy Gain Scaling and Driver Size: Burn Fraction Increases With Scale Size

For lower values of $\rho_f R_f$, f_b is approximately linearly proportional to $\rho_f R_f$. As $M_f \propto \rho_f R_f^3$, $R_f \propto \left(\frac{M_f}{\rho_f} \right)^{1/3}$ and this leads to

$$f_b^* \propto \rho_f^* R_f^* \propto \rho_f^{*2/3} M_f^{*1/3} \propto (\eta_T E_D)^{-2/10} (\eta_T E_D)^{(1.2/3)} \propto (\eta_T E_D)^{2/10} \tag{10.24}$$

where f_b^*, ρ_f^*, R_f^*, M_f^* are optimal values of the corresponding quantities.

Instinctively, larger targets driven would be expected to have a larger $\rho_f R_f$ from larger-scale drivers. Consequently, we will have a greater magnitude of f_b; in other words, burning will be more effective. f_b is expected to scale as $R_f \propto M_f^{1/3}$, which in turn will scale as $E_D^{1/3}$. The intuition of the aforesaid discussion with detailed derivation is that at larger scales, densities are lower and result in the variation in accordance to $f_b \propto E_D^{2/10}$. On summarization of drive scale and gain scaling, G is scaled as $f_b \rho_f^{-2/3}$; this, in turn, results in the following equation:

$$G \propto (\eta_T E_D)^{2/10} (\eta_T E_D)^{2/10} \propto (\eta_T E_D)^{4/10} \tag{10.25}$$

Hence, gain is naturally increased with about a factor of three by an order of magnitude in driver scale. We can get another benefit for the case of increment in coupling, $\eta_T = \eta_C \eta_H$, with scale as well. The gain is obviously enhanced with the modifications in the coupling.

A further increment in the gain with η_T can be achieved through the use of driver scale for optimal targets. We can also understand the additional benefit from another prospective. A relatively higher gain will be achieved with a smaller driver for the case of scaling of improved η_T compared with the case of scaling of unimproved η_T, where the relevant quantity is $\eta_T E_D$. The initial capital costs are reduced by the smaller sized drivers, and they provide a lower cost of electricity.

10.15 Conclusions

We have discussed various aspects of thermonuclear fusion, explored its feasibilities, and evaluated its challenges. In conclusion, the thermonuclear fusion reaction is an outstanding source of eco-friendly, safe, non–carbon-emitting, and unlimited output energy. Scientists are trying to generate enormous energy from fusion reactions. However, there are limitations and conditions to achieve fusion and that too in a controlled manner. For example, there is a specific confinement time of plasma (different for D-T and D-D plasmas) and the Lawson criterion should be satisfied. There are largely two schemes by which fusion can be achieved, ICF and MCF. In ICF the thermonuclear fusion is started by heating and squeezing the nuclear fuel comprising D and T elements, whereas in MCF the plasma is confined by strong magnetic fields at lower densities (about 10^{14} to 10^{15} cm^{-3}) for longer time. The DT fuel target is compressed so that it exceeds a critical value to very high density and very high temperatures. This is achieved by using a nanosecond laser (~10 ns) with extremely high energy ($E_{las} \geq 1MJ = 10^6$ J). This eventually creates central hot spot wherein ignition occurs. The CHSI approach needs to satisfy the high symmetry of the target's illumination and other various technical conditions. CHSI includes FI, SI, and II.

The ICF targets need to be imploded and burn for which there are direct and indirect drive implosion schemes that include linear, cylindrical, and spherical compressions. With respect to the physics of implosion, the important fields are laser-plasma interaction, electron energy transport, dynamics of implosion, hydrodynamic instability, radiation transport, and target design. Finally, the energy gain systematic, energy gain scaling, and driver size are important measures that need to be seen for an efficient fusion energy.

10.16 Selected Problems and Solutions

PROBLEM 10.1

In the case of ICF, if the radius is reduced by the factor of 2 and 10, i.e. r is reduced to $r/2$ and $r/10$, respectively, for the case of cylindrical and spherical compressions, then find the change in density for both the cases.

SOLUTION

The density in the case of ICF is given as

$$\rho = \left(\frac{r_0}{r}\right)^n$$

where n is the compression dimension and r_0 is the initial radius.

For the cylindrical case, $n = 2$ and the radius is reduced by a factor of 2, i.e. $r = \frac{r_0}{2}$. The change in density is calculated as follows:

$$\rho_{\text{cylindrical}} = \rho_0 (2)^2 = 4\rho_0$$

where ρ_0 is the initial density.

For the spherical case, $n = 3$ and the radius is reduced by the factor of 10, i.e. $r = \frac{r_0}{10}$. The change in density in this case is calculated as follows:

$$\rho_{\text{spherical}} = \rho_0 (10)^3 = 1000\rho_0$$

PROBLEM 10.2

What amount of energy is released in a D-T reaction? Given that $m(D) = 2.01410178$ amu, $m(T) = 3.0160492$ amu, $m(He) = 4.002602$ amu, $m(n) = 1.008664$ amu, and 1 amu = 931 MeV.

SOLUTION

The D-T reaction is stated as follows:

$${}_1^2D + {}_1^3T \rightarrow {}_2^4He + {}_0^1n + Q(?)$$

$$Q = (m_D + m_T - m_{He} - m_n) \times 931 \text{ MeV/amu}$$

Putting the values into the above relation, one gets

$$Q = 17.58 \text{ MeV}$$

PROBLEM 10.3

Evaluate Lawson criterion for the ICF and MCF. Given that for ICF, particle density n_e is $10^{26}\,\text{cm}^{-3}$ and confinement time τ is 10^{-11} s and for MCF, particle density n_e is $10^{14}\,\text{cm}^{-3}$ and confinement time τ is 10 s.

SOLUTION

In 1957, Lawson proposed the minimum condition for productive fusion, which is related to the product of confinement time and ion density. This product is designated as Lawson criterion.

For ICF, n_e is $10^{26}\,\text{cm}^{-3}$ and τ is 10^{-11} s, Lawson criterion is determined as follows:

$$\text{LC} = n_e\tau$$

$$= 10^{26} \times 10^{-11} = 10^{15} \text{ s/cm}^3$$

For MCF, n_e is $10^{14}\,\text{cm}^{-3}$ and τ is 10 s, Lawson criterion is determined as follows:

$$\text{LC} = n_e\tau$$

$$= 10^{14} \times 10 = 10^{15} \text{ s/cm}^3$$

ICF and MCF satisfy Lawson criterion in distinctive ways. In ICF, higher plasma densities are achieved for a smaller period of time, whereas, in MCF, lower plasma densities are confined for a longer period of time (about a few seconds).

PROBLEM 10.4

The net result of fusion of hydrogen $\left({}_{1}^{1}H\right)$ in the sun is that four ${}_{1}^{1}H$ atoms are fused into a single helium atom $\left({}_{2}^{4}He\right)$ with some junk. This reaction is given as follows:

$$4{}_{1}^{1}H \rightarrow {}_{2}^{4}He + 2\left({}_{+1}^{0}e + {}_{0}^{0}\gamma + {}_{0}^{0}v\right)$$

Given that mass of the sun and its output power, respectively, are 1.99×10^{30} kg and 3.85×10^{26} W, the mass of ${}_{1}^{1}H$ and ${}_{2}^{4}He$ are 1.007825 and 4.002602 amu, respectively, and the sun is composed of around 91% of ${}_{1}^{1}H$, calculate

i. the mass defect,
ii. the reduction rate for the mass of the sun,
iii. the distortion of the mass of the sun if all hydrogen present in the sun fused into helium, and
iv. the estimated life-time of the sun.

SOLUTION

i. The mass defect or the mass difference among a single helium atom and four hydrogen atoms is given as follows:

$$4 \times 1.007825 \text{ amu} - 4.002602 \text{ amu} = 0.028698 \text{ amu} = 26.718 \text{ MeV}$$

ii. The reduction rate for the mass of the sun is associated with the energy production rate. This is calculated as follows:

$$P = \frac{E}{t} = \frac{\Delta mc^2}{t} \rightarrow \frac{\Delta m}{t} = \frac{P}{c^2}$$

$$\frac{\Delta m}{t} = \frac{3.85 \times 10^{26} W}{(3.00 \times 10^8 m/s)^2}$$

$$\frac{\Delta m}{t} = 4.278 \times 10^9 \text{ kg/s}$$

It means around four times of the 10^9 kg mass of the sun is reduced per seconds.

iii. The distortion of the mass of the sun if all hydrogen present in the sun fused into helium is determined by the ratio of destroyed mass to initial mass, i.e.

$$\frac{\text{destroyed mass}}{\text{initial mass}} = \frac{m}{(0.91)\left(1.99 \times 10^{30} \text{ kg}\right)} = \frac{0.028698 \text{ amu}}{4 \times 1.007825 \text{ amu}}$$

$$\frac{m}{(0.91)\left(1.99 \times 10^{30} \text{ kg}\right)} = \frac{0.028698 \text{ amu}}{4.0313 \text{ amu}}$$

$$m = 1.289 \times 10^{28} \text{kg}$$

iv. The estimated lifetime of the sun is calculated as

$$\frac{\Delta m}{t} = 4.278 \times 10^9 \text{ kg/s} = \frac{1.289 \times 10^{28} \text{ kg}}{t}$$

$$t = 3.013 \times 10^{18} \text{s} = 95.54 \times 10^9 \text{ years}$$

This cannot be considered a good estimation. We know that only the central core of the sun is dense and hot enough; therefore, less than 10% of the hydrogen present in the sun will be feasible for thermonuclear fusion. Ten percent of the above calculated value is 9.5×10^9 years and this could be a good estimation. The Earth is around 4.5×10^9 years old, i.e. Earth is at the intermediate of the sun's life.

PROBLEM 10.5

In most of the high-yield thermonuclear weapons, the fuel used is solid lithium-6 deuteride (^{6}LiD) with a density of 820 kg/m^3. Such weapons are generally called 'hydrogen bombs' or 'H-bombs'. When the nucleus of deuterium (heavy hydrogen) and lithium are fused into two helium nuclei energy is released. This process is stated as follows:

$${}^6_3Li + {}^2_1H \rightarrow 2\ {}^4_2He$$

Calculate

i. the mass defect for aforesaid reaction; for the thermonuclear weapons whose explosive yield is around a few million tons (~10^6 tons) of TNT, determine
ii. the destroyed mass in this explosion,
iii. the amount of the fuel required, and
iv. the volume of the fuel required.

Given that 1000 kg (1 ton) of TNT is equivalent to 4.184×10^9 J.

SOLUTION

i. The mass defect will be calculated from the difference in the reactants' mass and products' mass, stated as follows:

$$\text{mass difference} = \text{reactant's mass} - \text{product's mass}$$
$$= (6.015121 \text{ a.m.u.} + 2.0140 \text{ a.m.u.}) - 2(4.0026) = 0.023921 \text{ a.m.u.} = 22.27 \text{ MeV.}$$

ii. The mass destroyed in such an explosion is calculated from the conversion relation of mass and energy, as follows:

$$E = mc^2 \rightarrow m = \frac{E}{c^2}$$

$$m = \frac{(10^6)(4.184 \times 10^9 \text{ J})}{\left(3 \times 10^8 \text{ m/s}\right)^2}$$

$$m = 0.04649 \text{ kg}$$

iii. The amount of the fuel required for the explosion will be determined from the ratio of original mass of ^{6}LiD to the destroyed mass, as follows

$$\frac{\text{intial mass}}{\text{destroyed mass}} = \frac{m}{0.04649\text{ kg}} = \frac{8.029121u}{0.023921u}$$

$$m = 15.604\text{ kg}$$

iv. The density and mass of the fuel will reveal the volume of fuel required for the explosion, as follows:

$$V = \frac{m}{\rho} = \frac{15.604\text{ kg}}{820\text{ kg/m}^3}$$

$$V = 0.01903\text{m}^3 = 19.029\text{ L}$$

PROBLEM 10.6

In the solar fusion reaction, given by ${}^1_1p + {}^1_1p \rightarrow {}^2_1H + {}^0_{+1}e + {}^0_0\nu$, calculate the neutrino's energy range, given that rest masses of the proton, electron, and deuteron, respectively, are 938.3, 0.511, and 1875.7 MeV/c^2. The B.E. of ${}^2_1H$ is 2.22 MeV and the kinetic energy of protons is negligible at the initial state.

SOLUTION

When deuteron and positron are produced in the opposite direction, the energy carried by the neutrino is zero. Therefore, the neutrino can have minimum zero energy. When deuteron moves in the opposite direction with respect to the neutrino, and the positron is at the rest, then the neutrino carries maximum energy. This maximum energy of the neutrino will be calculated from the law of energy and momentum conservations.

From the energy conservation,

$$T_d + T_v = Q.(T_{e+} = 0,\text{ because positron is at the rest}) \tag{10.26}$$

From the momentum conservation,

$$P_d^2 = P_v^2 \tag{10.27}$$

Also,

$$2m_dT_d = T_v^2 \tag{10.28}$$

Q value can be calculated as follows

$$Q = (2 \times 938.3 - 1875.7 - 0.511)\text{ MeV} = 0.389\text{ MeV}$$

From Eqs. (10.26) and (10.28), we have

$$T_v^2 + 2m_dT_v - 2m_dQ = 0$$

Positive value of this equation gives $(T_v)_{\text{max}} = 0.38895$ MeV. Therefore, the energy range for the neutrino will lie between zero and 0.38895 MeV.

PROBLEM 10.7

Q-value in the D-T reaction in a fusion reactor is 17.62 MeV. Let us consider that the density of deuteron is $7 \times 10^{18}/\text{m}^3$ and the experimental value of $<\sigma_{DT}.v>$ is $10^{-22}\,\text{m}^3/\text{s}$. The energy of tritons and deuterons, which are in equal number in plasma, is 10 keV. If the condition of the Lawson criterion is fulfilled, then determine the confinement time.

SOLUTION

The condition of Lawson criterion is fulfilled for

$$L = \frac{\text{energy output}}{\text{energy input}} = \frac{n_d < \sigma_{dt} v > t_c Q}{6kT}$$

$$= \frac{n_d < \sigma_{dt} v > t_c Q}{6kT} = 1$$

For $kT = 10$ keV $= 10^4$ eV, $n_d = 7 \times 10^{18}/m^3$, $Q = 17.62 \times 10^6$ eV, and $< \sigma_{dt}.v > = 10^{-22}\,\text{m}^3/\text{s}$, the confinement time t_c can be obtained as

$$t_c = \frac{6kT}{n_d \; < \sigma_{dt} v > Q}$$

$$= \frac{6 \times 10^4}{7 \times 10^{18} \times 10^{-22} \times 17.62 \times 10^6} = 4.86 \text{ s}$$

PROBLEM 10.8

Determine the confinement time for ICF when the radius of the spherical pellet is 200×10^{-4} cm and temperature is 10^8 K.

SOLUTION

We know that confinement time is given by $\tau = \frac{r_0}{v_s}$, where v_s is the thermal velocity of the ions, and is given by

$$v_s = \sqrt{\frac{KT_i}{m_i}}$$

where m_i is the average mass of ions = 2.515 amu

$$v_s = \sqrt{\frac{1.38 \times 10^{-23} \times 10^8}{2.515 \times 1.6726 \times 10^{-27}}}$$

$$= 0.5728 \times 10^6 \text{ m/s} = 0.5728 \times 10^8 \text{ cm/s}$$

$$\tau = \frac{200 \times 10^{-4}\,\text{cm}}{0.5728 \times 10^8\,\text{cm/s}} = 0.349 \text{ ns}$$

PROBLEM 10.9

Derive Lawson criterion for the D-T reaction and find its numerical value.

SOLUTION

Lawson criterion is about the product of confinement time and density of ions. The confinement time, τ_e, is defined as

$$\tau_e = \frac{\text{energy content of plasma}}{\text{power loss}} = \frac{W}{P_{\text{loss}}}$$

In plasma, thermal energy (W) is defined as

$$W = \int \frac{3}{2} K[n_e T_e + (n_D + n_T) T_{\text{ions}}] dV$$

where n_e, n_T, and n_D, respectively, are the densities of electrons, tritium, and deuterium, T_e is the electron temperature, and T_{ions} is temperature of tritium and deuterium. For simplicity temperature of all the species is assumed to be the same, and densities of deuterium and tritium are taken to be equal. Also, the sum of densities of tritium and deuterium is equal to the density of electrons. In other words, $n_D = n_T = \frac{n_e}{2}$ and $T_{\text{ions}} = T_e$. Then the result of the integration would be as follows:

$$\frac{W}{V} = 3 n_e KT$$

The number of fusions per unit time, N is defined as

$$N = n_D n_T < \sigma v > = \frac{1}{4} n_e^2 < \sigma v >$$

where relative velocity is v and fusion cross section is σ. The product of energy of charged fusion products (E_{ch}) with number of fusions per unit time (N) gives the heating rate per unit volume. Here, we have considered that in the plasma heating there is no contribution of neutron emission. Thus

$$N E_{ch} \geq P_{\text{loss}}$$

$$\frac{1}{4} n_e^2 < \sigma v > E_{ch} \geq \frac{3 n_e KT}{\tau_e}$$

$$n_e \tau_e \geq \frac{12KT}{E_{ch} < \sigma v >}$$

$$L = \frac{12KT}{E_{ch} < \sigma v >}$$

When $< \sigma v > = 1.3 \times 10^{-22}\ \text{m}^3/\text{sec}$ and $E_{ch} = 17.6\ \text{MeV}$, then

$$L = \frac{12 \times 1.38 \times 10^{-23} \times 10^8}{17.6 \times 10^6 \times 1.6 \times 10^{-19} \times 1.3 \times 10^{-22}} = 0.45 \times 10^{20}\ \text{s/m}^3$$

Suggested Reading Material

Aglitskiy, Y., Velikovich, A. L., & Obenschain, S. P. (2010). Basic hydrodynamics of Richtmyer–Meshkov-type growth and oscillations in the inertial confinement fusion-relevant conditions. *Philosophical Transactions of the Royal Society A: Mathematical, Physical and Engineering Sciences, 368*(1916), 1739–1768.

Atzeni, S., & Ciampi, M. L. (1997). Burn performance of fast ignited, tritium-poor ICF fuels. *Nuclear Fusion, 37*(12), 1665.

Atzeni, S., & Meyer-ter-Vehn, J. (2004). The physics of inertial fusion: Beam plasma interaction, *Hydrodynamics, Hot Dense Matter,* Vol. 125. Oxford, UK: Oxford University Press.

Azechi, H., Jitsuno, T., Kanabe, T., ... & Nishiguchi, A. (1991). High-density compression experiments at ILE, Osaka. *Laser and Particle beams, 9*(2), 193–207.

Azechi, H., Sakaiya, T., Watari, T., Karasik, M., Saito, H., Ohtani, K., ... & Shigemori, K. (2009). Experimental evidence of impact ignition: 100-fold increase of neutron yield by impactor collision. *Physical review letters, 102*(23), 235002.

Bar-Shalom, A., Oreg, J., Goldstein, W. H., Shvarts, D., & Zigler, A. (1989). Super-transition-arrays: a model for the spectral analysis of hot, dense plasma. *Physical Review A, 40*(6), 3183.

Bauche-Arnoult, C., Bauche, J., & Klapisch, M. (1979). Variance of the distributions of energy levels and of the transition arrays in atomic spectra. *Physical Review A, 20*(6), 2424.

Bell, A. R., Evans, R. G., & Nicholas, D. J. (1981). Electron energy transport in steep temperature gradients in laser-produced plasmas. *Physical Review Letters, 46*(4), 243.

Betti, R., Zhou, C. D., Anderson, K. S., Perkins, L. J., Theobald, W., & Solodov, A. A. (2007). Shock ignition of thermonuclear fuel with high areal density. *Physical Review Letters, 98*(15), 155001.

Bodner, S. E. (1981). Critical elements of high gain laser fusion. *Journal of Fusion Energy, 1*(3), 221–240.

Caruso, A., & Pais, V. A. (1996). The ignition of dense DT fuel by injected triggers. *Nuclear Fusion, 36*(6), 745.

Gardner, J. H., Bodner, S. E., & Dahlburg, J. P. (1991). Numerical simulation of ablative Rayleigh–Taylor instability. *Physics of Fluids B: Plasma Physics, 3*(4), 1070–1074.

Glendinning, S. G., Weber, S. V., Bell, P., ... & Wegner, P. J. (1992). Laser-driven planar Rayleigh-Taylor instability experiments. *Physical Review Letters, 69*(8), 1201.

Glenzer, S. H., MacGowan, B. J., Michel, P., Meezan, N. B., Suter, L. J., Dixit, S. N., ... & Dewald, E. L. (2010). Symmetric inertial confinement fusion implosions at ultra-high laser energies. *Science, 327*(5970), 1228–1231.

Kull, H. J. (1989). Incompressible description of Rayleigh–Taylor instabilities in laser-ablated plasmas. *Physics of Fluids B: Plasma Physics, 1*(1), 170–182.

Kull, H. J., & Anisimov, S. I. (1986). Ablative stabilization in the incompressible Rayleigh–Taylor instability. *The Physics of Fluids, 29*(7), 2067–2075.

Lawson, J. D. (1957). Some criteria for a power producing thermonuclear reactor. *Proceedings of the Physical Society. Section B, 70*(1), 6.

Lindl, J. (1995). Development of the indirect-drive approach to inertial confinement fusion and the target physics basis for ignition and gain. *Physics of Plasmas, 2*(11), 3933–4024.

Matte, J. P., & Virmont, J. (1982). Electron heat transport down steep temperature gradients. *Physical Review Letters, 49*(26), 1936.

Meade, D. (1987). Results and plans for the Tokamak fusion test reactor. *Fusion Energy Development, 7*(2–3), 107–114.

Murakami, M., & Nagatomo, H. (2005). A new twist for inertial fusion energy: Impact ignition. *Nuclear Instruments and Methods in Physics Research Section A: Accelerators, Spectrometers, Detectors and Associated Equipment, 544*(1–2), 67–75.

Nakai, S. (1989). Laser fusion experiment. *Laser and Particle Beams, 7*(3), 467–475.

Nakai, S., Watteau, J. P., Smirnov, V. P., & Rozanov, V. B. (1990). Inertial confinement. *Nuclear Fusion, 30*(9), 1863.

Remington, B. A., Weber, S. V., Haan, S. W., ... & Nash, J. K. (1993). Laser-driven hydrodynamic instability experiments. *Physics of Fluids B: Plasma Physics, 5*(7), 2589–2595.

Shiraga, H., Fujioka, S., Nakai, M., ... & Ishii, Y. (2011). Fast ignition integrated experiments with gekko and LFEX lasers. *Plasma Physics and Controlled Fusion, 53*(12), 124029.

Tabak, M., Munro, D. H., & Lindl, J. D. (1990). Hydrodynamic stability and the direct drive approach to laser fusion. *Physics of Fluids B: Plasma Physics, 2*(5), 1007–1014.

Takabe, H., Mima, K., Montierth, L., & Morse, R. L. (1985). Self-consistent growth rate of the Rayleigh–Taylor instability in an ablatively accelerating plasma. *The Physics of Fluids, 28*(12), 3676–3682.

Takabe, H., Montierth, L., & Morse, R. L. (1983). Self-consistent eigenvalue analysis of Rayleigh–Taylor instability in an ablating plasma. *The Physics of Fluids, 26*(8), 2299–2307.

Takabe, H., Yamanaka, M., Mima, K., Yamanaka, C., Azechi, H., Miyanaga, N., ... & Nishimura, H. (1988). Scalings of implosion experiments for high neutron yield. *The Physics of Fluids, 31*(10), 2884–2893.

Takabe, H., & Yamamoto, A. (1991). Reduction of turbulent mixing at the ablation front of fusion targets. *Physical Review A, 44*(8), 5142.

Theobald, W., Betti, R., Stoeckl, C., Anderson, K. S., Delettrez, J. A., Glebov, V. Y., ... & Meyerhofer, D. D. (2008). Initial experiments on the shock-ignition inertial confinement fusion concept. *Physics of Plasmas, 15*(5), 056306.

Yamanaka, C., & Nakai, S. (1986). High gain laser fusion. *Nature, 319*, 757.

11

Important Controlled Fusion Devices

11.1 Absorption of Laser in D-T Plasma

The most accessible fusion reaction is the reaction involving D-T where research on controlled fusion is being conducted. Under certain conditions, each fusion reaction of deuterium ($D = {}^2H$) and tritium ($T = {}^3H$) produces a helium nucleus (${}^4He$), a neutron, and 17.6 MeV of energy. D-T fusion is most suitable for inertial confinement fusion (ICF) and D-D or H-H fusion are not considered for an efficient fusion reactor. However, for fusion to occur, the nuclei need to overcome the Coulomb barrier and this requires very high energy input, more than 15 keV. At this very high temperature the fusion fuel is in the state of plasma called D-T plasma. In ICF, such a high temperature required for fusion is achieved by using laser or particle beams to compress and heat the small fuel pellet that is usually in a spherical capsule. Due to irradiation of the laser or particle beam, the surface of the spherical pellet ablates and shock is produced in the reaction as per Newton's third law. The interior fuel compresses and heats up due to the implosion and eventually burning occurs and fusion begins at the core. Because of a very high density, the inertia of the fuel protects it from escaping from the core, and the fusion continues. A situation is reached when pressure at the core becomes sufficiently high to stop implosion, then fuel cools down due to expansion, and the fusion reactions cease.

As the energy input to ignite the fuel to achieve fusion is very high, to ensure gain ICF aims to produce about 10^{20} D-T reactions for each shot from a laser. Moreover, for efficient energy production, the main objective is set to increase the repetition rate of fusion reactions. The helium nucleus carries 20% energy of the fusion yields and by virtue of its electrical charge it tries to remain within the plasma. The neutrons carry 80% of the kinetic energy produced in the fusion reaction and as they do not carry any charge, they can be absorbed easily by the walls of the fusion chamber. This kinetic energy carried by neutrons can be transformed into heat and can boil water into steam that drives a turbine for the production of electrical energy. The D-T fusion has the highest reaction rate of all fusion reactions; therefore, it is the central focus of worldwide fusion research.

11.2 Laser-Driven Thermonuclear Fusion: Present Status, Technical Challenges, and Current Research

Nuclear fusion is considered to be a secure, long-term source of clean and sustainable energy of the future. But to recreate the fusion reaction occurring in the sun and stars is not easy. The density at the core of the sun is 150 g/cc and its temperature is 15 million Kelvin. It is essential to mimic this condition to achieve fusion in a man-made environment. For several decades researchers have been trying to find methods for liberating fusion energy in a controlled manner. At the extremely high temperature required for fusion to occur, the fusion fuel becomes plasma. Thus, a powerful confinement of the plasma and ignition with extremely high temperature is essential to obtain fusion energy.

In terrestrial conditions like the sun and other stars 'gravitational confinement' occurs; the immense mass of a star causes the material at its core to be compressed or confined by the star's powerful gravity. As mentioned earlier, in the man-made conditions, there are two alternatives: MCF and ICF. Inertial fusion with a high energy gain was demonstrated in the form of a hydrogen bomb almost 60 years ago. Unfortunately, that was uncontrolled fusion, whereas to meet our energy needs we are looking for controlled fusion. The complexity involved with controlled fusion has resulted in various challenges in the development of ICF experimental facilities. The effectiveness of fast ignition depends on the effective coupling of the laser pulses in delivering its high energy to the core of the D-T pellets.

Recent advancement in laser technology has taken us closer to our goal of achieving controlled nuclear fusion. Researchers found that a laser-driven technique for creating thermonuclear fusion is now within our reach. The scientific challenges were overcome by advancements in technology resulting in the availability of powerful and high-intensity lasers (Moses 2003). At present laser fusion is a topic of prime interest worldwide and there is a global race for fusion power. Lawrence Livermore National Laboratory (LLNL) in Livermore, California, has been carrying out laser fusion experiments since the mid-1960s. The National Ignition Facility (NIF) in California is close to demonstrating a controlled, sustained fusion reaction with an array of high-powered lasers (Campbell et al. 2017). Several years ago NIF successfully compressed targets without D-T and experiments were carried out to put 1 MJ of laser energy into a 10-mm Au/U cylinder (hohlraum) producing X-rays that will crush a 2-mm diameter Beryllium sphere lined with D-T ice emitting 20 MJ in fusion products (Dunne et al. 2011; Landen et al. 2012). Laser Mégajoule (LMJ) in France is also intended to demonstrate inertial fusion. The Laboratory for Laser Energetics (LLE) in Rochester, New York, and the Institute of Laser Engineering (ILE) at Osaka, Japan, are other smaller but famous laboratories that have been working on laser fusion for the last few decades. LLE has demonstrated shock ignition with its OMEGA laser facility (Betti and Hurricane 2016). The Gekko XIII laser at ILE Osaka is capable of delivering more than 10 kJ in 2 ns with a 0.53 μm wavelength (green) plus one beam with 350 J, 1 ps with a 1 μm wavelength (IR). Here, the Fast Ignition Realization EXperiment (FIREX) has demonstrated the direct drive by using a gold cone to guide the ignition beam through the surrounding plasma corona with 20% coupling efficiency (Azechi et al. 2013). High Power laser Energy Research (HiPER) has been an ambitious pan-European project that explored the science of extreme conditions and developed the route to laser-driven fusion energy (Dunne 2006). The long-term goal of this project was to provide scientific basis for viable and efficient laser fusion and to demonstrate the detailed development of the fusion reactor. HiPER is based on the approach of direct drive, and the project studied the various aspects of fast ignition including the use of a laser channel in place of a gold cone and looked into the prospect of an energy deposit to the core of the D-T pellet by fast electrons generated by high-power laser beams. Looking at the research side, Singh et al. (2012) have demonstrated the formation of a millimeter-scale persistent laser channel relevant to fast ignition and investigated its dependence on laser polarization. In India, significant activities on the theory and modelling related to ICF are being carried out at the Institute of Plasma Research (IPR) at Ahmedabad. Raja Ramanna Centre for Advanced Technology (RRCAT) in Indore,

India, is also conducting some basic studies on laser-plasma interactions relevant to the ICF scheme, and a number of plasma diagnostics are operational at this centre.

11.3 Fuel Cycle and Economics

Ignition is beneficial only if we can extract the energy efficiently from the thermonuclear fuel. Neutron and a helium nuclei are produced from the D-T fusion reaction, stated as

$$D+T \rightarrow {}^4He(3.5 \times 10^6 eV) + n(14.1 \times 10^6 eV)$$

With D-T reaction, two practical problems occur. First, neutrons carry away most of the energy, which results in huge energy loss. Therefore, a significant neutron shielding is mandatory to any power plant for its efficient output. Second, while deuterium is rich and commonly available in nature, the existence of tritium is not so available. It appears that the first problem's solution results somewhat naturally in the second problem's solution.

A neutron blanket that absorbs the neutrons is surrounded in the reaction chamber by a 0.6-m-thick 'waterfall' of hot helium (900–1100 K). It is constantly cycled between a heat engine and the reactor. The heat engine produces electricity from the extraction of heat from the fusion reactions. All of the fuel debris and X-rays, and 95% of the neutrons, are absorbed by the blanket. This neutron blanket also results in the breeding of tritium along with the extraction of energy from the reactor and shielding of high energetic neutrons. The reactions, through which tritium is produced, are stated as follows

$$^6Li + n \rightarrow {}^4He + T + 4.8 \times 10^6 \text{ eV}$$
$$^7Li + n \rightarrow {}^4He + T + n - 2.46 \times 10^6 \text{ eV}$$

Therefore, from the lithium element, which is abundant in the Earth's crust, we can produce the essential and rare element of the fuel, i.e. tritium. Hence, the use of the neutron blanket resolves both problems stated above. Deuterium, the second fuel element, is available in significant amounts in seawater.

These neutrons can also be used in several applications. A single nucleus of uranium releases energy of around 200 MeV through the fission of each neutron. The energy generated per neutron in this process is approximately ten times more than that of fusion reactions. At LLNL a reactor concept of laser-initiated fusion-fission engine (LIFE) was introduced, which is based on this factor of 10 (Figure 11.1). The reactor is surrounded by a fissile fuel blanket, which absorbs its neutrons resulting in the fission products through further blanket heating in this process. In the LIFE reactor, it is the fusion neutrons that drive the fission and not the chain reaction. Fuel is subcritical always; therefore, the LIFE reactor is inherently stable and cannot go critical. The performance of the LIFE engines can be intensely good, as in the LIFE cycle, burned up fissile fuel could be around 99% of the initial value. The LIFE engine allows consumption of fuels through the fixed neutron source, which otherwise would have been inappropriate in standard nuclear reactors: spent fuel (transuranic waste), depleted uranium, thorium, weapons-grade fuel, and natural uranium can be adopted. Additionally, there would be a reduction in the quantity of the generated waste by a factor of 20.

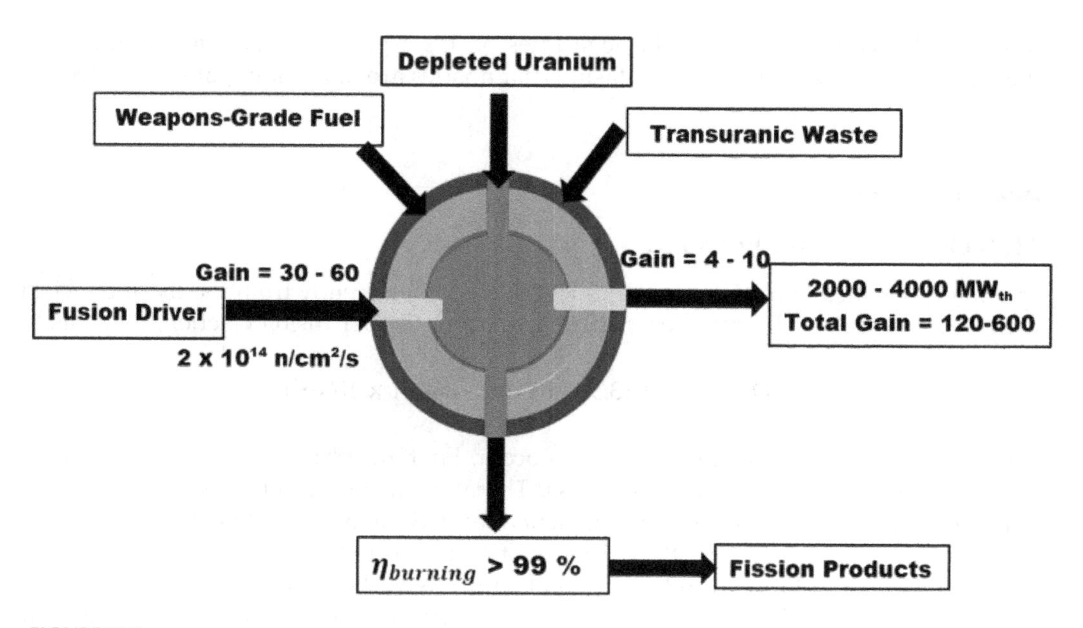

FIGURE 11.1
Fusion-fission engine provides more than 99% efficiency (η_{burning}) due to a complete burn-up closed fuel cycle once initiated by lasers.

Through the functioning of the LIFE one can see the usability of depleted fuel and nuclear waste with several other unwanted products, which may provide electricity. In the future, the development of fusion reactors will result in a milestone for society. However, questions arise like 'does it work' and 'how much does it cost' in the gathering of power executives while discussing the commercialization of the concept. Right now, there is no method to address such questions, but there are some clues given from a concise look at the efficiencies of the power plant (Figure 11.2).

The cost of the electricity strongly depends on the efficient operation of the plant, regardless of the type of plant. The following equation gives rise to the form of power generated by any type of fusion plant

$$P_{\text{out}} = E_{\text{pump}} f \left[G\eta_{th} - \frac{1}{\eta_{\text{pump}}} \right] \tag{11.1}$$

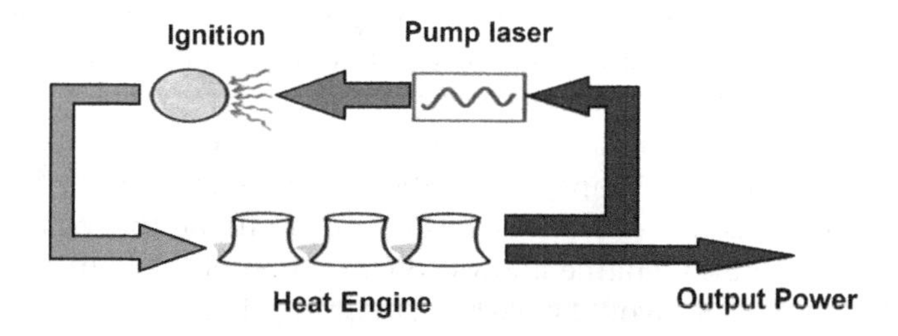

FIGURE 11.2
Power cycle of a fusion plant. The energy produced from the ignition must be greater than the energy required to trigger the ignition.

where the system frequency, gain (fusion to pump energy ratio), and pumped energy in the D-T pellet for every shot, respectively, are designated by f, G, and E_{pump}. The efficiencies of pump and heat engine are, respectively, represented by η_{pump} and η_{th}. Basically, the heat engines cannot provide better than 40% efficiency, so η_{th} may be calculated as ~0.3–0.4. However, there can be a wide variation in the pump efficiency. The magnitude of η_{pump} is 0.66% for NIF's flash lamp–driven lasers; however; this can be enhanced up to 5–13% with the help of diode technology from LLNL's Mercury laser. The power consumption by a power plant must be miniscule compared with its production to earn more money from the electricity supply. Practically, the output power is needed to be considerably larger than the input power (at least three times). The lower bond for the gain can be derived as follows:

$$G \gtrsim \frac{3}{\eta_{th}\eta_{\text{pump}}} \tag{11.2}$$

The magnitude of η_{pump} is enhanced by a factor of 10 while switching to laser diodes from the flash lamps, which usually gives the more acceptable gain of 65–70. NIF has a gain of around 20. LIFE technology and fast ignition have the ability to enhance this factor by 10, which helps us to reach the threshold. The main reason technologies such as LIFE, fast ignition, and diode lasers are adopted by scientists is for the advancement of the thermonuclear fusion process to achieve the efficient output of energy. Researchers are still hopeful to produce an energy gain by a factor of 10, which generates a considerable variation in the gain. There is always a small hope to have an estimate of the accurate price before we get the factors of 10. The technology of Inertial Fusion Energy (IFE) is in its early stage of development; therefore, there might be some unforeseen technical, economical, or scientific reasons that can create problems towards fruitful results. Besides these difficulties or challenges, there is also a huge possibility of a limitless, proliferation-free, and clear power source.

Looking at the inertial fusion energy (IFE) reactor, it consists of magnetic coils similar to the mirror machine of MCF. This magnetic field separates charged particles and neutrons along with protecting the first wall and solid blanket. This mirror magnet also induces efficient direct conversion of charged particles through end loss. An efficient laser can provide breakeven with low gain when a direct and efficient conversion takes place. Such a laser initiates efficient hydrogen production and high repetition rate. This reactor is expected to solve the following physics problems:

a. The plasma stability or stability of charged particles could be analyzed in the mirror magnetic field. This affects the structure of the reactor and the efficiency of direct conversion of energy.
b. There is the possibility of the estimation of thermal stress and radiation effects during the laser-induced fusion reaction. The thermal stress can affect the design parameters of the first wall and blanket.
c. In general, low-Z ceramics are a good candidate for making the wall and blanket because they have low activation properties. There is always a requirement of a detailed calculation of different properties of the required material.
d. Basically, the energy gain from the target fuel is assumed to be around 20 times, but the detailed design parameters of the target on the basis of physics experiments have yet to be solved.
e. The economics of a hydrogen fuel target radiated by laser beams having a higher repetition rate and higher efficiency need to be understood.

11.4 National Ignition Facility (NIF)

NIF is the world's largest laser facility and is located at LLNL. It is involved in the study of most fundamental questions pertinent to high energy density science in pursuit of ICF. NIF infrastructure is the size of a football stadium and is composed of two parallel laser bases, each containing 96 beamlines, i.e. a total of 192 laser beams of ultrahigh intensity designed to explore nuclear fusion. Each laser beam has a 351 nm wavelength and carries about 10 kJ of energy. NIF is capable of concentrating energy from all of the 192 beams into a D-T pellet of mm-size in a few nanoseconds (Moses and Wuest 2005). This results in a matter temperature of more than 10^8 K with a density of more than 10^3 g/cm^3 and more than 10^{11} atmospheric pressure.

NIF uses an indirect drive approach to laser fusion. The scheme to obtain ignition there is as follows:

i. The process starts by first energizing the laser amplifiers in the two laser bases by dumping electrical energy stored in the capacitors into flash lamps.
ii. The energy is converted into light and absorbed by laser glass in the amplifier.
iii. As the laser beam passes through this glass, the energy is extracted. Here a very low energy pulse created by an oscillator amplifies to very high energy.
iv. Further amplification takes place when it is passed several times through the laser cavity and power amplifier.
v. By the time the laser beam travels more than 1500 m from the master room to the target chamber, in total its energy increases 10^{15} times.
vi. The laser beams deliver about 1.8 MJ of energy into the cylindrical capsule called a hohlraum containing a spherical D-T fuel pellet. The size of the hohlraum is about 1 cm, whereas the size of the pellet is about 2 mm.
vii. About 70% (~1.3 MJ) of this energy is converted into X-rays. The remaining energy goes into creating low-density plasma.
viii. Most of the energy carried by the X-rays is used in heating the walls of the hohlraum, whereas 0.15 MJ of this energy is used to compress the fuel. Some of the energy also escapes through Laser Entrance Holes (LEHs). However, today a concept has evolved to apply an LEH 'storm window' to minimize such losses (Robey et al. 2012).
ix. Less than 1% (~12 kJ) of the laser energy gets into the fuel. As the fuel is compressed to a very high density (~100 times density of lead), this energy is sufficient to initiate burn. In a 300 ps hot spot of about 120 μm in size begins to form at the core of the fuel pellet.
x. The alpha particles produced in the D-T fusion reaction deposit their kinetic energy locally within the hot spot and, hence, increases its temperature further. Such heating is known as 'alpha particle self-heating' (Hurricane et al. 2016). The self-heating brings about two to three times yield enhancement and gives rise to 27–40 kJ of fusion energy.

For the production of fusion energy to the scale of a power plant, drivers with a high repetition rate with better efficiency are required. The efficiency η of a driver is defined as the ratio of electrical power E_{el} supplied to the capacitors to that which has been converted to driver energy E_{drive}.

$$\eta_{\text{drive}} = \frac{E_{el}}{E_{\text{drive}}} \tag{11.3}$$

Basically the lasers used for ICF are such that they can deliver more energy in a short-pulse time. The irradiance of the laser can be defined as

$$I_L = \frac{cE_{\max}}{8\pi}$$

$$E_{\max} = 2.75 \times 10^9 \left(\frac{I_L}{10^{16}} \right)^{1/2}$$

$$B_{\max} = 9.2 \times 10^6 \left(\frac{I_L}{10^{16}} \right)^{1/2} \tag{11.4}$$

In general, most common lasers for ICF have intensities of the order 10^{14} to 10^{15} W/cm^2. Since the absorption of laser light is inversely proportional to the intensity, the intensity of the laser cannot be much larger due to low absorption.

The construction of NIF was completed in March 2009, and immediately after this the 'National Ignition Campaign' (NIC) was started to achieve ignition by the end of 2012. The tests were started in October 2010 with about a 1-MJ laser energy fired at a target that was equivalent to 300-TW laser power. This was scaled up to 1.5 MJ of energy and 400 TW of power by 2011 and then reached to its design goal of 1.8 MJ of energy and 500 TW of power by July 2012.

With this facility, the fuel density $\rho_f \approx 600\ \text{g/cm}^3$, the fuel areal density $\rho_f r_f \approx 1\ \text{g/cm}^2$, and the average ion temperature of the fuel $T \approx 3.5$ keV were achieved. This leads to the production of fusion neutrons of the order 10^{15}. However, these parameters were still lower than required for ignition and significant energy gain. Along with the scaling up of laser parameters, there were a number of problems. The implosion symmetry required for hot-spot creation could not be maintained, which led to various laser-plasma instabilities (LPIs). Also, the desired ablation velocity could not be reached and most importantly the alpha self-heating did not occur as envisaged. As a result, even after 2 years of experimental and numerical efforts, the goal of NIC to achieve ignition by the year 2012 could not be realized. However, experiments supported by large-scale numerical simulations are ongoing and significant improvements have been reported in laser specifications, target fabrication and its symmetry, hohlraum design and X-rays parameters, implosion velocity, and so forth. It is believed that the goal to achieve ignition will be reached within the next few years. Once ignition in the laboratory is demonstrated, it would open the door for building a full-scale fusion power reactor.

Though the ignition at NIF will be a milestone for science and engineering, the energy yield would still be an order of magnitude less than the electrical energy pumped into the capacitors to drive the laser; thus, the device would be no closer to energetically break-even. To achieve significant gain for a fusion power reactor, laser shots with a very high repetition rate (~10 Hz) would be required. Presently, NIF is only able to fire one shot every few hours, whereas a fusion power reactor would require to ignite the fuel about ten times per second.

11.5 Steps towards an Inertial Fusion Power Plant

For over 100 years, we have known that nuclear fusion is the source of energy for the sun and other stars, and for more than six decades we have been trying to find methods to liberate fusion energy in a controlled manner. But to recreate the fusion reaction occurring in the sun and stars it is essential to mimic this condition in a man-made environment. Scientists around the world are looking forward to harnessing the energy released in fusion to use in a fusion power plant. They have invented two ways to make plasma hot enough to fuse: magnetic confinement and inertial confinement. The first type of reactor uses a magnetic field to squeeze the plasma into a doughnut-shaped chamber where the reaction takes place. These magnetic confinement reactors, such as the International Thermonuclear Experimental Reactor (ITER) in France, use a superconducting electromagnet cooled with liquid helium at –269 °C. The second type, called inertial confinement, uses pulses from a super power laser to heat the surface of a pellet of fuels, imploding it briefly and making the fuel hot and dense enough to fuse. In fact, one of the most powerful lasers in the world is used for fusion experiments at the NIF in the United States. These experiments and others like them around the world are today just experiments; scientists are still developing the technology.

In the United States, research and studies were done in the late 1970s to find out if building a power plant using the ICF technique was possible. Such power plants were called IFE plants or an IFE reactor. From several choices of drivers, the laser was found to be the most promising and it was decided to build a huge laser facility at the LLNL.

In the indirect drive approach to a fusion power plant, the electrical power is transformed into short pulses of energy through capacitors and lasing glasses and these are delivered to the D-T fuel pellet to cause implosion followed by hot-spot generation and ignition. The fuel pellets are manufactured in the pellet factory and are filled with D-T target fuel and placed into the fusion reaction chamber. To harness fusion energy at the scale of a fusion power plant, the driver beams need to be directed to the pellet to implode it with a repetition rate of a few times per second. The blanket captures the products of the thermonuclear explosion, and the energy is converted into thermal energy (heat), which gets converted into electricity, and a small portion is recirculated to power the driver. To understand the energetic efficiencies of such a reactor, the cycle discussed above is demonstrated in Figure 11.3. If f is a fraction of the electric power recirculated to supply the driver, then η_d and η_{th} are the driver and thermal cycle efficiencies, respectively, and G is the fusion target energy gain, then the condition for energy balance is

$$f\eta_d\eta_{th}G = 1 \tag{11.5}$$

Assuming $\eta_{th} = 40\%$ (typical value) and $f = 25\%$, we arrive at

$$G\eta_d = 0.1 \tag{11.6}$$

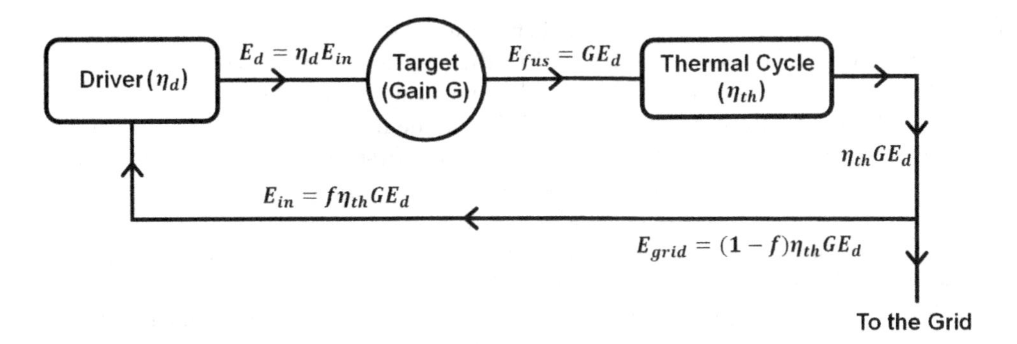

FIGURE 11.3
Energy flow in the fusion reactor.

Therefore, for the fusion energy gain G = 100, the driver efficiency has to be $\eta_d = 10\%$. Experimental results at NIF and LMJ have been promising and may be considered as a big technological step forward in the direction of a fusion power plant. But these are far from producing the high gain required by a fusion power plant. The target gain predicted for the NIF/LMJ facilities are $G \sim 10$, and the efficiency of lasers used in these devices are $\eta_d \sim 1\%$. These have to be increased by an order of magnitude to meet the conditions required for the fusion power plant. The high repletion rate of the driver (~10 Hz) may be obtained by using the diode-pumped solid-state lasers and KrF excimer lasers. Also, a great deal of research and technological advancement need to be done for the target fabrication for the fusion power plant. Currently, layered cryogenic targets are being considered for a full-scale ignition experiment. Recently, scientists have reported new techniques of target fabrication using Glow Discharge Polymer (GDP) shells and a Hollow Glass Microsphere (HGM) (Du et al. 2018).

Although fusion reactions have been successful during experiments, they are very far from commercial use as it costs more energy to do the experiments than the energy produced during fusion. The research has a long way to go and several serious technological challenges need to be met before making the process commercially viable. The fusion reactor requires an investment of more than $10 billion, which is of course a huge amount of money. On the budgetary front fusion is competing with other sources of clean energy such as solar, wind, and so forth, which have already been proven.

11.6 International Thermonuclear Experimental Reactor (ITER)

ITER, the world's largest tokamak, is a magnetic fusion device that has been designed to demonstrate the scientific and technological feasibility of nuclear fusion as a clean energy source of the future. It was designed to build the first fusion experiment, which would deliver 500 MW of fusion power with an input power of just 50 MW and hence opens the path for commercial exploitation of fusion power. The total cost of this project is more than €10 billion. The idea for ITER originated from the Geneva Superpower Summit in 1985 during which presidents of the former Soviet Union and the United States proposed an international effort to develop fusion energy. As mentioned earlier, the term 'tokamak' is a Russian acronym that means 'toroidal chamber with magnetic coil', and it was developed first by Soviet research in the late 1960s. ITER is an international project in which there are about 35 nations collaborating to build it in the southern France. The experimental campaign that will be carried out at ITER would contribute to one of the greatest challenges of mankind. There is a large possibility that ITER be the first thermonuclear fusion device in the world to produce the net energy. This will be the first fusion device to maintain fusion for long periods of time and to test the integrated technologies, materials, and physics regimes necessary for commercial production at a very large scale of fusion-based electricity. Thousands of engineers and scientists have participated in the ITER design. The journey of international joint experiments in the field of thermonuclear fusion reaction started in 1985. China, the European Union, India, Japan, Korea, Russia, and the United States are presently ITER members and have signed a Memorandum of Understanding for a 35-year collaboration to build and operate the ITER experimental devices. They are working hard to bring the fusion reaction to a level at which a fusion reactor can be designed.

The capability of producing the amount of fusion energy using a tokamak is a direct result of various fusion reactions taking place in the core. Scientists have found that the larger the vessel is, the larger the volume of plasma, and hence larger potential energy. The ITER tokamak is going to be a unique experimental facility that will acquire a ten times bigger size of plasma volume than the largest machine operational today. This will be able to confine the plasma more efficiently for longer durations. ITER is being designed mainly to make the world record in the field of production of fusion power. The European tokamak, JET, has been producing about 16 MW of fusion power from a total input power of 24 MW ($Q = 0.67$, where Q is the ratio of fusion power output to power injected into plasma) since 1997, whereas the design of ITER has been introduced to generate the ten times energy gain ($Q = 10$), i.e. 500 MW of power production with respect to 50 MW of input power. ITER will not capture the energy it produces as electricity, but it will create a condition to provide the power production efficiency $Q \geq 10$ (Aymar et al. 2002).

The researchers now will be able to study plasmas under critical conditions similar to those expected in a future power plant, and the ITER project will try to reduce the gap between today's smaller scale experimental fusion devices and the latest thermonuclear fusion power plants. They will also study the latest test technologies such as heating, diagnostics, cryogenics, control, and remote maintenance. Currently, fusion research is studying the burning of plasma, where the generated heat/temperature from the thermonuclear fusion reaction is confined within the plasma more efficiently. This helps in the sustainment of the fusion reaction for a long duration. The physicists have revealed that the confined plasmas in the ITER will not only generate enough fusion energy, but also the fusion reaction will be stable for longer durations. The other important objective of the ITER is to check the operational feasibility of producing tritium within the vacuum vessel, since the supply of tritium in the world is not sufficient to fulfill the requirements of future power plants. ITER will definitely provide a unique platform to cross-check the mock-up in-vessel tritium breeding blankets in a real fusion experimental environment.

ITER has reached a certain mark and achieved a higher position in fusion history. In 2012, the ITER Organization was given license as a nuclear operator in France. This was given on the basis of its thorough and non-discriminatory examination of safety and protection files. Today, power plants work either on nuclear fission, fossil fuels, or renewable energy sources but mainly water or wind. The plants generate electricity by converting mechanical power, such as the rotation of a turbine, into electrical power irrespective of energy sources. In a coal-fired steam station, the combustion of coal turns water into steam and the steam in turn drives turbine generators to produce electricity. The tokamak is a machine that is made to generate the energy of thermonuclear fusion. The energy inside a tokamak, produced through the fusion of atoms when the neutrons are absorbed by the walls and their energy is expected to convert into heat. A thermonuclear fusion power plant will use this heat to generate the steam and then electricity by using the turbines and generators similar to a conventional power plant. The main portion of a tokamak is a vacuum chamber like a doughnut-shaped chamber. Gaseous hydrogen fuel inside the vessel converts into plasma in the presence of extreme heat and pressure. This attains a critical condition in which hydrogen atoms can be brought to fuse and produce the energy. The confinement and controlling of charged particles is carried out by the massive magnetic coils placed around the vessel. Scientist use this innovative idea to confine the hot plasma away from the vessel walls.

11.7 Conclusions

To achieve controlled thermonuclear fusion, scientists are working hard and presently ICF and MCF are the two major schemes addressing our quest for clean energy. NIF and ITER are the two mega projects exploring these schemes, respectively. The objective of NIF is to perform with higher efficiency and higher energy gain, which are supposed to prove the physical and technical possibility of energy production from thermonuclear fusion reaction, showing the path for building the first prototype nuclear power plant. Through this, the research and study related to the behaviour of matter for nuclear weapons has been done. This is the largest and most energetic ICF tool and largest laser in the world. However, scientists faced various challenges but mainly the physical problems seen in the advanced ICF schemes. The problems are related to the production of ultraintense particle beams, relativistic laser-plasma interaction, transport and interaction of GA-current beams of particles with plasma, acceleration of micro-projectiles to hyper-velocities, or various kinds of instabilities in laser plasma.

The last 50 years of research on laser fusion reactions have resulted in the following scheme:

- Scheme to solve the physical challenges faced in ICF
- Formation of alternate and highly effective ICF schemes
- Generation of advanced technology, which enabled the construction of machines like LMJ, NIF and now ITER, for the mega-joule fusion
- Design and advancement program of the fusion reactor for ICF on the international level (FIREX, HiPER, and LIFE)

A heavy ion beam may also be applied to induce the fusion reaction in place of a laser beam. Most of the laser parameters, like the repetition rate and other important conditions, are met for the heavy ion beam, which are also similar to the laser required for the fusion reactions in the reactor. The mirror machine in the MCF device can act like an amplifier. For example, the energy of an injected neutral beam has been amplified by using a thermonuclear burning of heated plasma in a mirror machine. In principle, the laser is used to start the ignition process of the target fuel and helps in the heating of cold plasma to a certain temperature. The hot plasma is confined in a mirror machine, which amplifies the energy of the original inertial fusion reaction. In this case, there is a possibility of an additional energy gain, which may be added to the target gain after the implementation of some modifications in the simple mirror magnet. These parameters are still under research and investigation.

In summary, there is always considerable further progress in fusion technology, handling, and production of technologies of the D-T reactor and target, which is necessary for the generation of commercial energy through the laser fusion reaction (specifically in the drivers or lasers' term). In general, there are several common difficulties for both MCF and ICF, but the cooperation between the communities that are on the track to resolve these difficulties are immensely preferable. The study of laser-induced fusion is constructing a safe, potentially unlimited and efficient energy source irrespective of its main objective, and it is a 'driving-horse' of the development of several areas of science, particularly in the fields of laboratory astrophysics, material research, physics of high-energy density, laser technology, acceleration of particle and dense matter, the interaction of high-intensity light with matter, and several others.

This is also worth pointing out that despite the long and dedicated efforts of the scientists and the availability of funds, several facilities could not achieve their main goals due to unexpected problems and hence, the objectives were diverted and these are now being used for other research purposes. For example, as a fusion energy effort, LIFE could run at LLNL between 2008 and 2013 only, which started just after NIF was reaching completion in 2008. This was focussed to establish a commercial power plant design using the NIF technologies as basis for lowering the costs using mass-produced fuel elements along with simplified maintenance and the use of diode lasers in view of their larger electrical efficiency. During 2009 to 2012, another program named 'National Ignition Campaign (NIC)' was run in parallel by the NIF with aim of reaching ignition by September 30, 2012. The campaign ultimately failed due to unpredicted performance problems, though the system could yield best-case shots by the end of 2012, which were further enhanced in couple of years but remained far from the condition of achieving ignition. After several reviews, the LIFE effort was ultimately called off in early 2013.

The efforts appear to be inactive concerning HiPER also, which was designed to study the fast ignition approach based on lasers with much smaller capacity but producing fusion power outputs of almost the same magnitude. The points of attraction were its construction cost of about $1/10^{\text{th}}$ and a total fusion gain much higher compared to the case of NIF. This led to the hope for a small machine to be constructed promptly and where ignition can be achieved before NIF; the Japanese FIREX design was also intended to employ the same concept. Since number of problems encountered in the Omega laser (small machine) in the US while conducting the research on the fast ignition, the search for an alternative approach began and around 2012 study on shock ignition took over future development. However, since 2012 HiPER and FIREX both appear to have seen no additional development and the US National Academy of Science concluded that HiPER was no longer a worthwhile research direction.

11.8 Selected Problems and Solutions

PROBLEM 11.1

Determine the laser power required for the fusion for $n\tau = 10^{14}\,\text{s/cm}^3$, $\upsilon_s = 0.5728 \times 10^8$ cm/s and temperature 10^8 K.

SOLUTION

Laser power is given by

$$P_{\text{laser}} = \frac{E_{\text{laser}}}{\tau}$$

$$E_{\text{laser}} = \text{density of ions} \times \text{volume of spherical pellet} \times \text{energy of ions}$$

$$E_{\text{laser}} = n \times \frac{4}{3}\pi r_0^3 \times \frac{3}{2}KT = 2\pi r_0^3 KTn$$

$$P_{\text{laser}} = \frac{E_{\text{laser}}}{\tau} = \frac{2\pi r_0^3 KTn}{r_0/v_s} = 2\pi n(10^{14}v_s/n)^2 KTv_s$$

Since $n\tau = 10^{14}\,\text{s/cm}^3$, $r_0 = v_s\tau = \frac{10^{14}v_s}{n}$,

$$P_{\text{laser}} = \frac{2\pi(10^{14})^2 v_s^3}{n} \times KT$$

$$= \frac{2\times 3.14\times 10^{28}\times(0.5728\times 10^8)^3\times 1.38\times 10^{-23}\times 10^8}{n}$$

$$= \frac{1.6287\times 10^{37}}{n}\times\frac{n_s}{n_s} = \frac{1.6287\times 10^{37}}{n_s}\times\frac{n_s}{n} \approx \frac{10^{15}n_s}{n}$$

where n_s is the density of the D-T pellet and is of the order 10^{22} cm^{-3}. If $n_s = n$, then

$$P_{\text{laser}} \approx 1000 \text{ TW}$$

PROBLEM 11.2

Determine the power generated by a D-T fusion plant for a frequency (f) of 2 MHz, gain (G) of 20, efficiency of pump (η_{pump}) of 66%, efficiency of heat engine (η_{th}) of 30%, and energy pumped into DT pellet (E_{pump}) of 0.88 MeV.

SOLUTION

The power generated by a D-T fusion plant is given as

$$P_{\text{out}} = E_{\text{pump}} f\left[G\eta_{th} - \frac{1}{\eta_{\text{pump}}}\right]$$

For the given data, $f = 2$ MHz, $G = 20$, $\eta_{\text{pump}} = 0.66$, $\eta_{th} = 0.3$, and $E_{\text{pump}} = 0.88$ MeV, one gets

$$P_{\text{out}} = 0.88\times 2\times 10^6\times\left[20\times 0.3 - \frac{1}{0.66}\right]$$

$$P_{\text{out}} = 7.89\times 10^6 \text{ W} = 7.89 \text{ MW}$$

Suggested Reading Material

Aymar, R., Barabaschi, P., & Shimomura, Y. (2002). The ITER design. *Plasma Physics and Controlled Fusion, 44*(5), 519.

Azechi, H., Mima, K., Shiraga, S., ... & Kondo, K. (2013). Present status of fast ignition realization experiment and inertial fusion energy development. *Nuclear Fusion, 53*(10), 104021.

Betti, R., & Hurricane, O. A. (2016). Inertial-confinement fusion with lasers. *Nature Physics, 12*(5), 435–448.

Campbell, E. M., Goncharov, V. N., Sangster, T. C., Regan, S. P., Radha, P. B., Betti, R., ... & Seka, W. (2017). Laser-direct-drive program: Promise, challenge, and path forward. *Matter and Radiation at Extremes, 2*(2), 37–54.

Du, K., Liu, M., Wang, T., He, X., Wang, Z., & Zhang, J. (2018). Recent progress in ICF target fabrication at RCLF. *Matter and Radiation at Extremes, 3*(3), 135–144.

Dunne, M. (2006). A high-power laser fusion facility for Europe. *Nature Physics, 2*(1), 2–5.

Dunne, M., Moses, E. I., Amendt, P., ... & Farmer, J. C. (2011). Timely delivery of laser inertial fusion energy (LIFE). *Fusion Science and Technology, 60*(1), 19–27.

Hurricane, O. A., Callahan, D. A., Casey, D. T., ... & Kritcher, A. L. (2016). Inertially confined fusion plasmas dominated by alpha-particle self-heating. *Nature Physics, 12*(8), 800–806.

Landen, O. L., Benedetti, R., Bleuel, D., ... & Collins, G. W. (2012). Progress in the indirect-drive National Ignition Campaign. *Plasma Physics and Controlled Fusion, 54*(12), 124026.

Moses, E. I. (2003). The National Ignition Facility: the world's largest laser. In *20th IEEE/NPSS Symposium on Fusion Engineering, 2003*, IEEE, pp. 413–418.

Moses, E. I., & Wuest, C. R. (2005). The National Ignition Facility: laser performance and first experiments. *Fusion Science and Technology, 47*(3), 314–322.

Robey, H. F., Boehly, T. R., Celliers, P. M., ... & Munro, D. H. (2012). Shock timing experiments on the National Ignition Facility: Initial results and comparison with simulation. *Physics of Plasmas, 19*(4), 042706.

Singh, D. K., Davies, J. R., Sarri, G., Fiuza, F., & Silva, L. O. (2012). Dynamics of intense laser propagation in underdense plasma: Polarization dependence. *Physics of Plasmas, 19*(7), 073111.

Index

9780367651077